Aspects of Semidefinite Programming

Applied Optimization

Volume 65

Series Editors:

Panos M. Pardalos
University of Florida, U.S.A.

Donald Hearn
University of Florida, U.S.A.

The titles published in this series are listed at the end of this volume.

Aspects of Semidefinite Programming

Interior Point Algorithms and Selected Applications

by

Etienne de Klerk

Delft University of Technology,
Delft, The Netherlands

KLUWER ACADEMIC PUBLISHERS
DORDRECHT / BOSTON / LONDON

A C.I.P. Catalogue record for this book is available from the Library of Congress.

ISBN 1-4020-0547-4

Published by Kluwer Academic Publishers,
P.O. Box 17, 3300 AA Dordrecht, The Netherlands.

Sold and distributed in North, Central and South America
by Kluwer Academic Publishers,
101 Philip Drive, Norwell, MA 02061, U.S.A.

In all other countries, sold and distributed
by Kluwer Academic Publishers,
P.O. Box 322, 3300 AH Dordrecht, The Netherlands.

Printed on acid-free paper

Printed in the Netherlands.

Contents

Part II Selected Applications

Acknowledgments

I would like to thank Támas Terlaky for his encouragement to publish a revised version of my PhD thesis in the Kluwer Applied Optimization series. Several people have assisted me by proof-reading preliminary versions of the manuscript. In particular, I am indebted to Aharon Ben-Tal, Maria Gonzalez, Margareta Halická, Dorota Kurowicka, Hans van Maaren, Renato Monteiro, Dima Pasechnik and Kees Roos in this regard.

I also thank Jos Sturm for patiently answering many questions about the convergence behaviour of the central path.

I thank Lucie Aarts and Margareta Halická for sharing their notes on which I based Appendix E and Appendix D respectively. Florin Barb I thank for moral and technical support.

Much of the material presented here is based on joint research, and I gratefully acknowledge the contributions of my co-authors: Immanuel Bomze, Margareta Halická, Bingsheng He, Dick den Hertog, Hans van Maaren, Dima Pasechik, Kees Roos, Támas Terlaky, and Joost Warners.

Finally, I thank John Martindale, Angela Quilici and Edwin Beschler at Kluwer Academic Publishers for their patience and assistance.

Foreword

This monograph has grown from my PhD thesis *Interior point Methods for Semidefinite Programming* [39] which was published in December 1997. Since that time, Semidefinite Programming (SDP) has remained a popular research topic and the associated body of literature has grown considerably. As SDP has proved such a useful tool in many applications, like systems and control theory and combinatorial optimization, there is a growing number of people who would like to learn more about this field.

My goal with this monograph is to provide a personal view on the theory and applications of SDP in such a way that the reader will be equipped to read the relevant research literature and explore new avenues of research. Thus I treat a selected number of topics in depth, and provide references for further reading. The chapters are structured in such a way that the monograph can be used for a graduate course on SDP.

With regard to algorithms, I have focused mainly on methods involving the so-called Nesterov–Todd (NT) direction in some way. As for applications, I have selected interesting ones — mainly in combinatorial optimization — that are not extensively covered in the existing review literature.

In making these choices I hasten to acknowledge that much of the algorithmic analysis can be done in a more general setting (*i.e.,* working with self-concordant barriers, self-dual cones and Euclidean Jordan algebras). I only consider real symmetric positive semidefinite matrix variables in this book; this already allows a wealth of applications.

Required background

The reader is expected to have some background on the following topics:

- linear algebra and multivariate calculus;
- linear and non-linear programming;
- basic convex analysis;
- combinatorial optimization or complexity theory.

I have provided appendices on the necessary matrix analysis (Appendix A), matrix calculus (Appendix C) and convex analysis & optimization (Appendix B), in an attempt to make this volume self–contained to a large degree. Nevertheless, when reading this book it will be handy to have access to the following textbooks, as I refer to them frequently.

- *Convex Analysis* by Rockafellar [160];
- *Matrix Analysis* and *Topics in Matrix Analysis* by Horn and Johnson [85, 86];
- *Nonlinear Programming: Theory and Algorithms* by Bazarraa, Sherali and Shetty [16];
- *Theory and Algorithms for Linear Optimization: An interior point approach*, by Roos, Terlaky and Vial [161];
- *Randomized Algorithms* by Motwani and Raghavan [128].

In particular, the analysis of interior point algorithms presented in this monograph owes much to the analysis in the book by Roos, Terlaky and Vial [161].

Moreover, I do not give an introduction to complexity theory or to randomized algorithms; the reader is referred to the excellent text by Motwani and Raghavan [128] for such an introduction.

Matrix notation

A^T	:	transpose of $A \in \mathbf{R}^{m\times n}$;
A^{-T}	:	$\left(A^T\right)^{-1}$;
a_{ij}	:	ijth entry of $A \in \mathbf{R}^{m\times n}$;
$A \sim B$	:	$A = T^{-1}BT$ for some nonsingular T
	$\equiv$	the matrices A and B are similar;
$A \succeq B$ $(A \succ B)$	:	$A - B$ is symmetric positive semidefinite (positive definite);
$A \preceq B$ $(A \prec B)$	:	$A - B$ is symmetric negative semidefinite (negative definite);
$\mathcal{R}(A)$	:	range (column space) of $A \in \mathbf{R}^{n\times n}$;

Special vectors and matrices

I_r	:	$r \times r$ identity matrix;
I	:	identity matrix of size depending on the context;
$0_{m\times n}$	:	$m \times n$ zero matrix;
0_n	:	zero vector in $\mathbf{R}^n$;
0	:	zero vector/matrix of size depending on the context;
e_n	:	vector of all-ones in $\mathbf{R}^n$;
e	:	vector of all-ones of size depending on the context;

Sets of matrices and vectors

$\mathbb{R}^n$	:	n-dimensional real Euclidian vector space;
$\mathbb{R}^n_+$	:	positive orthant of $\mathbf{R}^n$;
$\mathbb{Z}^n$	:	n-tuples of integers
$\mathbb{Z}^n_+$	:	n-tuples of *nonnegative* integers
$\mathbb{R}^{n\times n}$	:	space of real $(n \times n)$ matrices;
$\mathcal{S}_n$	$=$	$\left\{X \mid X \in \mathbb{R}^{n\times n},\ X = X^T\right\}$;
$\mathcal{S}_n^+$	$=$	$\{X \mid X \in \mathcal{S}_n,\ X \succeq 0\}$;
$\mathcal{S}_n^{++}$	$=$	$\{X \mid X \in \mathcal{S}_n,\ X \succ 0\}$;
$\mathcal{C}_n$	$=$	$\left\{A \in \mathcal{S}_n \;\middle|\; x^TAx \geq 0 \quad \forall x \in \mathbf{R}^n_+\right\}$ (Copositive matrices);
$\mathcal{N}_n$	$=$	$\{A \in \mathcal{S}_n \mid a_{ij} \geq 0 \quad \forall i,j = 1,\ldots,n\}$ (Nonnegative matrices);
$\mathbf{svec}\left(\mathcal{S}_n^+\right)$	$=$	$\left\{x \in \mathbf{R}^{\frac{1}{2}n(n+1)} \;\middle|\; \mathbf{smat}(x) \in \mathcal{S}_n^+\right\}$;
$\{-1,1\}^n$	$=$	$\{x \in \mathbf{R}^n \mid x_i \in \{-1,1\},\ i = 1,\ldots,n\}$;

Functions of matrices

$$
\begin{aligned}
\lambda_i(A) &: && i\text{th largest eigenvalue of } A, \text{ if } \lambda_j(A) \in \mathbb{R} \quad \forall j;\\
\lambda_{\max}(A) &= && \max_i \lambda_i(A), \text{ if } \lambda_i(A) \in \mathbb{R} \quad \forall i;\\
\lambda_{\min}(A) &= && \min_i \lambda_i(A), \text{ if } \lambda_i(A) \in \mathbb{R} \quad \forall i;\\
\mathbf{Tr}\,(A) &= && \sum_i a_{ii} = \sum_i \lambda_i(A) \quad \text{(trace of } A \in \mathbb{R}^{n\times n});\\
\langle A, B\rangle &= && \mathbf{Tr}\left(AB^T\right);\\
\det(A) &: && \text{determinant of } A \in \mathbb{R}^{n\times n} = \prod_i \lambda_i(A);\\
\|A\|^2 &= && \mathbf{Tr}\,(AA^T) = \sum_i \sum_j a_{ij}^2 \quad \text{(Frobenius norm)}\\
&= && \sum_i \lambda_i^2(A) \text{ if } A \in \mathcal{S}_n;\\
\|A\|_2 &= && (\lambda_{\max}(A^TA))^{\frac{1}{2}} \quad \text{(spectral norm)}\\
&= && \lambda_{\max}(A) \text{ if } A \succeq 0;\\
\rho(A) &= && \max_i |\lambda_i(A)| \quad \text{(spectral radius of } A);\\
\kappa(A) &= && \frac{\lambda_{\max}(A)}{\lambda_{\min}(A)} \text{ if } \lambda_i(A) > 0\ \forall i\\
&= && \text{condition number of } A \text{ if } A \succ 0;\\
A^{\frac{1}{2}} &: && \text{unique symmetric square root factor of } A \succeq 0;\\
\mathrm{Diag}\,(x) &: && n \times n \text{ diagonal matrix with components of } x \in \mathbb{R}^n \text{ on diagonal};\\
\mathrm{diag}\,(X) &: && n\text{-vector obtained by extracting diagonal of } X \in \mathbb{R}^{n\times n};\\
\mathbf{vec}(A) &= && [a_{11}, a_{21}, \ldots a_{n1}, a_{12}, a_{22}, \ldots, a_{nn}]^T \text{ for } A \in \mathbb{R}^{n\times n};\\
\mathbf{svec}(A) &= && [a_{11}, \sqrt{2}a_{12}, \ldots \sqrt{2}a_{1n}, a_{22}, \sqrt{2}a_{23}, \ldots, a_{nn}]^T \text{ for } A \in \mathcal{S}_n;\\
\mathbf{smat} &: && \text{inverse operator of } \mathbf{svec};
\end{aligned}
$$

Set notation

$$
\begin{aligned}
\mathrm{ri}\,(\mathcal{C}) &: && \text{relative interior of a convex set } C;\\
\dim(\mathcal{L}) &: && \text{dimension of a subspace } \mathcal{L};\\
\mathcal{C}^* &: && \text{Dual cone of a cone } \mathcal{C} \subset \mathbb{R}^n\\
&= && \left\{x \in \mathbb{R}^n \mid x^Ty \geq 0\ \forall y \in \mathcal{C}\right\};
\end{aligned}
$$

SDP problems in standard form

$$
\begin{aligned}
(P) &: \text{ primal problem in standard from;}\\
(D) &: \text{ Lagrangian dual problem of } (P);\\
\mathcal{P} &: \text{ feasible set of problem } (P);\\
\mathcal{D} &: \text{ feasible set of problem } (D);\\
\mathcal{P}^* &: \text{ optimal set of problem } (P);\\
\mathcal{D}^* &: \text{ optimal set of problem } (D);\\
(P_{gf}) &: \text{ ELSD dual of } (D);\\
(D_{cor}) &: \text{ Lagrangian dual of } (P_{gf});\\
\mathcal{L} &= \text{span}\,\{A_1, \ldots, A_m\}\,;
\end{aligned}
$$

Interior point analysis

$$
\begin{aligned}
\log(t) &: \text{ natural logarithm of } t > 0;\\
\psi(t) &= t - \log(1+t) \quad t > -1;\\
f_p^{\mu}(X) &= \frac{1}{\mu}\mathbf{Tr}\,(CX) - \log\det X;\\
f_d^{\mu}(S, y) &= \frac{1}{\mu}b^T y + \log\det(S);\\
f_{pd}^{\mu}(X, S) &= \mathbf{Tr}\left(\frac{XS}{\mu}\right) - \log\det\left(\frac{XS}{\mu}\right) - n;\\
D &= \left[X^{\frac{1}{2}}\left(X^{\frac{1}{2}}SX^{\frac{1}{2}}\right)^{-\frac{1}{2}}X^{\frac{1}{2}}\right]^{\frac{1}{2}} \quad \text{where } (X,S) \in \text{ri}\,(\mathcal{P}\times\mathcal{D})\\
&\equiv \text{ Nesterov–Todd scaling matrix;}\\
V &= D^{-\frac{1}{2}}XD^{-\frac{1}{2}} = D^{\frac{1}{2}}SD^{\frac{1}{2}};\\
\delta(X, S, \mu) &= \frac{1}{2}\left\|\sqrt{\mu}V^{-1} - \frac{1}{\sqrt{\mu}}V\right\|;\\
D_X &= D^{-\frac{1}{2}}\Delta X D^{-\frac{1}{2}} \quad (\Delta X \in \mathcal{L}^{\perp});\\
D_S &= D^{\frac{1}{2}}\Delta S D^{\frac{1}{2}} \quad (\Delta S \in \mathcal{L});\\
\Psi(X, S) &= -\log\det(XS) + n\log\mathbf{Tr}\,(XS) - n\log n \text{ (Centrality function);}\\
\Phi(X, S) &= (n + \nu\sqrt{n})\log\mathbf{Tr}\,(XS) - \log\det(XS) - n\log n\\
&\equiv \text{ Tanabe–Todd–Ye potential function;}
\end{aligned}
$$

Graph theory

$$
\begin{aligned}
G = (V, E) &: \text{ simple undirected graph with vertex set } V \text{ and edge set } E;\\
\alpha(G) &: \text{ stability number of } G;
\end{aligned}
$$

$\chi(G)$	:	chromatic number of G;
$\omega(G)$	:	clique number of G;
$\bar{G}$	:	complementary graph (complement) of G;
$\vartheta(G)$	:	Lovász ϑ-function of G;
$\Theta(G)$	:	Shannon capacity of G;

Notation for asymptotic analysis Let $f(n), g(n) : \mathbf{R} \mapsto \mathbf{R}_+$. We say that

$f(n) = O(g(n))$	:	$f(n)/g(n)$ is bounded from above;
$f(n) = \tilde{O}(g(n))$	:	$f(n) = O\left([\log(n)]^p \, g(n)\right)$ for some $p > 0$ (independent of n);
$f(n) = \Omega(g(n))$	:	$g(n)/f(n)$ is bounded from above;
$f(n) \sim g(n)$	:	$\lim_{n\to\infty} f(n)/g(n) = 1$.

1
INTRODUCTION

Preamble

This monograph deals with algorithms for a subclass of nonlinear, convex optimization problems, namely semidefinite programs, as well as selected applications of these problems. To place the topics which are dealt with in perspective, a short survey of the field of semidefinite programming[1] is presented in this chapter. Applications in combinatorial optimization and engineering are reviewed, after which interior point algorithms for this problem class are surveyed.

1.1 PROBLEM STATEMENT

Semidefinite programming (SDP) is a relatively new field of mathematical programming, and most papers on SDP were written in the 1990's, although its roots can be traced back a few decades further (see *e.g.* Bellman and Fan [19]). A paper on semidefinite programming from 1981 is descriptively named *Linear Programming with Matrix Variables* (Craven and Mond [37]), and this apt title may be the best way to introduce the problem.

The goal is to minimize the inner product

$$\langle C, X \rangle := \mathbf{Tr}\,(CX),$$

[1]Some authors prefer the more descriptive term 'optimization' to the historically entrenched 'programming'.

of two $n \times n$ symmetric matrices, namely a constant matrix C and a variable matrix X, subject to a set of constraints, where '**Tr**' denotes the trace (sum of diagonal elements) of a matrix.[2] The first of the constraints are linear:

$$\mathbf{Tr}\,(A_i X) = b_i, \quad i = 1, \ldots, m,$$

where the A_i's are given symmetric matrices, and the b_i's given scalars. Up to this point, the stated problem is merely a linear programming (LP) problem with the entries of X as variables. We now add the convex, nonlinear constraint that X must be symmetric positive semidefinite[3], denoted by $X \succeq 0$.[4]

The convexity follows from the convexity of the cone of positive semidefinite matrices. (We recommend that the reader briefly review the properties of positive semidefinite matrices in Appendix A.)

The problem under consideration is therefore

$$(P): \;\; p^* := \inf_X \{\mathbf{Tr}\,(CX) \; : \mathbf{Tr}\,(A_i X) = b_i \;(i = 1, \ldots, m), \; X \succeq 0\},$$

which has an associated Lagrangian dual problem:

$$(D): \;\; d^* := \sup_{y,S} \left\{ b^T y \; : \; \sum_{i=1}^m y_i A_i + S = C, \; S \succeq 0, \; y \in \mathbb{R}^m \right\}.$$

The duality theory of SDP is weaker than that of LP. One still has the familiar *weak duality* property: Feasible X, y, S satisfy

$$\mathbf{Tr}\,(CX) - b^T y = \mathbf{Tr}\left(\left(S + \sum_{i=1}^m y_i A_i\right) X\right) - \sum_{i=1}^m y_i \mathbf{Tr}\,(A_i X) = \mathbf{Tr}\,(SX) \geq 0,$$

where the inequality follows from $X \succeq 0$ and $S \succeq 0$ (see Theorem A.2 in Appendix A). In other words, the *duality gap* is nonnegative for feasible solutions.

Solutions (X, y, S) with zero duality gap

$$\mathbf{Tr}\,(CX) - b^T y = \mathbf{Tr}\,(SX) = 0$$

are optimal. For LP, if either the primal or the dual problem has an optimal solution, then both have optimal solutions, and the duality gap at optimality is zero. This is the

[2] This inner product corresponds to the familiar Euclidean inner product of two vectors – if the columns of the two matrices C and X are stacked to form vectors $\mathbf{vec}(X)$ and $\mathbf{vec}(C)$, then $\mathbf{vec}(C)^T \mathbf{vec}(X) = \mathbf{Tr}\,(CX)$. The inner product induces the so-called *Frobenius norm*:

$$\|A\|^2 := \langle A, A \rangle = \mathbf{Tr}\left(AA^T\right) = \sum_{i,j} A_{ij}^2.$$

See Appendix A for more details.

[3] By definition, a symmetric matrix X is positive semidefinite if $z^T X z \geq 0 \; \forall z \in \mathbb{R}^n$, or equivalently, if all eigenvalues of X are nonnegative.

[4] The symbol '$\succeq$' denotes the so-called Löwner partial order on the symmetric matrices: $A \succeq B$ means $A - B$ is positive semidefinite; see also Appendix A.

strong duality property. The SDP case is more subtle: One problem may be solvable and its dual infeasible, or the duality gap may be positive at optimality, *etc.* The existence of primal and dual optimal solutions is guaranteed if both (P) and (D) allow positive definite solutions, *i.e.* feasible $X \succ 0$ and $S \succ 0$. This is called the *Slater constraint qualification* (or Slater regularity condition). These duality issues will be discussed in detail in Chapter 2.

1.2 THE IMPORTANCE OF SEMIDEFINITE PROGRAMMING

SDP problems are of interest for a number of reasons, including:

- SDP contains important classes of problems as special cases, such as linear and quadratic programming (LP and QP);
- important applications exist in combinatorial optimization, approximation theory, system and control theory, and mechanical and electrical engineering;
- Loosely speaking, SDP problems can be solved to ϵ-optimality in polynomial time by *interior point algorithms* (see Section 1.9 for a more precise discussion of the computational complexity of SDP); interior point algorithms for SDP have been studied intensively in the 1990's, explaining the resurgence in research interest in SDP.

Each of these considerations will be discussed briefly in the remainder of this chapter.

1.3 SPECIAL CASES OF SEMIDEFINITE PROGRAMMING

If the matrix X is restricted to be diagonal, then the requirement $X \succeq 0$ reduces to the requirement that the diagonal elements of X must be nonnegative. In other words, we have an LP problem. Optimization problems with convex quadratic constraints are likewise special cases of SDP.[5] This follows from the well-known *Schur complement* theorem (Theorem A.9 in Appendix A). Thus we can represent the quadratic constraint

$$(Ax+b)^T(Ax+b) - (c^Tx+d) \leq 0, \quad x \in \mathbb{R}^n,$$

by the semidefinite constraint

$$\begin{bmatrix} I & Ax+b \\ (Ax+b)^T & c^Tx+d \end{bmatrix} \succeq 0.$$

In the same way, we can represent the *second order cone* (or 'ice cream cone'):

$$\left\{ (t,x) \;\middle|\; t \geq \sqrt{\sum_{i=1}^{n} x_i^2} \right\},$$

[5]This includes the well-known convex quadratic programming (QP) problem.

by requiring that a suitable arrow matrix be positive semidefinite:

$$\begin{bmatrix} tI & x \\ x^T & t \end{bmatrix} \succeq 0.$$

Another nonlinear example is

$$\min_x \left\{ \frac{(c^T x)^2}{d^T x} \;\middle|\; Ax \geq b \right\},$$

where it is known that $d^T x > 0$ if $Ax \geq b$. An equivalent SDP problem is:[6]

$$\min_{t,x} \left\{ t \;\middle|\; \begin{bmatrix} t & c^T x & 0 \\ c^T x & d^T x & 0 \\ 0 & 0 & \mathrm{Diag}\,(Ax - b) \end{bmatrix} \succeq 0 \right\}.$$

Several problems involving matrix norm or eigenvalue minimization may be stated as SDP's. A list of such problems may be found in Vandenberghe and Boyd [181]. A simple example is the classical problem of finding the largest eigenvalue $\lambda_{\max}(A)$ of a symmetric matrix A. The key observation here is that $t \geq \lambda_{\max}(A)$ if and only if $tI - A \succeq 0$. The SDP problem therefore becomes

$$\min_t \{t \mid tI - A \succeq 0, \quad t \in \mathbb{R}\}.$$

An SDP algorithm for this problem is described by Jansen *et al.* [88, 90].

1.4 APPLICATIONS IN COMBINATORIAL OPTIMIZATION

In this section we give a short review of some of the most important and successful applications of SDP in combinatorial optimization.

The Lovász ϑ-function

The most celebrated example of application of SDP to combinatorial optimization is probably the Lovász ϑ-function [115].

The Lovász ϑ-function maps a graph $G = (V, E)$ to $\mathbb{R}_+$, in such a way that

$$\omega(G) \leq \vartheta(\bar{G}) \leq \chi(G), \tag{1.1}$$

where $\omega(G)$ denotes the clique number[7] of G, $\chi(G)$ the chromatic number[8] of G, and $\bar{G}$ the complement of G.[9]

[6] We use the following notation: for a matrix X, diag (X) is the vector obtained by extracting the diagonal of X; for a vector x, Diag (x) is the diagonal matrix with the coordinates of x as diagonal elements.

[7] A maximum clique (or completely connected subgraph) is a subset $C \subset V$ with $\forall i, j \in C (i \neq j)$: $(i, j) \in E$, such that $|C|$ is as large as possible. The cardinality $|C|$ is called the clique number.

[8] The chromatic number is the number of colours needed to colour all vertices so that no two adjacent vertices share the same colour.

[9] The complementary graph (or complement) of $G = (V, E)$ is the graph $\bar{G} = (V, \bar{E})$ such that for each pair of vertices $i \neq j$ one has $(i, j) \in \bar{E}$ if and only if $(i, j) \notin E$.

The ϑ-function value is given as the optimal value of the following SDP problem:

$$\vartheta\left(\bar{G}\right) := \max_{X} \mathbf{Tr}\left(ee^T X\right) = e^T X e,$$

where $e \in \mathbb{R}^{|V|}$ denotes the vector of all-ones, subject to

$$\begin{aligned} x_{ij} &= 0, \ (i,j) \notin E \ (i \neq j) \\ \mathbf{Tr}\,(X) &= 1 \\ X &\succeq 0. \end{aligned}$$

The relation (1.1) is aptly known as the 'sandwich theorem' (a name coined by Knuth [103]). The sandwich theorem implies that $\vartheta(\bar{G})$ can be seen as a polynomial time approximation to both $\omega(G)$ and $\chi(G)$, and that the approximation cannot be off by more than a factor $|V|$. This may seem like a rather weak approximation guarantee, but some recent in-approximability results suggest that neither $\omega(G)$ nor $\chi(G)$ can be approximated within a factor $|V|^{1-\epsilon}$ for any $\epsilon > 0$ in polynomial time (see Håstad [81] and Feige and Kilian [57]).

We will review some of the properties of the ϑ-function in Chapter 10, and will give a proof of the sandwich theorem there; moreover, we will look at some alternative definitions of this function. We will also review how the ϑ-function may be used to estimate the Shannon capacity[10] of a graph.

The MAX–CUT problem and extensions

Another celebrated application of SDP to combinatorial optimization is the MAX-CUT problem. Consider a clique $G = (V, E)$ where each edge (i, j) has a given weight $w_{ij} \geq 0$ $(i \neq j)$.

The goal is to colour all the vertices of G using two colours (say red and blue), in such a way that the total weight of edges where the endpoints have different colours (called *non–defect edges*) is as large as possible. The non–defect edges define a 'cut' in the graph — if one 'cuts' all the non–defect edges, then the blue and red vertices are separated. The total weight of the non–defect edges is therefore also called the weight of the cut.

Goemans and Williamson [66] derived a randomized 0.878-approximation algorithm[11] for this problem using SDP. The first step was to write the MAX-CUT problem as a Boolean quadratic optimization problem, *i.e.* an optimization problem with quadratic objective function and Boolean variables. For each vertex in V we introduce

[10]The Shannon capacity is a graph theoretical property that arises naturally in applications in coding theory; see Chapter 10 for details.

[11]Consider any class of NP-complete maximization problems. An α-approximation algorithm $(0 < \alpha \leq 1)$ for this problem class is then defined as follows. For any problem instance from this class, the algorithm terminates in time bounded by a polynomial in the size of the instance, and produces a feasible solution with objective value at least α times the optimal value for the instance.

a $\{-1,1\}$ variable:

$$x_i = \begin{cases} 1 & \text{if vertex } i \text{ is coloured red} \\ -1 & \text{if vertex } i \text{ is coloured blue} \end{cases} \qquad i = 1,\ldots,|V|.$$

Note that, for a given edge $(i,j) \in E$ we have

$$x_i x_j = \begin{cases} 1 & \text{if } (i,j) \text{ is defect} \\ -1 & \text{if } (i,j) \text{ is non–defect.} \end{cases}$$

The weight of the maximum cut is therefore given by

$$OPT := \max_{x\in\{-1,1\}^{|V|}} \left\{ \frac{1}{2}\sum_{i<j} w_{ij}\left(1 - x_i x_j\right) \right\} = \max_{x\in\{-1,1\}^{|V|}} \frac{1}{4} x^T L x, \tag{1.2}$$

where L is the matrix

$$L = -W + \text{Diag}(We),$$

where W is the matrix with zero diagonal and the nonnegative edge weights as off-diagonal entries, and e the all-ones vector. If all weights are zero or one, then L is simply the Laplacian of the graph.[12] Note that L is a diagonally dominant matrix, and is therefore positive semidefinite (see Appendix A).

We can relax problem (1.2) to an SDP problem by noting four things:

1. $x^T Lx = \mathbf{Tr}\left(Lxx^T\right)$, by the properties of the trace operator (see Appendix A);
2. the matrix $X := xx^T$ is positive semidefinite;
3. the ith diagonal element of $X := xx^T$ is given by $x_i^2 = 1$;
4. the matrix $X := xx^T$ has rank one.

If we drop the last (rank one) requirement, we arrive at the SDP relaxation

$$OPT \le \overline{OPT} := \max_X \left\{ \frac{1}{4}\mathbf{Tr}\left(LX\right) \mid \text{diag}\left(X\right) = e,\ X \succeq 0 \right\}. \tag{1.3}$$

Goemans and Williamson [66] devised a randomized rounding scheme that uses the optimal solution of (1.3) to generate cuts in the graph. Their algorithm produces a cut of weight at least $0.878\overline{OPT} \ge 0.878 OPT$.

This seminal result has been extended to MAX-k-CUT (where we allow k colours instead of two) by Frieze and Jerrum [59], and their results were further refined by De Klerk *et al.* [42]. We will review all these results in Chapter 11. A variation on

---

[12]Recall that if A is the adjacency matrix of G, and $d \in \mathbf{R}^{|V|}$ is the vector with degrees of the vertices as components, then the Laplacian matrix is given by $L = \text{Diag}(d) - A$.

the MAX-CUT problem is MAX-BISECTION where the cut must partition the vertex set into two sets of equal cardinality. A 0.65-approximation algorithm was given for MAX-BISECTION by Frieze and Jerrum [59] and this approximation guarantee was improved by Ye [190] to 0.699. Another related problem is the *approximate graph colouring* problem. The goal is to give a legal colouring[13] of a κ-colourable[14] graph in polynomial time, using as few colours as possible. An SDP-based approximation algorithm for this problem is given by Karger *et al.* [98]. We will derive their approximation results in Chapter 11, Section 11.8.

The (maximum) satisfiability problem

An instance of the maximum satisfiability (MAX-SAT) problem is defined by a collection of Boolean clauses $\{\mathbf{C}_1, \ldots, \mathbf{C}_k\}$, where each clause is a disjunction of literals drawn from a set of variables $\{p_1, \ldots, p_n\}$. A literal is either a variable p_i or its negation $\neg p_i$ for some i. An example of a three literal clause (or 3-clause) is

$$\mathbf{C}_1 := p_1 \vee p_2 \vee \neg p_3,$$

where '$\vee$' denotes the logical OR operator. This clause evaluates to TRUE if p_1 is true, p_2 is TRUE, or p_3 is false. Each clause has an associated nonnegative weight, and an optimal solution to a MAX-SAT instance is an assignment of truth values to the variables which maximizes the total weight of the satisfied clauses. MAX-k-SAT is a special case of MAX-SAT where each clause contains at most k literals. The satisfiability (SAT) problem is the special case where we only wish to know if all the clauses can be satisfied simultaneously.

Goemans and Williamson [66] showed that the MAX-2-SAT problem — like the MAX-CUT problem — allows a 0.878-approximation algorithm. Feige and Goemans [56] have shown that the addition of valid inequalities (so-called triangle inequalities) improves the approximation guarantee of the SDP relaxation from 0.878 to 0.931.

Karloff and Zwick [99] extended the analysis to MAX-3-SAT and proved a 7/8-approximation guarantee (for satisfiable instances). This bound is tight in view of a recent in-approximability result by Håstad [80]. Zwick [194], and Halperin and Zwick [79] have continued this line of research, and have given approximation guarantees for MAX-4-SAT and other related problems.

For the SAT problem, De Klerk *et al.* [50] have investigated a simple SDP relaxation and showed that it can be used to detect unsatisfiability of several polynomially solvable subclasses of SAT problems. Most of these results on SAT and MAX-SAT are derived in Chapter 13.

Approximating the stable set polytope

For a graph $G = (V, E)$, a subset $V' \subset V$ is called a *stable set* (or co-clique or independent set) of G if the induced subgraph on V' contains no edges. The cardinality of the largest stable set is called the *stability number* (or co-clique number or

[13] A legal colouring is a colouring of the vertices so that there are no defect edges.
[14] We call a graph G κ-colourable if $\chi(G) \leq \kappa$.

independence number) of G. The *incidence vector* $x_{V'}$ of a stable set V' is defined as

$$(x_{V'})_i = \begin{cases} 1 & \text{if } i \in V' \\ 0 & \textit{otherwise.} \end{cases}$$

The *stable set polytope* STAB(G) of G is defined as the convex hull of the incidence vectors of all stable sets.

Sherali and Adams [164] first showed how to describe the stable set polytope as the projection of a polytope in a higher dimension. Lovász and Schrijver [116] presented a similar idea and also gave a description of STAB(G) using SDP. These methods are called lift-and-project methods. Anjos and Wolkowicz [11] and Laserre [111, 110] have introduced other lift-and-project schemes, and Laurent [113, 112] has recently investigated the relationships between all these approaches.

In Chapter 12 we will give a somewhat unusual perspective on lift-and-project methods. We will show how the stability number of a graph can be computed by solving a suitably large LP or SDP problem, following the methodology by De Klerk and Pasechnik [41].

Other combinatorial applications

SDP relaxations have also been employed for the traveling salesman problem (Cvetković *et al.* [38]), quadratic assignment problem (Zhao *et al.* [193]), machine scheduling (Yang *et al.* [186]), and other problems. A recent survey was done by Goemans and Rendl [64].

1.5 APPLICATIONS IN APPROXIMATION THEORY

Non-convex quadratic optimization

The randomized MAX-CUT algorithm by Goemans and Williamson [66] has been extended to general Boolean quadratic optimization, as well as to some related problems. Recall from (1.2) that the weight of a maximum cut in a graph can be obtained by solving a problem of the form

$$q^{\max} = \max\left\{x^T Q x \mid x_i \in \{-1, 1\} \ (\forall i)\right\}, \tag{1.4}$$

where Q corresponds to the 'weighted Laplacian' of a graph. In the same way as for MAX-CUT we can derive the SDP relaxation for general symmetric matrices Q, to obtain

$$\bar{q} = \max\left\{\mathbf{Tr}\,(QX) \mid \operatorname{diag}(X) = e,\ X \succeq 0\right\}. \tag{1.5}$$

The results by Goemans and Williamson [66] show that $0.878\bar{q} \le q^{\max} \le \bar{q}$ if Q corresponds to the 'weighted Laplacian' of a graph with nonnegative edge weights. Nesterov [136] proved that $\frac{2}{\pi}\bar{q} \le q^{\max}$ if $Q \succeq 0$, and for general symmetric Q proved that

$$\bar{q} - \underline{q} \ge q^{\max} - q^{\min} \ge \frac{4-\pi}{\pi}\left(\bar{q} - \underline{q}\right),$$

where $(q^{\min}, q^{\max})$ is the range of feasible objective values in (1.4), and $(\underline{q}, \bar{q})$ is the range of feasible values in the relaxation problem (1.5). As for the MAX-CUT problem, one can use an optimal solution of (1.5) to generate a feasible solution, say x, of (1.4) via a randomized rounding scheme. The expected objective value of x, say $E(x)$, satisfies

$$\frac{q^{\max} - E(x)}{q^{\max} - q^{\min}} = \frac{\pi}{2} - 1 < \frac{4}{7}.$$

The same bounds were obtained by Ye [189] for the 'box-constrained' problem where $x_i \in \{-1, 1\}$ is replaced by $-1 \leq x_i \leq 1$ in problem (1.4), as well as for problems with *simple quadratic constraints* of the form: $\sum_{i=1}^n a_i x_i^2 = b$. For a more detailed review of all these results and some further extensions, see Nesterov *et al.* [139].

Bomze and De Klerk [27] have considered SDP relaxations of the so-called *standard quadratic optimization* problem:

$$\min_{x \in \Delta} x^T Q x,$$

where Δ is the standard simplex in $\mathbb{R}^n$, namely

$$\Delta := \left\{ x \;\middle|\; \sum_{i=1}^n x_i = 1,\ x_i \geq 0\ (i = 1, \ldots, n) \right\}.$$

The authors showed that this problem allows a polynomial time ϵ-approximation for each $\epsilon > 0$. This improves an earlier result by Nesterov [134] who showed that a 2/3-approximation is always possible. These results are reviewed in Section 12.7.

Quadratic least squares and logarithmic Chebychev approximation

Den Hertog *et al.* [51] showed that the quadratic least squares approximation of a multivariate convex function in a finite number of points is not necessarily a convex quadratic approximation. The best *convex* quadratic approximation (in the least squares sense) can be found using SDP. This is discussed in detail in Chapter 9.

Another class of approximation problems that can be modeled as SDP are *logarithmic Chebychev approximation problems.* Given vectors $a_1, \ldots, a_p \in \mathbb{R}^k$ and $b \in \mathbb{R}^p$, the problem is defined by

$$\min_x \max_{i=1,\ldots,p} \left| \log a_i^T x - \log b_i \right|. \tag{1.6}$$

By observing that

$$\left| \log a_i^T x - \log b_i \right| = \log \left(\max \left\{ \frac{a_i^T x}{b_i}, \frac{b_i}{a_i^T x} \right\} \right),$$

we see that problem (1.6) is equivalent to

$$\min \left\{ t \;\middle|\; \frac{1}{t} \leq \frac{a_i^T x}{b_i} \leq t, \quad i = 1, \ldots, p \right\},$$

which in turn is equivalent to

$$\min \left\{ t \;\middle|\; \begin{bmatrix} t - \frac{a_i^T x}{b_i} & 0 & 0 \\ 0 & \frac{a_i^T x}{b_i} & 1 \\ 0 & 1 & t \end{bmatrix} \succeq 0, \quad i = 1, \ldots, p \right\}.$$

The resulting problem is an SDP problem of dimension $n = 3p$, $m = k + 1$; for more details, see Vandenberghe and Boyd [181].

Nonnegative polynomials

It is well known that a univariate homogeneous polynomial is nonnegative on $\mathbb{R}$ if and only it can be written as a sum of squares. The same is not true in the multivariate case. SDP can be used to decide whether a given homogeneous polynomial has a sum of squares decomposition.

If a multivariate homogeneous polynomial $p(x)$ is positive on $\mathbb{R}^n$, then

$$p(x)\left(\sum_{i=1}^{n} x_i^2\right)^r \tag{1.7}$$

can be written as a sum of squares for r sufficiently large, even if $p(x)$ cannot. This is a celebrated theorem by Polya [146], and related to the famous $17th$ problem by Hilbert.[15]

For a given value of r we can therefore use SDP to decide if the polynomial in (1.7) can be written as a sum of squares. As an application one can give sufficient conditions for copositivity of a matrix. A symmetric matrix A is called *copositive* if

$$x^T Ax \geq 0 \quad \forall x \in \mathbb{R}^n_+,$$

or, equivalently

$$p_A(x) := \sum_{i,j} a_{ij} x_i^2 x_j^2 \geq 0 \quad \forall x \in \mathbb{R}^n.$$

One can therefore give a sufficient condition for copositivity by deciding if the homogeneous polynomial $p_A(x)\left(\sum_{i=1}^n x_i^2\right)^r$ has a sum of squares decomposition for some given $r \geq 1$.

The procedure for finding a sum of squares decomposition is explained on page 193 of this monograph, based on the work by Parillo [141].

1.6 ENGINEERING APPLICATIONS

One of the richest fields of application of SDP is currently *system and control theory*, where SDP has become an established tool. The standard reference for these problems is Boyd *et al.* [31]; a more recent survey was done by Balakrishnan and Wang [13]. Introductory examples are given by Vandenberghe and Boyd [181] and Olkin [140]; the latter reference deals with a problem in *active noise control*: The noise level inside a dome is reduced by emitting sound waves at the same frequency but with a suitable phase shift. The underlying control problem involves optimization over the second order cone, which was shown in Section 1.3 to be a special case of SDP.

Other engineering applications of SDP include *VLSI transistor sizing* and *pattern recognition* using ellipsoids (see Vandenberghe and Boyd [181]).

[15] For a history of Hilbert's 17th problem, see Reznick [159].

An application which receives less attention is *structural design*, where the best known SDP problem involves optimal truss[16] design. Two variants are:

1. minimize the weight of the structure such that its fundamental frequency[17] remains above a critical value;

2. minimize the worst-case compliance ('stored energy') of the truss given a set of forces which the structure has to withstand.

For a recent review, see Ben-Tal and Nemirovski [21].

An important consideration in optimal design is that of *robustness*, *i.e.* one should often allow for uncertainty in the data of the optimization problem and compute the best worst-case scenario. A survey of robust optimization problems that can be formulated as SDP's is given by Ben-Tal *et al.* in [20].

1.7 INTERIOR POINT METHODS

Bearing the links between LP and SDP in mind, it may come as little surprise that interior point algorithms for LP have been successfully extended to SDP.

The field of interior point methods for LP more or less started with the *ellipsoid algorithm* of Khachiyan [101] in 1979, that allowed a polynomial bound on the worst-case iteration count. This resolved the question whether linear programming problems are solvable in polynomial time, but practical experiences with the ellipsoid method were disappointing. The next major development was the famous paper by Karmarkar [100] in 1984, who introduced an algorithm with an improved complexity bound that was also accompanied by claims of computational efficiency. In the following decade several thousand papers appeared on this topic. A major survey of interior point methods for LP was done by Gonzaga [70] (up to 1992). Several new books on the subject have appeared recently, including Roos *et al.* [161], Wright [184] and Ye [188]. It has taken nearly ten years to substantiate the claims of the computational efficiency of interior point methods; several studies have now indicated that these methods have superior performance to state-of-the-art Simplex algorithms on large scale problems (see *e.g.* Lustig *et al.* [119] and more recently Andersen and Andersen [9]).

The first extension of interior point algorithms from LP to SDP was by Nesterov and Nemirovski [137], and independently by Alizadeh [3] in 1991. This explains the resurgence of research interest in SDP in the 1990's. Nesterov and Nemirovski actually considered convex optimization problems in the generic *conic formulation*:

$$\inf_{x} \left\{ c^T x \mid x \in (\mathcal{L} + b) \cap \mathcal{C} \right\}, \tag{1.8}$$

where $\mathcal{L}$ denotes a linear subspace of $\mathbb{R}^n$, $b, c \in \mathbb{R}^n$, and $\mathcal{C}$ is a closed and pointed[18] convex cone with nonempty interior. The associated dual problem is

$$\sup_{y} \left\{ b^T y \mid y \in (\mathcal{L}^{\perp} + c) \cap \mathcal{C}^* \right\},$$

[16] A truss is a structure of bars which connect a fixed ground structure of nodes. (A famous example is the Eiffel tower!)

[17] Frequency at which the structure resonates.

[18] A cone is called pointed if it contains no lines.

where $\mathcal{L}^{\perp}$ is the orthogonal complement of $\mathcal{L}$ in $\mathbf{R}^n$, and $\mathcal{C}^*$ is the *dual cone* of $\mathcal{C}$:

$$\mathcal{C}^* := \{x \mid \langle x, y\rangle \geq 0 \,\forall\, y \in \mathcal{C}\}.$$

Note that the nonlinearity in the problem is 'banished' to a convex cone. In the SDP case this cone is the cone of semidefinite matrices:

$$\mathcal{S}_n^+ := \{X \mid X \in \mathcal{S}_n,\ X \succeq 0\},$$

where $\mathcal{S}_n$ denotes the space of symmetric $n \times n$ matrices. Nesterov and Nemirovski showed that such conic optimization problems can be solved by sequential minimization techniques, where the conic constraint is discarded and a barrier term is added to the objective. Suitable barriers are called *self-concordant.* These are smooth convex functions with second derivatives which are Lipschitz continuous with respect to a local metric (the metric induced by the Hessian of the function itself).

Self-concordant barriers go to infinity as the boundary of the cone is approached, and can be minimized efficiently by Newton's method.[19] Each convex cone C possesses a self-concordant barrier, although such barriers are only computable for some special cones. The function $f(x) = -\sum_{i=1}^n \log(x_i)$ is such a barrier for the positive orthant of $\mathbf{R}^n$, and is instrumental in designing interior point methods for LP. Likewise, the function

$$f_{bar}(X) = -\log\det(X)$$

is a self-concordant barrier for the cone of semidefinite matrices (see Nesterov and Nerovski [137]). Using this barrier, several classes of algorithms may be formulated which have polynomial worst-case iteration bounds for the computation of ϵ-optimal solutions, *i.e.* feasible (X^*, S^*) with duality gap $\mathbf{Tr}\,(X^*S^*) \leq \epsilon$, where $\epsilon > 0$ is a given tolerance; see Section 1.9 for a more precise statement of the complexity results.

LOGARITHMIC BARRIER METHODS

Primal methods Primal log-barrier methods use the projected Newton method to solve a sequence of problems of the form

$$\min_X \{\mathbf{Tr}\,(CX) - \mu \log\det(X) \mid \mathbf{Tr}\,(A_iX) = b_i\ (i = 1, \ldots, m)\},$$

where the parameter μ is sequentially decreased to zero. Such algorithms were analysed by Faybusovich in [54, 55] and later by other authors in He *et al.* [82] and Anstreicher and Fampa [12]. Note that the condition $X \succeq 0$ has been replaced by adding a 'barrier term' to the objective.[20] The condition $X \succeq 0$ is maintained by controlling the Newton process carefully — large decreases of μ necessitate damped Newton steps, while small updates allow full Newton steps.

These results will be reviewed in Chapter 5.

---

[19]The definition of self-concordant barriers will not be used here and is omitted; for an excellent introductory text dealing with self-concordance, see Renegar [158].

[20]This idea actually dates back to the 1960's and the work of Fiacco and McCormick [58]; the implications for complexity theory only became clear two decades later, when Gill *et al.* [63] showed that the method of Karmarkar could be interpreted as a logarithmic barrier method.

Dual methods Dual logarithmic barrier methods are analogous to primal ones, and one solves a sequence of problems

$$\min_{y,S} \left\{ b^T y - \mu \log\det S \;\middle|\; \sum_{i=1}^{m} y_i A_i + S = C \right\},$$

where the parameter μ is again sequentially decreased to zero.

There has been a resurgence in interest in these methods recently. The reason is that one can exploit the sparsity structure of problems like the MAX-CUT relaxation (1.3) more efficiently than with other interior point methods. Benson *et al.* [23, 22] and Choi and Ye [34] have implemented the dual method and report results for the MAX-CUT relaxation (1.3) of sparse graphs with up to 14000 vertices. These developments are reviewed in Chapter 5, Section 5.10.

Primal–dual methods Following the trend in LP, so-called *primal-dual* methods for SDP have received a great deal of attention. These methods minimize the duality gap

$$\mathbf{Tr}\,(CX) - b^T y = \mathbf{Tr}\,(XS),$$

and employ the combined primal-dual barrier function

$$-\left(\log\det(X) + \log\det(S)\right) = -\log\det(XS).$$

This means that a sequence of problems of the following form are solved

$$\min_{X,y,S} \left\{ \mathbf{Tr}\,(XS) - \mu \log\det(XS) \;:\; \mathbf{Tr}\,(A_i X) = b_i \;\; \forall\, i,\; \sum_{i=1}^{m} y_i A_i + S = C \right\}. \tag{1.9}$$

The minimizers of problem (1.9) satisfy

$$\left.\begin{array}{rcl} \mathbf{Tr}\,(A_i X) & = & b_i, \quad i = 1, \ldots, m \\ \sum_{i=1}^{m} y_i A_i + S & = & C \\ XS & = & \mu I \\ X, S & \succ & 0. \end{array}\right\} \tag{1.10}$$

These equations can be viewed as a perturbation of the optimality conditions of (P) and (D), where $\mu = 0$. System (1.10) has a unique solution under the assumptions that the A_i's $(i = 1, \ldots, m)$ are linearly independent, and that there exist positive definite feasible solutions of (P) and (D). This solution will be denoted by $X(\mu), S(\mu), y(\mu)$, and can be interpreted as the parametric representation of a smooth curve (the *central path*) in terms of the parameter μ. The properties of the central path are reviewed in detail in Chapter 3.

Logarithmic-barrier methods are also called *path-following* methods, due to the relation between the central path and the log-barrier function.

Primal-dual log-barrier methods solve the system (1.10) approximately, followed by a reduction in μ. Ideally, the goal is to obtain primal and dual directions ΔX and ΔS, respectively, that satisfy $X + \Delta X \succeq 0$, $S + \Delta S \succeq 0$ as well as

$$\left.\begin{array}{rcl} \mathbf{Tr}\,(A_i \Delta X) & = & 0, \quad i = 1, \ldots, m \\ \sum_{i=1}^{m} \Delta y_i A_i + \Delta S & = & 0 \\ (X + \Delta X)(S + \Delta S) & = & \mu I \\ \Delta X & = & \Delta X^T. \end{array}\right\} \tag{1.11}$$

The third equation in (1.11) is nonlinear, and the system is overdetermined.

Perhaps the most straightforward solution approach is to linearize the system (1.11) and subsequently find a least squares solution of the resulting overdetermined linear system (the Gauss–Newton method). This approach was explored by Kruk *et al.* [109], and a variation thereof by De Klerk *et al.* [43].

Another approach was taken by Zhang [192], who suggested to replace the nonlinear equation in (1.11) by

$$H_P\,(\Delta X S + X \Delta S) = \mu I - H_P(XS), \tag{1.12}$$

where H_P is the linear transformation given by

$$H_P(M) := \frac{1}{2}\left[PMP^{-1} + P^{-T} M^T P^T\right],$$

for any matrix M, and where the *scaling matrix* P determines the symmetrization strategy. Some popular choices for P are listed in Table 1.1. If one replaces the

P	Reference
$\left[X^{\frac{1}{2}} \left(X^{\frac{1}{2}} S X^{\frac{1}{2}}\right)^{-\frac{1}{2}} X^{\frac{1}{2}}\right]^{\frac{1}{2}}$	Nesterov and Todd [138];
$X^{-\frac{1}{2}}$	Monteiro [124], Kojima *et al.* [108];
$S^{\frac{1}{2}}$	Monteiro [124], Helmberg *et al.* [84], Kojima *et al.* [108];
I	Alizadeh, Haeberley and Overton [6];

Table 1.1. Choices for the scaling matrix P.

nonlinear equation in (1.11) by (1.12), and drops the requirement $\Delta X = \Delta X^T$, then we obtain a square linear system. Moreover, if this system has a solution, then ΔX is necessarily symmetric. The proof of the existence and uniqueness of each of the resulting search directions from Table 1.1 was done by Shidah *et al.* in [165].[21]

[21] For $P = I$ existence and uniqueness is not always guaranteed; a sufficient condition is $XS + SX \succeq 0$.

The conspicuous entry $P = \left[X^{\frac{1}{2}}\left(X^{\frac{1}{2}}SX^{\frac{1}{2}}\right)^{-\frac{1}{2}}X^{\frac{1}{2}}\right]^{\frac{1}{2}}$ in Table 1.1 warrants some comment. Nesterov and Todd [138] showed[22] that for each pair $X \succ 0$, $S \succ 0$ there exists a matrix D such that

$$\nabla^2 f_{bar}(D)X = S.$$

It is shown in Appendix C that the Hessian $\nabla^2 f_{bar}(D)$ is the linear operator which satisfies

$$\nabla^2 f_{bar}(D) : X \mapsto D^{-1}XD^{-1}.$$

It follows that $X = DSD$, from which it easily follows that $D = P^2$. In this way we obtain the *symmetric primal-dual scaling* $P^{-1}XP^{-1} = PSP$.

The list of primal–dual search directions given here is by no means complete; for a more detailed survey, see Todd [172].

Algorithms differ in how μ is updated, and how the symmetrized equations are solved. Methods that use large reductions of μ followed by several damped Newton steps are called long step (or large update) methods. These are analysed in Jiang [96], Monteiro [124], and Sturm and Zhang [169].

Methods that use dynamic updates of μ include the popular *predictor-corrector* methods. References include [6, 107, 148, 170]. Superlinear convergence properties of predictor-corrector schemes are studied in [148, 105, 117].

There have been several implementations of predictor–corrector algorithms for SDP, including SeDuMi by Sturm [167] and SDPT3 by Toh *et al.* [175].

We will review some primal–dual path–following methods that use the direction due to Nesterov and Todd [138] in Chapter 7.

AFFINE–SCALING METHODS

Affine–scaling algorithms for LP have been of interest since it became clear that Karmarkar's algorithm was closely related to the primal affine–scaling method of Dikin [53] (from 1967!). In fact, modifications of Karmarkar's algorithm by Vanderbei *et al.* [182] and Barnes [15] proved to be a rediscovery of the primal affine–scaling method.

The primal affine–scaling direction for SDP minimizes the primal objective over an ellipsoid that is inscribed in the primal feasible region. Surprisingly, Muramatsu [130] has shown that an SDP algorithm using this search direction may converge to a non-optimal point, regardless of which step length is used. This is in sharp contrast to the LP case, and shows that extension of algorithms from LP to SDP cannot always be taken for granted.

Two primal-dual variants of the affine–scaling methods were extended by De Klerk *et al.* in [48] from LP to SDP. These algorithms minimize the duality gap over ellipsoids in the scaled primal-dual space, where the matrix $P = D^{\frac{1}{2}}$ is used for the scaling. One of the two methods is the *primal-dual affine–scaling method*, where the

[22] This result was actually proved in the more general setting of conic optimization problems where the cone $\mathcal{C}$ in (1.8) is self-dual, *i.e.* $\mathcal{C} = \mathcal{C}^*$. The interested reader is referred to Nesterov and Todd [138].

search direction is obtained by using $\mu = 0$ in (1.12). The primal-dual affine–scaling method can fail if the scaling $P = I$ from Table 1.1 is used (instead of $P = D^{\frac{1}{2}}$). This was recently proved by Muramatsu and Vanderbei [131]. We review these results in Chapter 6.

PRIMAL-DUAL POTENTIAL REDUCTION METHODS

These algorithms are based on the so-called Tanabe–Todd–Ye potential function

$$\Phi(X,S) = (n + \nu\sqrt{n}) \log(\mathbf{Tr}(XS)) - \log(\det(XS)) - n \log n,$$

where $\nu \geq 1$. In order to obtain a polynomial complexity bound it is sufficient to show that Φ can be reduced by a constant at each iteration. A survey of algorithms which achieve such a reduction is given by Vandenberghe and Boyd [181]. We will review this methodology in Chapter 8, and give the analysis of a potential reduction method due to Nesterov and Todd [138]. An implementation of this method was done by Vandenberghe and Boyd [179] (the SP software package).

INFEASIBLE–START METHODS

Algorithms that do not require a feasible starting point are usually called *infeasible–start methods*. Traditional big-M initialization strategies are reviewed in Vandenberghe and Boyd [181]; other references for infeasible-start methods include [105, 117, 148].

An elegant way of avoiding big-M parameters is to embed the SDP problem in a larger problem that is essentially its own dual, and for which a feasible starting point is known. A solution of this *self-dual embedding* problem then gives information about the solution of the original problem. The idea of self-dual embeddings for LP dates back to the 1950's and the work of Goldman and Tucker [69]. With the arrival of interior point methods for LP, the embedding idea was revived to be used in infeasible–start algorithms by Ye *et al.* [191].

The idea of embedding the SDP problem in an extended self-dual problem with a known starting point on the central path was investigated for SDP by De Klerk *et al.* [44] and independently by Luo *et al.* [118], as extensions of the work by Potra and Sheng [148], who used so-called *homogeneous embeddings*.

The embedding approach will be described in detail in Chapter 4.

1.8 OTHER ALGORITHMS FOR SDP

Although the mathematical analysis of interior point methods for LP and SDP is quite similar, the implementational issues are different. In particular, in the LP case one can exploit sparsity in the data very efficiently, and sparse problems with hundreds of thousands of variables and constraints can be solved routinely using primal–dual predictor–corrector methods. In the SDP case this is much more difficult, and the largest problems that can be solved by primal–dual predictor–corrector methods involve matrices of size of the order of a thousand rows, and a thousand constraints (even if the data matrices are very sparse). For a detailed discussion on this issue, see Fujisawa *et al.* [60].

For this reason there has been interest in methods that do not require computation of the Hessian (or even the gradient) of the barrier function $f_{bar}(X) = \log\det(X)$.

- *Bundle methods* (zero–order methods) for eigenvalue optimization and related problems are surveyed by Helmberg and Oustry [83];

- *First order methods* (that use only gradient information) have been implemented by Burer and Monteiro [32] for the MAX-CUT relaxation (1.3). These methods are used to solve a (nonconvex) nonlinear programming reformulation of the SDP problem. The drawback with the nonconvex formulation is that one cannot guarantee convergence to an optimal solution.

1.9 THE COMPLEXITY OF SDP

In this section we review known results on the computational complexity of SDP, based on the review by Ramana and Pardalos [156].

We first consider the bit (Turing machine) model of computation.[23]

Consider an SDP instance in the standard primal form (P) (see page 2) with integer data and feasible solution set $\mathcal{P}$, and let a rational $\epsilon > 0$ be given.

- If an integer $R > 0$ is known a priori such that either $\mathcal{P} = \emptyset$ or $\|X\| \leq R$ for some $X \in \mathcal{P}$, then one can find an X^* at distance at most ϵ from $\mathcal{P}$, such that $|\mathbf{Tr}\ (CX^*) - p^*| \leq \epsilon$, or a certificate that $\mathcal{P}$ does not contain a ball of radius ϵ. The complexity of the procedure is polynomial in n, m, $\log(R)$, $\log\left(\frac{1}{\epsilon}\right)$, and the bit length of the input data. This result follows from the complexity of the ellipsoid algorithm of Khachiyan [101] (see also Grötschel *et al.* [73]);

- If a rational $X \in \mathcal{P}$ is given such that $X \succ 0$, and an integer $R > 0$ is known a priori such that $\|X\| \leq R$ for all $X \in \mathcal{P}$, then one can compute a rational $X^* \in \mathcal{P}$ such that $\mathbf{Tr}\ (CX^*) - p^* \leq \epsilon$ using interior point methods. The complexity of this procedure is polynomial in n, m, $\log(R)$, $\log\left(\frac{1}{\epsilon}\right)$, the bit length of the input data, and the bit length of X (Nesterov and Nemirovski [137]).[24]

We can also ask what the complexity is of obtaining an exact optimal solution (as opposed to an ϵ-optimal solution). Consider the SDP instance

$$\max y$$

subject to

$$\begin{bmatrix} 2 & y \\ y & 1 \end{bmatrix} \succeq 0.$$

[23] See *e.g.* Garey and Johnson [61] for a discussion of the bit model of computation.

[24] As mentioned by Ramana and Pardalos [156], a polynomial bound has not been established for the bit lengths of the intermediate numbers occurring in the interior point algorithms. Strictly speaking, this should be done when deriving complexity results in the bit model of computation.

Note that the data is integer, but the unique optimal solution $y = \sqrt{2}$ is irrational. It is therefore not meaningful to speak of the complexity of SDP in the bit model of computation, since the output is not always representable in this model.

However, we can still consider the semidefinite feasibility problem (SDFP):

Does there exist a $y \in \mathbb{R}^m$ such that

$$C - \sum_{i=1}^{m} y_i A_i \succeq 0?$$

(Here the data matrices are integer.) The following is known about the complexity of SDFP (Ramana [153]):

- In the bit model of computation: Either SDFP $\in$ NP $\cap$ co-NP or SDFP $\notin$ NP $\cup$ co-NP;

- In the real number model of computation of Blum, Shub, and Smale [24]: SDFP $\in$ NP $\cap$ co-NP.

In other words, the complexity of SDFP is not known, but it cannot be an NP-complete problem, unless NP = co-NP. Porkolab and Khachiyan [147] have proven that SDFP $\in$ P (in the bit model) if either m or n is a fixed constant. The complexity question for SDFP is recognized as one of the most important open problems in the field of semidefinite programming.

1.10 REVIEW LITERATURE AND INTERNET RESOURCES

The presentation in this chapter was loosely based on the short survey by De Klerk *et al.* [45]. The seminal work of Nesterov and Nemirovski [137] contains a section on special cases of SDP problems (§6.4), as well as the development of an entire interior point methodology. An excellent survey by Vandenberghe and Boyd [181] deals with basic theory, diverse applications, and potential reduction algorithms (up to 1995). Three more recent surveys that focus more on applications of SDP in combinatorial optimization are by Alizadeh [4], Ramana and Pardalos [156] and Goemans [65]. The paper [4] also deals with interior point methodology, while [156] contains a survey of geometric properties of the SDP feasible set (so-called spectrahedra), as well as complexity and duality results. Lewis and Overton [114] give a nice historical perspective on the development of SDP and focus on the relation with eigenvalue optimization. Most recently, the *Handbook of Semidefinite Programming* [183] contains an extensive review of both the theory and applications of SDP up to the year 2000.

Christoph Helmberg maintains a web-site for SDP with links to the latest papers and software at

`http://www.zib.de/helmberg/semidef.html`

A web-site with benchmarks of different SDP solvers is maintained by Hans Mittelmann at

`http://plato.la.asu.edu/bench.html`

I THEORY AND ALGORITHMS

2
DUALITY, OPTIMALITY, AND DEGENERACY

Preamble

All convex optimization problems can in principle be restated as so-called *conic linear programs* (conic LP's for short); these are problems where the objective function is linear, and the feasible set is the intersection of an affine space with a convex cone. For conic LP's, all nonlinearity is therefore hidden in the definition of the convex cone. Conic LP's also have the strong duality property under a constraint qualification: if the affine space intersects the relative interior of the cone, it has a solvable dual with the same optimal value (if the dual problem is feasible).

A special subclass of conic LP's is formed if we consider cones which are self-dual. There are three such cones over the reals: the positive orthant in $\mathbf{R}^n$, the Lorentz (or ice-cream or second order) cone, and the positive semidefinite cone. These cones respectively define the conic formulation of linear programming (LP) problems, second order cone (SOC) programming problems, and semidefinite programming (SDP) problems. The self-duality of these cones ensures a perfect symmetry between primal and dual problems, *i.e.* the primal and dual problem can be cast in exactly the same form. As discussed in Chapter 1, LP and SCO problems may be viewed as special cases of SDP.

Some fundamental theoretical properties of semidefinite programs (SDP's) will be reviewed in this chapter. We define the standard form for SDP's and derive the associated dual problem. The classical weak and strong duality theorems are proved to obtain necessary and sufficient optimality conditions for the standard form SDP.

Subsequently we review the concepts of degeneracy and maximal complementarity of optimal solutions.

2.1 PROBLEMS IN STANDARD FORM

We say that a problem (P) and its Lagrangian dual (D) are in *standard form* if they are in the form

$$(P): \; p^* := \inf_X \left\{ \mathbf{Tr}\,(CX) \mid \mathbf{Tr}\,(A_i X) = b_i \; (i = 1, \ldots, m), \; X \in \mathcal{S}_n^+ \right\},$$

and

$$(D): \; d^* := \sup_{y,S} \left\{ b^T y \;\middle|\; \sum_{i=1}^m y_i A_i + S = C, \;\; S \in \mathcal{S}_n^+, \; y \in \mathbf{R}^m \right\}.$$

We will refer to X and (y, S) as *feasible solutions* as they satisfy the primal and dual constraints respectively. The primal and dual feasible sets (*i.e.* sets of feasible solutions) will be denoted by $\mathcal{P}$ and $\mathcal{D}$ respectively:

$$\mathcal{P} := \left\{ X \mid \mathbf{Tr}\,(A_i X) = b_i \; (i = 1, \ldots, m), \; X \succeq 0 \right\},$$

and

$$\mathcal{D} := \left\{ (y, S) \;\middle|\; \sum_{i=1}^m y_i A_i + S = C, \;\; S \succeq 0, \; y \in \mathbf{R}^m \right\}.$$

Similarly, $\mathcal{P}^*$ and $\mathcal{D}^*$ will denote the respective optimal sets (*i.e.* sets of optimal solutions):

$$\mathcal{P}^* := \left\{ X \in \mathcal{P} \mid \mathbf{Tr}\,(CX) = p^* \right\} \text{ and } \mathcal{D}^* := \left\{ (S, y) \in \mathcal{D} \mid b^T y = d^* \right\}.$$

The values p^* and d^* will be called the *optimal values* of (P) and (D), respectively. We use the convention that $p^* = -\infty$ if (P) is *unbounded* and $p^* = \infty$ if (P) is *infeasible* ($\mathcal{P} = \emptyset$), with the analogous convention for (D).

A problem (P) (resp. (D)) is called *solvable* if $\mathcal{P}^*$ (resp. $\mathcal{D}^*$) is nonempty.

It is not hard to see that (D) is indeed the Lagrangian dual of (P). Note that

$$\begin{aligned} p^* &= \inf_{X \succeq 0} \sup_{y \in \mathbf{R}^m} \left\{ \mathbf{Tr}\,(CX) - \sum_{i=1}^m y_i \left(\mathbf{Tr}\,(A_i X) - b_i \right) \right\} \qquad (2.1) \\ &= \inf_{X \succeq 0} \sup_{y \in \mathbf{R}^m} \left\{ \mathbf{Tr}\left(\left(C - \sum_{i=1}^m y_i A_i \right) X \right) + b^T y \right\}. \end{aligned}$$

The Lagrangian dual of (P) is obtained by interchanging the 'sup' and 'inf', to obtain the problem:

$$(D'): \quad \sup_{y \in \mathbf{R}^m} \left\{ b^T y + \inf_{X \succeq 0} \left\{ \mathbf{Tr}\left(\left(C - \sum_{i=1}^m y_i A_i \right) X \right) \right\} \right\}. \qquad (2.2)$$

The inner minimization problem in (D') will be bounded from below if and only if

$$\mathbf{Tr}\left(\left(C-\sum_{i=1}^{m}y_iA_i\right)X\right)\geq 0 \quad \forall\, X\succeq 0,$$

i.e. if and only if

$$C-\sum_{i=1}^{m}y_iA_i\succeq 0.$$

In this case the inner minimization problem in (D') has optimal value zero. Problem (D') can therefore be rewritten as

$$\sup_{y\in\mathbb{R}^m}\left\{b^Ty \;\middle|\; C-\sum_{i=1}^{m}y_iA_i\succeq 0\right\}.$$

Defining $S:=C-\sum_{i=1}^{m}y_iA_i$ we obtain problem (D).

ASSUMPTIONS

The following assumption will be made throughout this monograph.

Assumption 2.1 *The matrices A_i $(i=1,\ldots,m)$ are linearly independent.*

Under Assumption 2.1, y is uniquely determined for a given dual feasible S. Assumption 2.1 therefore allows us to use the shorthand notation: $y\in\mathcal{D}$ or $S\in\mathcal{D}$ instead of $(y,S)\in\mathcal{D}$. It is essentially the same assumption as the assumption in LP that the constraint matrix must have full rank. To see this, note that the linear independence of A_i $(i=1,\ldots,m)$ is equivalent to the linear independence of $\mathbf{vec}(A_i)$ $(i=1,\ldots,m)$. The assumption can therefore be made without loss of generality.[1]

Another assumption which will be used when specified is the assumption of *strict feasibility*.

Assumption 2.2 (Strict feasibility) *There exist $X\in\mathcal{P}$ and $S\in\mathcal{D}$ such that $X\succ 0$ and $S\succ 0$.*

Strict feasibility is also called *strong feasibility*, *Slater's constraint qualification* or *Slater's regularity condition*. Assumption 2.2 is sometimes also referred to as the *interior point assumption*.

Note that the assumption is consistent with the usual definition of Slater regularity for convex optimization (Assumption B.1 in Appendix B), if we use the fact that the (relative) interior of $\mathcal{S}_n^+$ is given by the $n\times n$ symmetric positive definite matrices. Note that if Assumption 2.2 holds, then the relative interior of the primal–dual feasible set is given by (see Theorem B.3 on page 239):

$$\mathrm{ri}\,(\mathcal{P}\times\mathcal{D})=\left\{(X,S)\in\mathcal{P}\times\mathcal{D}\mid X\succ 0,\; S\succ 0\right\}.$$

We will see in the next section that Assumption 2.2 guarantees that $\mathcal{P}^*\neq\emptyset$, $\mathcal{D}^*\neq\emptyset$ and $p^*=d^*$. For the time being we do not make this assumption, however.

[1] Practical algorithms for ensuring full row rank of a matrix are described by Andersen [8].

ORTHOGONALITY OF FEASIBLE DIRECTIONS

The following *orthogonality property* for (P) and (D) is easily proven, and will be used extensively.

Lemma 2.1 (Orthogonality) *Let $(X,(y,S)) \in \mathcal{P} \times \mathcal{D}$ and $(X^0,(y^0,S^0)) \in \mathcal{P} \times \mathcal{D}$ be two pairs of feasible solutions. Denoting $\Delta X = X - X^0$ and $\Delta S = S - S^0$, one has:* $\mathbf{Tr}\,(\Delta X \Delta S) = 0$.

Proof:
By the definition of ΔS and $\mathcal{D}$:

$$\Delta S = \sum_{i=1}^{m} (y_i^0 - y_i) A_i,$$

i.e.

$$\Delta S \in \text{span}\,\{A_1, \ldots, A_m\}.$$

Similarly, by the definition of ΔX and $\mathcal{P}$ one has:

$$\mathbf{Tr}\,(A_i \Delta X) = \mathbf{Tr}\,(A_i X) - \mathbf{Tr}\,(A_i X^0) = b_i - b_i = 0, \quad i = 1, \ldots, m,$$

which implies

$$\Delta X \in \text{span}\,\{A_1, \ldots, A_m\}^{\perp}.$$

This shows that ΔS lies in the subspace of $\mathcal{S}_n$ spanned by the matrices A_i $(i = 1, \ldots, m)$, and ΔX lies in the orthogonal complement of this subspace. □

Now let

$$\mathcal{L} := \text{span}\,\{A_1, \ldots, A_m\}. \tag{2.3}$$

Given a strictly feasible $X \in \text{ri}\,(\mathcal{P})$ we call ΔX a *feasible direction* at X if $\Delta X \in \mathcal{L}^{\perp}$. Similarly ΔS is a feasible direction at a strictly feasible $S \in \text{ri}\,(\mathcal{D})$ if $\Delta S \in \mathcal{L}$. The idea is that there exists an $\alpha > 0$ (called the *step length*) in this case so that $X + \alpha \Delta X \in \mathcal{P}$ and $S + \alpha \Delta S \in \mathcal{D}$. For this reason we also refer to $X + \alpha \Delta X$ and $S + \alpha \Delta S$ as *feasible steps*.

SYMMETRIC PROBLEM REFORMULATION

It is sometimes useful to reformulate (P) and (D) in in a symmetric *conic formulation*, where both problems have exactly the same form. To this end, let $M \in \mathcal{S}_n$ be such that $\mathbf{Tr}\,(A_i M) = b_i$ $(i = 1, \ldots, m)$ and $\mathbf{Tr}\,(CM) = 0$. Then (P) has the alternative formulation

$$(P'): \quad p^* = \inf_X \left\{ \mathbf{Tr}\,(CX) \mid X \in \mathcal{L}^{\perp} + M,\ X \succeq 0 \right\}, \tag{2.4}$$

and the Lagrangian dual of this problem is the conic reformulation of (D):

$$(D'): \quad d^* = \sup_S \left\{ \mathbf{Tr}\,(-MS) \mid S \in \mathcal{L} + C,\ S \succeq 0 \right\}, \tag{2.5}$$

where $\mathcal{L}$ is defined in (2.3).

Note that $X \in \mathcal{P}$ if and only if X is feasible for (P'). Similarly, one has $S \in \mathcal{D}$ if and only if S is feasible for (D'), and $b^T y = \mathbf{Tr}\,(-MS)$ if $S = C - \sum_{i=1}^{m} y_i A_i$.

2.2 WEAK AND STRONG DUALITY

The difference between the primal and dual objective values at feasible solutions of (P) and (D) is called the *duality gap.*

Definition 2.1 (Duality gap) *Let $X \in \mathcal{P}$ and $(y, S) \in \mathcal{D}$. The quantity*

$$\mathbf{Tr}\,(CX) - b^T y$$

is called the duality gap of (P) and (D) at (X, y, S).

By the definitions of (P) and (D) one has for feasible (X, y, S):

$$\mathbf{Tr}\,(CX) - b^T y = \mathbf{Tr}\left(\left(\sum_{i=1}^{m} y_i A_i + S\right) X\right) - \sum_{i=1}^{m} y_i \mathbf{Tr}\,(A_i X) = \mathbf{Tr}\,(SX) \geq 0,$$

where the inequality follows from $X \succeq 0$ and $S \succeq 0$ (see Theorem A.8 in Appendix A). The nonnegativity of the duality gap is called the *weak duality* property, which we can state as a theorem.

Theorem 2.1 (Weak duality) *Let $X \in \mathcal{P}$ and $(y, S) \in \mathcal{D}$. One has*

$$\mathbf{Tr}\,(CX) - b^T y = \mathbf{Tr}\,(SX) \geq 0, \tag{2.6}$$

i.e. *the duality gap is nonnegative at feasible solutions.*

Of course, the weak duality relation between (P) and (D) also follows from the fact that (D) is the Lagrangian dual of (P). Indeed, one can easily prove that for any function $f : \mathcal{S}_1 \times \mathcal{S}_2 \mapsto \mathbf{R}$ where $\mathcal{S}_1$ and $\mathcal{S}_2$ are arbitrary subsets of $\mathbf{R}^n$, there holds

$$\inf_{x \in \mathcal{S}_1} \sup_{y \in \mathcal{S}_2} f(x, y) \geq \sup_{y \in \mathcal{S}_2} \inf_{x \in \mathcal{S}_1} f(x, y).$$

Applying this inequality to (2.1) and (2.2) yields that $p^* \geq d^*$. However, the direct proof of Theorem 2.1 has the additional advantage that we get the useful expression (2.6) for the duality gap.

Definition 2.2 (Perfect duality) *The problems (P) and (D) are said to be in* perfect duality *if $p^* = d^*$.*

Note that this definition does not imply that $\mathcal{P}^*$ and $\mathcal{D}^*$ are nonempty. If $\mathcal{D}^*$ is also nonempty, then we say that *strong duality* holds for (P) and its dual (D).

Example 2.1 (Adapted from Vandenberghe and Boyd [181]) *This example shows a pair of dual problems for which perfect duality holds but $\mathcal{P}^* = \emptyset$. Consider the following problem in the standard dual form:*

$$(D): \quad \sup_{y \in \mathbf{R}^2} \; y_2$$

subject to

$$y_1 \begin{bmatrix} 1 & 0 \\ 0 & 0 \end{bmatrix} + y_2 \begin{bmatrix} 0 & 0 \\ 0 & 1 \end{bmatrix} \preceq \begin{bmatrix} 0 & 1 \\ 1 & 1 \end{bmatrix}.$$

This problem is not solvable but $\sup_{y \in \mathcal{D}} y_2 = 1$. *Its dual problem (which is in the standard primal form) takes the form:*

$$(P): \quad \min_X \mathbf{Tr} \left(\begin{bmatrix} 0 & 1 \\ 1 & 1 \end{bmatrix} X \right)$$

subject to

$$X = \begin{bmatrix} 0 & x_{12} \\ x_{12} & 1 \end{bmatrix} \succeq 0.$$

Note that $X \succeq 0$ *implies* $x_{12} = 0$ *so that*

$$X^* = \begin{bmatrix} 0 & 0 \\ 0 & 1 \end{bmatrix} \succeq 0,$$

is the unique optimal solution with optimal value 1. □

If (P) (resp. (D)) is strictly feasible, and (D) (resp. (P)) is feasible, then perfect duality holds *and* $\mathcal{D}^* \neq \emptyset$ (resp. $\mathcal{D}^* \neq \emptyset$). The proof of this theorem requires a fundamental result from convex analysis, namely the *separation theorem for convex sets* (Theorem B.4 in Appendix B). This theorem essentially states that two convex sets in $\mathbf{R}^n$ can be separated by a hyperplane if and only if their relative interiors are disjoint.

Theorem 2.2 (Strong duality) *Assume that* $d^* < \infty$ *(resp.* $p^* > -\infty$*). Further assume that* (D) *(resp.* (P)*) is strictly feasible. It now holds that* $\mathcal{P}^* \neq \emptyset$ *(resp.* $\mathcal{D}^* \neq \emptyset$*) and* $p^* = d^*$.

Proof:
We will first consider the case where $d^* < \infty$ and (D) is strictly feasible. The proof is trivial if $b = 0$ since then $X^* := 0$ is optimal for (P). We can therefore assume $b \neq 0$.

Let us define the (nonempty) convex set

$$\mathcal{M} = \left\{ S \in \mathcal{S}_n \;\middle|\; S = C - \sum_{i=1}^{m} y_i A_i, \; b^T y \geq d^*, \; y \in \mathbf{R}^m \right\}.$$

The relative interiors of $\mathcal{M}$ and $\mathcal{S}_+^n$ are disjoint, by construction. To see this, recall that the (relative) interior of $\mathcal{S}_n^+$ consists of all symmetric positive definite matrices. If there exists a positive definite $S \in \mathcal{M}$ then d^* obviously cannot be the optimal value of (D).

We can now apply Theorem B.4 by associating $\mathcal{S}_n$ with $\mathbb{R}^{\frac{1}{2}n(n+1)}$ in the usual way, and using that $\mathbf{svec}(A)^T\mathbf{svec}(B) = \mathbf{Tr}\,(AB)$ for $A, B \in \mathcal{S}_n$. Thus we find that there exists a (nonzero) $\Lambda \in \mathcal{S}_n$ such that

$$\sup_{S \in \mathcal{M}} \mathbf{Tr}\,(S\Lambda) \leq \inf_{U \in \mathcal{S}_n^+} \mathbf{Tr}\,(U\Lambda). \tag{2.7}$$

Since $\mathcal{S}_n^+$ is a cone, the infimum on the right-hand side must be either $-\infty$ or zero. Since $\mathcal{M} \neq \emptyset$, the latter case applies, *i.e.*

$$\inf_{U \in \mathcal{S}_n^+} \mathbf{Tr}\,(U\Lambda) = 0,$$

which shows that $\Lambda \succeq 0$, and therefore by (2.7) one has:

$$\sup_{S \in \mathcal{M}} \mathbf{Tr}\,(S\Lambda) \leq 0. \tag{2.8}$$

Each $y \in \mathbf{R}^m$ satisfying $b^T y \geq d^*$ defines an $S \in \mathcal{M}$, and

$$\begin{aligned} -\sum_{i=1}^m y_i \mathbf{Tr}\,(A_i\Lambda) &= \mathbf{Tr}\,((S - C)\Lambda) \\ &= \mathbf{Tr}\,(S\Lambda) - \mathbf{Tr}\,(C\Lambda) \\ &\leq -\mathbf{Tr}\,(C\Lambda), \end{aligned} \tag{2.9}$$

where the inequality follows from (2.8). In other words, the linear function

$$f(y) = \sum_{i=1}^m y_i \mathbf{Tr}\,(A_i\Lambda)$$

is bounded from below on the half space defined by $b^T y \geq d^*$. This is only possible if

$$\mathbf{Tr}\,(A_i\Lambda) = \beta b_i, \quad i = 1, \ldots, m, \tag{2.10}$$

for some $\beta \geq 0$. If $\beta = 0$, then

$$\mathbf{Tr}\,(A_i\Lambda) = 0, \quad i = 1, \ldots, m,$$

and consequently $\mathbf{Tr}\,(C\Lambda) \leq 0$ by (2.9). At this point we use the assumption of strict feasibility: there exist $(y^0, S^0) \in \mathcal{D}$ with $S^0 \succ 0$. Taking the inner product of S^0 and Λ yields:

$$\begin{aligned} \mathbf{Tr}\,\left(S^0\Lambda\right) &\equiv \mathbf{Tr}\,(C\Lambda) - \sum_{i=1}^m y_i^0 \mathbf{Tr}\,(A_i\Lambda) \\ &= \mathbf{Tr}\,(C\Lambda) \leq 0. \end{aligned}$$

This is a contradiction: $\Lambda \succeq 0$ and $S^0 \succ 0$ imply that $\mathbf{Tr}(\Lambda S^0) > 0$. Thus $\beta > 0$ and we can define

$$X^* := \frac{1}{\beta}\Lambda \succeq 0.$$

It therefore holds that $\mathbf{Tr}(A_i X^*) = b_i$ $(i = 1, \ldots, m)$ by (2.10), and

$$\mathbf{Tr}(CX^*) \leq b^T y \quad \forall y \text{ such that } b^T y \geq d^*,$$

by (2.9), which implies $\mathbf{Tr}(CX^*) \leq d^*$. Consequently, $\mathbf{Tr}(CX^*) = d^*$ by the weak duality theorem (Theorem 2.1). We conclude that $X^* \in \mathcal{P}^*$, which completes the first part of the proof.

It remains to prove the analogous result where (P) is assumed bounded and strictly feasible. The easiest way to do this is to use the symmetric reformulation of (P) and (D) as described in Section 2.1. The symmetry implies that we can apply the result for (D) in the first part of this proof to (P) as well; thus we conclude that strict feasibility and boundedness of one of (P) or (D) imply solvability of the other and $p^* = d^*$. This completes the proof. □

As a consequence of Theorem 2.2 we have the following useful result, which gives a sufficient condition for strong duality.

Corollary 2.1 *Let (P) and (D) be strictly feasible. Then $p^* = d^*$ and both optimal sets $\mathcal{P}^*$ and $\mathcal{D}^*$ are nonempty.*

Proof:
Since (P) and (D) are feasible, both p^* and d^* are finite, by Theorem 2.1 (weak duality). The boundedness and strict feasibility of both problems imply that the optimal sets $\mathcal{P}^*$ and $\mathcal{D}^*$ are nonempty and $p^* = d^*$, by Theorem 2.2 (strong duality). □

THE CONIC DUALITY THEOREM FOR MORE GENERAL CONES

The proof of Theorem 2.2 can actually be extended to cover more general conic LP's. In other words, we can replace the positive semidefinite cone by a more general convex cone. To this end, let $\mathcal{K}$ be a closed convex cone with dual cone $\mathcal{K}^*$, and define the primal and dual pair of conic linear programs:

$$(P_{\mathcal{K}}) \quad p^* := \inf_X \left\{ \mathbf{Tr}(CX) \mid \mathbf{Tr}(A_i X) = b_i \ (i = 1, \ldots, m),\ X \in \mathcal{K} \right\}$$

$$(D_{\mathcal{K}}) \quad d^* := \sup_{y \in \mathbf{R}^m} \left\{ b^T y \;\middle|\; \sum_{i=1}^m y_i A_i + S = C,\ S \in \mathcal{K}^* \right\}.$$

If $\mathcal{K} = \mathcal{S}_n^+$ we return to the SDP case.

Theorem 2.3 (General conic duality theorem) *If there exists a feasible solution $X^0 \in \mathrm{ri}(\mathcal{K})$ of $(P_{\mathcal{K}})$, and a feasible solution of $(D_{\mathcal{K}})$, then $p^* = d^*$ and the supremum in (D) is attained. Similarly, if there exist feasible y^0, S^0 for $(D_{\mathcal{K}})$ where $S^0 \in \mathrm{ri}(\mathcal{K}^*)$, and a feasible solution of $(P_{\mathcal{K}})$, then $p^* = d^*$ and the infimum in $(P_{\mathcal{K}})$ is attained.*

STRONGER DUALITY THEORIES

It is also possible to formulate strong duality results without requiring any regularity condition, via a procedure known as *regularization*. It is important, however, to notice that a feasible problem that does not satisfy the Slater condition is ill-posed in the sense that an arbitrary small perturbation of the problem data can change its status from feasible to infeasible. We give a brief review of regularized duals in Appendix G.

2.3 FEASIBILITY ISSUES

To detect possible infeasibility and unboundedness of the problems (P) and (D) we need the following definition.

Definition 2.3 (Primal and dual improving rays) *We say that the primal problem (P) has an improving ray if there is a symmetric matrix $\bar{X} \succeq 0$ such that $\mathbf{Tr}\,(A_i\bar{X}) = 0,\ (i = 1,\dots,m)$ and $\mathbf{Tr}\,(C\bar{X}) < 0$. Analogously, the dual problem (D) has an improving ray if there is a vector $\bar{y} \in \mathbb{R}^m$ such that $\bar{S} := -\sum_{i=1}^m \bar{y}_i A_i \succeq 0$ and $b^T\bar{y} > 0$.*

Primal improving rays cause infeasibility of the dual problem, and *vice versa*. Formally one has the following result.

Lemma 2.2 *If there is a dual improving ray $\bar{y}$, then (P) is infeasible. Similarly, a primal improving ray $\bar{X}$ implies infeasibility of (D).*

Proof:
Let a dual improving ray $\bar{y}$ be given. By assuming the existence of a primal feasible X one has

$$0 < b^T\bar{y} = \sum_{i=1}^m \mathbf{Tr}\,(A_iX)\bar{y}_i = -\mathbf{Tr}\,(X\bar{S}) \le 0,$$

which is a contradiction. The proof in case of a primal improving ray proceeds similarly. □

Definition 2.4 (Strong infeasibility) *Problem (P) (resp. (D)) is called strongly infeasible if (D) (resp. (P)) has an improving ray.*

Every infeasible LP problem is strongly infeasible, but in the SDP case so-called *weak infeasibility* is also possible.

Definition 2.5 (Weak infeasibility) *Problem (P) is weakly infeasible if $\mathcal{P} = \emptyset$ and yet for each $\epsilon > 0$ there exists an $X \succeq 0$ such that*

$$|\mathbf{Tr}\,(A_iX) - b_i| \le \epsilon, \quad \forall i.$$

Similarly, problem (D) is called weakly infeasible if $\mathcal{D} = \emptyset$ and for every $\epsilon > 0$ there exist $y \in \mathbb{R}^m$ and $S \succeq 0$ such that

$$\left\| \sum_{i=1}^m y_iA_i + S - C \right\| \le \epsilon.$$

Example 2.2 *An example of weak infeasibility is given if (D) is defined by $n = 2$, $m = 1$, $b = [1\ 0]^T$,*

$$A_1 = \begin{bmatrix} 1 & 0 \\ 0 & 0 \end{bmatrix}, \text{ and } C = \begin{bmatrix} 0 & 1 \\ 1 & 0 \end{bmatrix},$$

so that problem (D) becomes:

$$\sup y_1$$

subject to

$$y_1 \begin{bmatrix} 1 & 0 \\ 0 & 0 \end{bmatrix} + S = \begin{bmatrix} 0 & 1 \\ 1 & 0 \end{bmatrix}.$$

We can construct an 'ϵ-infeasible solution' by setting

$$S = \begin{bmatrix} 1/\epsilon & 1 \\ 1 & \epsilon \end{bmatrix}, \quad y_1 = -\frac{1}{\epsilon},$$

so that

$$\|y_1 A_1 + S - C\| = \left\| -\frac{1}{\epsilon} \begin{bmatrix} 1 & 0 \\ 0 & 0 \end{bmatrix} + \begin{bmatrix} 1/\epsilon & 1 \\ 1 & \epsilon \end{bmatrix} - \begin{bmatrix} 0 & 1 \\ 1 & 0 \end{bmatrix} \right\| = \epsilon.$$

□

It is not difficult to show that an infeasible SDP problem is either weakly infeasible or strongly infeasible.

Theorem 2.4 *Assume (D) (resp. (P)) has no improving ray. Then (P) (resp. (D)) is either feasible or weakly infeasible.*

Proof:
Let problems (P) and (D) be given. We will show that (P) is either feasible or weakly infeasible if (D) has no improving ray. The proof where (P) has no improving ray then follows from the symmetric problem reformulation.

Define the two (nonempty) convex cones

$$\mathcal{K}_1 := \left\{ y \in \mathbb{R}^m \;\middle|\; -\sum_{i=1}^m y_i A_i \succeq 0 \right\}, \quad \mathcal{K}_2 := \left\{ y \in \mathbb{R}^m \mid b^T y > 0 \right\}.$$

Since there is no dual improving ray, the relative interiors of these two cones are disjoint. If we apply the separation theorem for convex sets (Theorem B.4) we find that there exists a nonzero $r \in \mathbb{R}^m$ such that

$$\sup_{y \in \mathcal{K}_1} r^T y \leq \inf_{y \in \mathcal{K}_2} r^T y.$$

Since $0 \in \mathcal{K}_1$ one has $\sup_{y \in \mathcal{K}_1} r^T y \geq 0$, and since $\mathcal{K}_2$ is a nonempty cone one has $\inf_{y \in \mathcal{K}_2} r^T y \in \{0, -\infty\}$. It follows that

$$\sup_{y \in \mathcal{K}_1} r^T y = \inf_{y \in \mathcal{K}_2} r^T y = 0.$$

Thus we find that the function $f(y) = r^T y$ is bounded from below on the half-space $b^T y \geq 0$. This is only possible if $r = \alpha b$ for some $\alpha > 0$. Obviously, we can take $\alpha = 1$ without loss of generality. Thus we have shown that

$$\sup_{y \in \mathbf{R}^m} \left\{ b^T y \;\middle|\; -\sum_{i=1}^{m} y_i A_i \succeq 0 \right\} = 0. \tag{2.11}$$

We can now show that (P) is either feasible or weakly infeasible. To this end, define auxiliary variables $t^+ \in \mathbf{R}^m_+$ and $t^- \in \mathbf{R}^m_+$ and consider the problem:

$$\inf_{X \succeq 0, t^+ \in \mathbf{R}^m_+, t^- \in \mathbf{R}^m_+} \left\{ \sum_{i=1}^{m} (t_i^+ + t_i^-) \;\middle|\; \mathbf{Tr}\,(A_i X) + t_i^+ - t_i^- = b_i \ (i = 1, \ldots, m) \right\}. \tag{2.12}$$

Note that the optimal value of this problem is zero if and only if (P) is either feasible or weakly infeasible. If the optimal value of problem (2.12) is zero, then (P) is:

- feasible if the optimal set of problem (2.12) is nonempty;
- weakly infeasible if the optimal set of problem (2.12) is empty.

The dual problem of (2.12) is given by

$$\sup_{y \in \mathbf{R}^m} \left\{ b^T y \;\middle|\; -\sum_{i=1}^{m} y_i A_i \succeq 0, \ -1 \leq y_i \leq 1 \ (i = 1, \ldots, m) \right\},$$

which clearly has optimal value zero because problem (2.11) does. Moreover, problem (2.12) is clearly strictly feasible, and we can apply the strong duality theorem (Theorem 2.2) to conclude that the optimal value of problem (2.12) is indeed zero. □

We can give an alternative characterization of weak infeasibility by introducing the concept of a weakly improving ray. Whereas an improving ray in (P) causes strict infeasibility in (D) (and *vice versa*), weakly improving rays cause weak infeasibility.

Definition 2.6 (Weakly improving ray) *We say that the primal problem (P) has a weakly improving ray if there exists a sequence $\bar{X}^{(k)} \succeq 0$ ($k = 1, 2, \ldots$) such that*

$$\liminf_{k \to \infty} \left| \mathbf{Tr}\left(A_i \bar{X}^{(k)}\right) \right| = 0 \ (i = 1, \ldots, m), \ \limsup_{k \to \infty} \mathbf{Tr}\,(C\bar{X}^{(k)}) = -1.$$

Analogously, the dual problem (D) has a weakly improving ray if there are sequences $\bar{y}^{(k)} \in \mathbf{R}^m$ and $\bar{S}^{(k)} \succeq 0$ such that

$$\liminf_{k \to \infty} \left\| \bar{S}^{(k)} + \sum_{i=1}^{m} \bar{y}_i^{(k)} A_i \right\| = 0, \ \liminf_{k \to \infty} b^T \bar{y}^{(k)} = 1.$$

Theorem 2.5 *The problem (P) (resp. (D)) is weakly infeasible if and only if (D) (resp. (P)) has a weakly improving ray.*

Proof:
We will show that (D) is weakly infeasible if and only if (P) has a weakly improving ray. The proof where (P) and (D) are interchanged then follows from the symmetric problem reformulation as before.

Consider the problem

$$\inf_{X\succeq 0, t^+\in\mathbf{R}_+^{m+1}, t^-\in\mathbf{R}_+^{m+1}} \sum_{i=1}^{m+1}(t_i^+ + t_i^-) \tag{2.13}$$

subject to

$$\begin{aligned} \mathbf{Tr}\,(A_iX) + t_i^+ - t_i^- &= 0 \quad (i = 1,\dots,m) \\ \mathbf{Tr}\,(CX) + t_{m+1}^+ - t_{m+1}^- &= -1. \end{aligned}$$

Note that problem (2.13) has optimal value zero if and only if (P) has either an improving or a weakly improving ray. If the optimal value is zero, it is attained if (P) has an improving ray and it is not attained if (P) only has a weakly improving ray.

The dual of problem (2.13) is given by

$$\sup_{y\in\mathbf{R}^{m+1}} \left\{ -y_{m+1} \;\middle|\; y_{m+1}C - \sum_{i=1}^{m} y_iA_i \succeq 0,\ -1 \le y_i \le 1\ (i = 1,\dots,m+1) \right\}. \tag{2.14}$$

Note that problem (2.13) is strictly feasible, and the zero solution is feasible for its dual problem (2.14). We can therefore apply the strong duality theorem (Theorem 2.2) to conclude that the optimal value of problem (2.14) is attained, and that it is positive if and only if (P) has neither an improving ray nor a weakly improving ray. On the other hand, it is easy to see that problem (D) is feasible if and only if the optimal value of problem (2.14) is positive: if the optimal value in (2.14) is positive, then we can divide $y_{m+1}C - \sum_{i=1}^m y_iA_i \succeq 0$ by y_{m+1} to obtain a feasible solution to (D). Conversely, if there is a feasible $y^0 \in \mathbf{R}^m$ such that $C - \sum_{i=1}^m y_i^0 A_i \succeq 0$, then we can construct a feasible solution to (2.14) with positive objective value as follows: if $\max_{i=1,\dots,m} |y_i^0| \le 1$ set $y_i = y_i^0$ $(i = 1,\dots,m)$ and $y_{m+1} = 1$; otherwise set

$$y_{m+1} = \frac{1}{\max_{i=1,\dots,m}|y_i^0|},\quad y_i = \frac{y_i^0}{\max_{j=1,\dots,m}|y_j^0|} \quad (i = 1,\dots,m).$$

This completes the proof. □

2.4 OPTIMALITY AND COMPLEMENTARITY

By weak duality (Theorem 2.1) we know that $X^* \in \mathcal{P}$ and $S^* \in \mathcal{D}$ will be optimal if the duality gap at (X^*, S^*) is zero, *i.e.* $\mathbf{Tr}\,(X^*S^*) = 0$. The condition $\mathbf{Tr}\,(X^*S^*) =$

0 is equivalent to $X^*S^* = 0$, since $X^* \succeq 0$ and $S^* \succeq 0$ (see Lemma A.2 in Appendix A). It follows that sufficient optimality conditions for (P) and (D) are:

$$\left.\begin{aligned} \mathbf{Tr}\,(A_iX) &= b_i, \quad X \succeq 0, \quad i = 1, \ldots, m \\ \sum_{i=1}^{m} y_iA_i + S &= C, \quad S \succeq 0 \\ XS &= 0. \end{aligned}\right\} \tag{2.15}$$

The condition $XS = 0$ is called the *complementarity condition*, and optimal solutions that satisfy this condition are called *complementary*.

The strong duality result of Corollary 2.1 implies that these optimality conditions are also necessary if (P) and (D) are strictly feasible.

Theorem 2.6 (Necessary and sufficient optimality conditions) *Under Assumption (2.2) (strict feasibility), (2.15) is a system of necessary and sufficient optimality conditions for (P) and (D).*

In what follows, we will assume strict feasibility of (P) and (D), unless otherwise indicated. Also, the range (or column) space of any primal (resp. dual) feasible $X \in \mathcal{P}$ ($S \in \mathcal{D}$) will be denoted by $\mathcal{R}(X)$ (resp. $\mathcal{R}(S)$).

In the special case of linear programming (LP) one always has *strict complementarity, i.e.* there exists an optimal solution pair $(X^*, S^*) \in \mathcal{P}^* \times \mathcal{D}^*$ such that $X^* + S^* \succ 0$.[2] For general SDP this is not the case, as the next example shows.

Example 2.3 (Alizadeh *et al.* [5]) *Let $n = m = 3$, $b = [1\ 0\ 0]^T$ and*

$$C = \begin{bmatrix} 0 & 0 & 0 \\ 0 & 0 & 0 \\ 0 & 0 & 1 \end{bmatrix} \quad A_1 = \begin{bmatrix} 1 & 0 & 0 \\ 0 & 0 & 0 \\ 0 & 0 & 0 \end{bmatrix} \quad A_2 = \begin{bmatrix} 0 & 0 & 1 \\ 0 & 1 & 0 \\ 1 & 0 & 0 \end{bmatrix} \quad A_3 = \begin{bmatrix} 0 & 1 & 0 \\ 1 & 0 & 0 \\ 0 & 0 & 1 \end{bmatrix}.$$

The optimal solutions of (P) and (D) are given by

$$X^* = \begin{bmatrix} 1 & 0 & 0 \\ 0 & 0 & 0 \\ 0 & 0 & 0 \end{bmatrix}, \quad y_i^* = 0\ (i = 1, 2, 3), \quad S^* = \begin{bmatrix} 0 & 0 & 0 \\ 0 & 0 & 0 \\ 0 & 0 & 1 \end{bmatrix}.$$

The solution X^* is clearly optimal, since $C \succeq 0$ and therefore $\mathbf{Tr}\,(CX) \geq 0\ \forall X \in \mathcal{P}$. It is also easy to see that the optimal solutions (X^*, S^*) are unique, and therefore strict complementarity does not *hold for this example.* □

[2] The strict complementarity property for LP is known as the Goldman–Tucker theorem[69].

It is therefore natural to introduce the weaker concept of *maximal complementarity*, that corresponds to optimal solutions of maximal rank.

Definition 2.7 (Strict and maximal complementarity) *We call* $X^* \in \mathcal{P}^*$ *a* maximally complementary primal optimal solution *if*

$$\mathcal{R}(X) \subset \mathcal{R}(X^*) \quad \forall X \in \mathcal{P}^*,$$

and, similarly, $S^* \in \mathcal{D}^*$ *is called a* maximally complementary dual solution *if*

$$\mathcal{R}(S) \subset \mathcal{R}(S^*) \quad \forall S \in \mathcal{D}^*.$$

If a maximally complementary solution pair (X^*, S^*) *satisfies* $X^* + S^* \succ 0$, *then we call* (X^*, S^*) *a strictly complementary solution pair.*

It may not be immediately clear that maximally complementary solutions exist, but any optimal solution in the relative interior of the optimal set is maximally complementary, as the following lemma shows.

Lemma 2.3 *The following statements are equivalent for a given* $X^* \in \mathcal{P}^*$*:*

(i) X^* *is maximally complementary,* i.e. $\mathcal{R}(X) \subset \mathcal{R}(X^*) \quad \forall X \in \mathcal{P}^*$;

(ii) $X^* \in \text{ri}\,(\mathcal{P}^*)$;

(iii) $\textit{rank}\,(X^*) \geq \textit{rank}\,(X) \quad \forall X \in \mathcal{P}^*$.

Proof:
We only show that (ii) implies (i). The other relationships can be shown easily.[3] Let $X^* \in \text{ri}\,(\mathcal{P}^*)$ and $X \in \mathcal{P}^*$ be given. We will show that $\mathcal{R}(X) \subset \mathcal{R}(X^*)$. Since $X^* \in \text{ri}\,(\mathcal{P}^*)$, we can find a $\bar{X} \in \mathcal{P}^*$ such that $X^* = \lambda X + (1-\lambda)\bar{X}$ for some $\lambda \in (0,1)$. Now let $x \in \mathbf{R}^n$ be in the null-space of X^*. One has

$$0 = x^T X^* x = x^T\left(\lambda X + (1-\lambda)\bar{X}\right)x = \lambda x^T X x + (1-\lambda)x^T \bar{X} x.$$

Since $X \succeq 0$ and $\bar{X} \succeq 0$, we have

$$\lambda x^T X x = (1-\lambda)x^T \bar{X} x = 0.$$

Thus

$$0 = x^T X x = x^T X^{\frac{1}{2}} X^{\frac{1}{2}} x = \left\|X^{\frac{1}{2}} x\right\|^2,$$

and therefore $X^{\frac{1}{2}} x = 0$ where $X^{\frac{1}{2}}$ denotes the symmetric square root factorization of X. This implies $X^{\frac{1}{2}}\left(X^{\frac{1}{2}} x\right) = Xx = 0$, as required. □

[3] Lemma 2.3 actually holds for any face of the positive semidefinite cone (it is stated here for the optimal face, but we do not make use of the optimality property). For detailed proofs of Lemma 2.3, see Barker and Carlson [14] and Pataki [145]. A more detailed analysis of the faces of the semidefinite cone, and results on the rank of optimal solutions may be found in Pataki [143].

The lemma shows that all $X^* \in \text{ri}(\mathcal{P}^*)$ have the same range space, which we will call $\mathcal{B}$, and that for all other optimal $X \in \mathcal{P}^*$ one has $\mathcal{R}(X) \subset \mathcal{B}$. Similarly we can define the subspace $\mathcal{N}$ such that $\mathcal{R}(S^*) \subset \mathcal{N}$ for all $S^* \in \mathcal{D}^*$ and $\mathcal{R}(S^*) = \mathcal{N}$ for all $S^* \in \text{ri}(\mathcal{D}^*)$.

Since optimal $X^* \in \mathcal{P}^*$ and $S^* \in \mathcal{D}^*$ commute ($X^*S^* = S^*X^* = 0$), the spectral (eigenvector-eigenvalue) decompositions of an optimal pair X^* and S^* take the form:

$$X^* = Q\Lambda Q^T, \quad S^* = Q\Sigma Q^T, \tag{2.16}$$

where Q is orthogonal and the diagonal matrices Λ and Σ have the (nonnegative) eigenvalues of X^* and S^* on their respective diagonals. Obviously $X^*S^* = 0$ if and only if $\Lambda\Sigma = 0$. Furthermore, $\mathcal{R}(X^*) = \mathcal{R}(Q\Lambda)$ and $\mathcal{R}(S^*) = \mathcal{R}(Q\Sigma)$, *i.e.* the range spaces are spanned by those eigenvectors corresponding to strictly positive eigenvalues.

Now let Q_B be an orthogonal matrix whose columns form an orthonormal basis for the subspace $\mathcal{B}$. For example, if $X^* = Q\Lambda Q^T \in \text{ri}(\mathcal{P}^*)$ we can choose Q_B as the submatrix of Q obtained by taking the columns corresponding to positive eigenvalues.

Note that the spaces $\mathcal{B}$ and $\mathcal{N}$ are orthogonal to each other. In the case of strict complementarity they span $\mathbb{R}^n$; if strict complementarity does not hold, then we can define the subspace $\mathcal{T}$ which is the orthogonal complement of $\mathcal{B} \cup \mathcal{N}$. Note that $\mathcal{B}$, $\mathcal{N}$ and $\mathcal{T}$ partition $\mathbb{R}^n$ into three mutually orthogonal subspaces.[4] Thus we also define the matrices Q_N and Q_T as orthogonal matrices such that $\mathcal{R}(Q_N) = \mathcal{N}$ and $\mathcal{R}(Q_T) = \mathcal{T}$. The matrices Q_B, Q_N and Q_T are not uniquely defined, of course, but since an orthonormal basis of a given subspace of $\mathbb{R}^n$ is unique up to a rotation, we have the following.

Lemma 2.4 *Let Q_B and $\bar{Q}_B$ be orthogonal matrices such that*

$$\mathcal{R}(Q_B) = \mathcal{R}(\bar{Q}_B) = \mathcal{B}.$$

Now there exists a square orthogonal matrix U such that $Q_BU = \bar{Q}_B$.

Using this lemma, it is easy to derive the following useful representation of optimal solutions.

Theorem 2.7 *Let Q_B be given such that $\mathcal{R}(Q_B) = \mathcal{B}$ and $Q_B^TQ_B = I$. Any optimal solution $X \in \mathcal{P}^*$ can be written as*

$$X = Q_BU_XQ_B^T \tag{2.17}$$

where U_X is a suitable positive semidefinite matrix of size $\dim(\mathcal{B}) \times \dim(\mathcal{B})$. If $X \in \text{ri}(\mathcal{P}^*)$, then $U_X \succ 0$. Similarly, any dual optimal solution $S\mathcal{P}^*$ can be written as

$$S = Q_NU_SQ_N^T \tag{2.18}$$

[4]This is the natural extension of the concept of the *optimal partition* in LP, and the optimal tri-partition in LCP.

where U_S is a suitable positive semidefinite matrix of size dim($\mathcal{N}$) $\times$ dim($\mathcal{N}$).

Note that U_X in (2.17) is uniquely determined by Q_B and is given by $U_X = Q_B^T U_X Q_B$, since $Q_B^T Q_B = I$. Thus it can easily be understood that the matrices X and U_X in (2.17) have the same spectrum, except that the multiplicity of zero in the spectrum of X may be larger than in the spectrum of U_X.

2.5 DEGENERACY

A feasible solution x^0 of an optimization problem

$$\left\{\min_x f(x), \quad g_i(x) \leq 0, \; i = 1, \ldots, m\right\}$$

is called degenerate if all the gradients of the active constraints[5] at the point x^0 are linearly dependent. It is well-known in linear programming that degeneracy can cause *cycling* of the Simplex algorithm, unless suitable pivoting rules are used. It is also known that the absence of degeneracy ensures that optimal solutions are unique.

The concept of degeneracy can be extended to SDP.[6] Of course, SDP problems are in conic form, and therefore the concept of active constraints is not well-defined.

The gradient of a constraint at some point is orthogonal to the level set of the constraint function at that point. In the SDP case, we can replace the level set by the smallest face of $\mathcal{S}_n^+$ that contains the given point $X_0 \succeq 0$. We can then replace the gradient by the orthogonal complement of this face.

Let us denote

$$\mathcal{L} = \text{span}\left\{A_1, \ldots, A_m\right\},$$

as before. For a given $\hat{X} \in \mathcal{P}$ of rank r we further define the subspace:

$$\mathcal{T}_{\hat{X}} := \left\{ X \in \mathcal{S}_n \;\middle|\; X = Q \begin{bmatrix} U & V \\ V^T & 0 \end{bmatrix} Q^T \right\},$$

where $U \in \mathcal{S}_r$, $V \in \mathbf{R}^{r \times (n-r)}$ and $Q \in \mathbf{R}^{n \times n}$ is an orthogonal matrix with the eigenvectors of $\hat{X}$ corresponding to positive eigenvalues as the first r columns. This subspace is simply the tangent space to the smallest face of $\mathcal{S}_n^+$ that contains $\hat{X}$ (see Pataki [145], §3.2).

Definition 2.8 (Primal degeneracy) *We call $X \in \mathcal{P}$ primal nondegenerate if*

$$\mathcal{S}_n = \mathcal{T}_X + \mathcal{L}^{\perp}. \tag{2.19}$$

Otherwise, we call X primal degenerate.

---

[5]A constraint $g_i(x) \leq 0$ is active at the point x^0 if $g_i(x^0) = 0$.

[6]The approach in this section closely follows that by Alizadeh *et al.* [5], where the concept degeneracy was first extended to SDP.

Dual degeneracy can be defined in a similar way. As before, for a given $\hat{S} \in \mathcal{D}$ of rank s we define the subspace:

$$\mathcal{T}_{\hat{S}} := \left\{ S \in \mathcal{S}_n \;\middle|\; S = Q \begin{bmatrix} 0 & V \\ V^T & W \end{bmatrix} Q^T \right\}$$

where $W \in \mathcal{S}_s$, $V \in \mathbb{R}^{(n-s)\times s}$ and $Q \in \mathbb{R}^{n\times n}$ is an orthogonal matrix with the eigenvectors of $\hat{S}$ corresponding to positive eigenvalues as the last s columns.

Definition 2.9 (Dual degeneracy) *We call $S \in \mathcal{D}$ dual nondegenerate if*

$$\mathcal{S}_n = \mathcal{T}_S + \mathcal{L}. \tag{2.20}$$

Otherwise, we call S dual degenerate.

Problem (P) (resp. (D)) is called nondegenerate if all $X \in \mathcal{P}$ (resp. $S \in \mathcal{D}$) are nondegenerate.

Example 2.4 (Alizadeh *et al.* [5]) *We show that the optimal solution X^* in Example 2.3 is nondegenerate. Note that we have $Q = I$,*

$$\mathcal{T}_X = \begin{bmatrix} u & v_1 & v_2 \\ v_1 & 0 & 0 \\ v_2 & 0 & 0 \end{bmatrix}, \quad \mathcal{L}^\perp = \begin{bmatrix} 0 & -\frac{1}{2}x_1 & -\frac{1}{2}x_2 \\ -\frac{1}{2}x_1 & x_2 & x_3 \\ -\frac{1}{2}x_2 & x_3 & x_1 \end{bmatrix},$$

where $u, v_2, v_2 \in \mathbb{R}$, and $x_1, x_2, x_3 \in \mathbb{R}$, so that

$$\mathcal{S}_3 = \mathcal{T}_{X^*} + \mathcal{L}^\perp,$$

i.e. *X^* is nondegenerate. Similarly, one can show that S^* is dual nondegenerate.* □

The example illustrates a general result: an optimal primal nondegenerate solution implies a unique dual optimal solution and vice versa.

Theorem 2.8 (Alizadeh *et al.* [5]) *Let (P) and (D) be strictly feasible. If there exists a nondegenerate $X^* \in \mathcal{P}^*$ (resp. $S^* \in \mathcal{D}^*$), then (D) (resp. (P)) has a unique optimal solution.*

Proof:
Let $S_1 \in \mathcal{D}^*$, $S_2 \in \mathcal{D}^*$, and let $X^* \in \mathcal{P}^*$ be primal nondegenerate. We can assume that the orthogonal matrix X^* with eigenvectors of X^* as columns takes the form $Q = [\bar{Q}, Q_N]$ where $\bar{Q}$ is an orthogonal matrix, whose columns include those eigenvectors of X^* that correspond to positive eigenvalues, and where Q_N is an orthogonal matrix whose columns span the space $\mathcal{N}$. To fix our ideas, we assume that $\dim(\mathcal{N}) = s$ such that Q_N is an $n \times s$ matrix.

By (2.18), we can write

$$S_1 = Q_N U_1 Q_N^T, \quad S_2 = Q_N U_2 Q_N^T$$

where U_1 and U_2 are suitable positive semidefinite matrices of size $s \times s$.

We will prove that $\Delta S := S_1 - S_2 = 0$. Note that $\Delta S \in \mathcal{L}$ and

$$\Delta S = Q_N(U_1 - U_2)Q_N^T.$$

Let $Z \in \mathcal{S}_n$ now be given, and recall that by nondegeneracy

$$\mathcal{S}_n = \mathcal{T}_{X^*} + \mathcal{L}^\perp.$$

We can therefore decompose Z as

$$Z = Z_\mathcal{T} + Z_{\mathcal{L}^\perp},$$

say, where $Z_\mathcal{T} \in \mathcal{T}_{X^*}$ and $Z_{\mathcal{L}^\perp} \in \mathcal{L}^\perp$. We now show that the inner product of Z and ΔS is zero:

$$\begin{aligned}
\mathbf{Tr}\,(\Delta S Z) &\equiv \mathbf{Tr}\,(\Delta S(Z_\mathcal{T} + Z_{\mathcal{L}^\perp})) \\
&= \mathbf{Tr}\,(\Delta S Z_\mathcal{T}) \\
&= \mathbf{Tr}\left(Q_N(U_1 - U_2)Q_N^T Q \begin{bmatrix} U & V \\ V^T & 0 \end{bmatrix} Q^T\right) \\
&= \mathbf{Tr}\left(\left(Q_N^T Q\right)^T (U_1 - U_2) Q_N^T Q \begin{bmatrix} U & V \\ V^T & 0 \end{bmatrix}\right) \\
&= \mathbf{Tr}\left(\begin{bmatrix} 0 & 0 \\ 0 & U_1 - U_2 \end{bmatrix} \begin{bmatrix} U & V \\ V^T & 0 \end{bmatrix}\right) = 0,
\end{aligned}$$

where we have used the fact that

$$Q_N^T Q = Q_N^T[\bar{Q}, Q_N] = [0_{s\times(n-s)}, I_s].$$

We conclude that $\Delta S = 0$, since $Z \in \mathcal{S}_n$ was arbitrary. □

If strict complementarity holds for (P) and (D), then the converse of Theorem 2.8 is also true. In other words, if we assume strict complementarity, then the concepts of primal nondegeneracy and unique dual optimal solutions coincide.

Theorem 2.9 (Alizadeh *et al.* [5]) *If strict complementarity holds for (P) and (D), then nondegeneracy of (P) (resp. (D)) is a necessary and sufficient condition for an optimal solution of (D) (resp. (P)) to be unique.*

Proof:
First note that the nondegeneracy condition (2.19) for some $X^* \in \mathcal{P}^*$ can be restated as:

$$\mathcal{T}_{X^*}^{\perp} \cap \mathcal{L} = \{0\}.$$

Also note that

$$\mathcal{T}_{X^*}^{\perp} = \left\{ X \in \mathcal{S}_n \;\middle|\; X = Q \begin{bmatrix} 0 & 0 \\ 0 & W \end{bmatrix} Q^T \right\} = Q_N W {Q_N}^T,$$

where Q and Q_N are defined as in the previous lemma. It follows that a linear equation of the form

$$Q_N \bar{W} Q_N^T + \sum_{i=1}^{m} \bar{y}_i A_i = 0 \tag{2.21}$$

in the variables $\bar{W}$ and $\bar{y}$ only has the zero solution in case of primal nondegeneracy. Assume that primal nondegeneracy does not hold; in this case the linear equation

$$Q_N \bar{W} Q_N^T + \sum_{i=1}^{m} \bar{y}_i A_i = C \tag{2.22}$$

has an affine solution set with positive dimension, since the homogeneous system (2.21) has multiple solutions. Any maximally complementary dual solution, say $\bar{S} \in$ ri $(\mathcal{D}^*)$ corresponds to a solution of (2.22) with $\bar{W} \succ 0$. Since any $S^* = Q_N W {Q_N}^T$ which satisfies (2.22) is dual optimal if $W \succeq 0$, there must be an open neighbourhood of $\bar{S}$ containing other optimal solutions of (D). The proof where (D) is nondegenerate is similar. □

In the absence of strict complementarity, nondegeneracy is *not* necessary to ensure unique optimal solutions, as the following example shows.

Example 2.5 (Alizadeh *et al.* [5]) *If we replace the matrix A_3 in the previous example by*

$$A_3 = \begin{bmatrix} 0 & 0 & 0 \\ 0 & 0 & 0 \\ 0 & 0 & 1 \end{bmatrix},$$

then it is easy to show that dual nondegeneracy no longer holds, and there are still no strictly complementarity optimal solutions. The primal optimal solution remains unique and unchanged, however. □

3
THE CENTRAL PATH

Preamble

If the system of necessary and sufficient optimality conditions for (P) and (D) is perturbed by introducing a parameter $\mu > 0$ in a special way, then the solution of the perturbed system defines an analytic curve (parameterized by μ) through the feasible region, which leads to the optimal set as $\mu \downarrow 0$. This curve is called the central path and most interior point methods 'follow' the central path approximately to reach the optimal set. We will review various properties of the central path.

3.1 EXISTENCE AND UNIQUENESS OF THE CENTRAL PATH

We now perturb the optimality conditions (2.15) for (P) and (D) to

$$\left.\begin{array}{rcl} \mathbf{Tr}\,(A_i X) & = & b_i, \quad X \succeq 0, \quad i = 1, \ldots, m \\ \sum_{y=1}^{m} y_i A_i + S & = & C, \quad S \succeq 0 \\ XS & = & \mu I, \end{array}\right\} \tag{3.1}$$

for some $\mu > 0$. Note that if $\mu = 0$ we regain the optimality conditions (2.15). We will show that system (3.1) — which we sometimes refer to as the *centrality conditions* — has a unique solution for any $\mu > 0$. This solution will be denoted by

$$X(\mu), S(\mu), y(\mu),$$

and can be seen as the parametric representation of an analytic curve (the *central path*) in terms of the parameter μ. The existence and uniqueness of the central path can be proved in the following way: consider the problem

$$\min_{X \succ 0} \left\{ f_p^\mu(X) := \frac{1}{\mu}\mathbf{Tr}\,(CX) - \log\det X \;\middle|\; \mathbf{Tr}\,(A_iX) = b_i \; (i = 1, \ldots, m) \right\},$$

i.e. the minimization of the *primal log-barrier function* over the relative interior of $\mathcal{P}$. The function f_p^μ is strictly convex (see Theorem C.1 in Appendix C). The KKT (first order) optimality conditions for this problem are therefore necessary and sufficient, and are given by:[1]

$$\begin{aligned} \nabla f_p^\mu(X) = \frac{1}{\mu}C - X^{-1} &= \sum_{i=1}^m \hat{y}_i A_i \\ \mathbf{Tr}\,(A_iX) &= b_i \quad i = 1, \ldots, m \\ X &\succ 0. \end{aligned}$$

Defining $S = C - \sum_{i=1}^m y_i A_i$ where $y_i = \mu\hat{y}_i$, this system becomes identical to system (3.1). In other words, the existence and uniqueness of the central path is equivalent to the existence of a unique minimizer of f_p^μ in $\mathrm{ri}(\mathcal{P})$ for each $\mu > 0$. Since the function f_p^μ is strictly convex, any minimizer of f_p^μ is unique.[2] One way to prove existence of the central path is therefore to show that the level sets of f_p^μ are compact if (D) is strictly feasible. This ensures the existence of a minimizer, say X_p^*. We can use this minimizer to construct a solution of (3.1) as follows:

$$X(\mu) := X_p^*, \quad S(\mu) := \mu \left(X_p^*\right)^{-1}. \tag{3.2}$$

Note that $S(\mu)$ as defined in (3.2) is dual feasible, as it should be.

One can also approach the proof from the 'dual side', by maximizing the dual barrier

$$f_d^\mu(S, y) := \frac{1}{\mu}b^T y + \log\det(S), \quad (y, S) \in \mathcal{D},$$

and proving that its level sets are compact if (P) is strictly feasible.

We will give a proof below, where we consider the combined problem of minimizing the difference between the primal barrier f_p^μ and the dual barrier f_d^μ. To this end, we define the primal-dual barrier function $f_{pd}^\mu : \mathcal{P} \times \mathcal{D} \mapsto \mathbf{R}_+$:

$$\begin{aligned} f_{pd}^\mu(X, S) &:= f_p^\mu(X) - f_d^\mu(S, y) - n - n\log(\mu) \\ &= \frac{1}{\mu}\mathbf{Tr}\,(CX) - \frac{1}{\mu}b^T y - \log\det(X) - \log\det(S) - n - n\log(\mu) \\ &= \mathbf{Tr}\left(\frac{XS}{\mu}\right) - \log\det\left(\frac{XS}{\mu}\right) - n \end{aligned}$$

[1] The gradient of f_p^μ is derived in Appendix C.

[2] For a proof that a strictly convex function has a unique minimizer over a compact convex set, see *e.g.* Bazaraa *et al.* [16], Theorem 3.4.2.

$$= \sum_{i=1}^{n} \left(\frac{\lambda_i(XS)}{\mu} - \log\left(\frac{\lambda_i(XS)}{\mu} \right) \right) - n.$$

Note that (X^*, S^*) is a minimizer of f_{pd}^{μ} if and only if X^* and S^* are minimizers of f_p^{μ} and $-f_d^{\mu}$ respectively. Also note that $f_{pd}^{\mu}(X, S) = 0$ if and only if $XS = \mu I$. We therefore aim to prove that a unique minimizer of f_{pd}^{μ} exists and satisfies (3.1).

We can rewrite $f_{pd}^{\mu}(X, S)$ as

$$f_{pd}^{\mu}(X, S) = \sum_{i=1}^{n} \psi\left(\frac{\lambda_i(XS)}{\mu} - 1 \right) \tag{3.3}$$

where $\psi(t) := t - \log(1 + t)$ (see Figure 3.1). Note that f_{pd}^{μ} is given as the sum of two

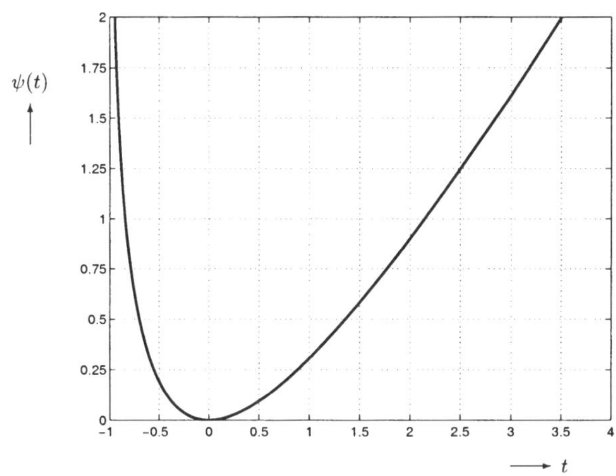

Figure 3.1. The graph of ψ.

strictly convex functions (f_p^{μ} and $-f_d^{\mu}$) up to a constant, and is therefore also strictly convex. We therefore only have to prove that its level sets are compact in order to establish existence and uniqueness of the central path. We will do this in two steps:

1. First we will show that the level sets of the duality gap are compact (Lemma 3.1);

2. Subsequently we will show that the compactness of the level sets of the duality gap implies that the level sets of the primal-dual barrier function f_{pd}^{μ} are also compact (Theorem 3.1).

Lemma 3.1 (Compact level sets of duality gap) *Assume that (P) and (D) are strictly feasible. The set*

$$G_\alpha := \{(X, S) \in \mathcal{P} \times \mathcal{D} \mid \mathbf{Tr}\,(XS) \leq \alpha\}$$

is compact, for each $\alpha \geq 0$.

Proof:
Let (X^0, S^0) be any strictly feasible primal-dual solution, and $(X, S) \in G_\alpha$ for some given $\alpha \geq 0$. By orthogonality, Lemma 2.1, one has

$$\mathbf{Tr}\left((X - X^0)(S - S^0)\right) = 0. \tag{3.4}$$

Using $\mathbf{Tr}\,(XS) \leq \alpha$ simplifies (3.4) to

$$\mathbf{Tr}\,(XS^0) + \mathbf{Tr}\,(X^0S) \leq \alpha + \mathbf{Tr}\,(X^0S^0). \tag{3.5}$$

The left-hand side terms of the last inequality are nonnegative by feasibility. One therefore has

$$\mathbf{Tr}\left(XS^0\right) \leq \alpha + \mathbf{Tr}\,(X^0S^0),$$

which implies

$$\mathbf{Tr}\,(X) \leq \frac{\alpha + \mathbf{Tr}\,(X^0S^0)}{\lambda_{\min}(S^0)},$$

where $\lambda_{\min}(S^0)$ denotes the smallest eigenvalue of S^0. Now using the fact that any positive semidefinite matrix X satisfies $\|X\| \leq \mathbf{Tr}\,(X)$ for the Frobenius norm (see Appendix A), one has

$$\|X\| \leq \frac{\alpha + \mathbf{Tr}\,(X^0S^0)}{\lambda_{\min}(S^0)}.$$

A similar bound can be derived for $\|S\|$. It remains to show that G_α is closed; this follows from the closedness of $\mathcal{P}$ and $\mathcal{D}$ and the linearity of the duality gap function $\mathbf{Tr}\,(XS) = \mathbf{Tr}\,(CX) - b^Ty$ on $\mathcal{P} \times \mathcal{D}$. □

Theorem 3.1 (Existence of the central path) *The central path exists if (P) and (D) are strictly feasible.*

Proof:
Let $X^0 \succ 0$ and $S^0 \succ 0$ be strictly feasible solutions to (P) and (D) respectively and let $\mu > 0$ be given. Choose $\alpha > 0$ such that $f_{pd}^{\mu}\left(X^0, S^0\right) < \alpha$ and denote the α-level set of f_{pd}^{μ} by

$$L_\alpha := \left\{(X, S) \in \mathcal{P} \times \mathcal{D} \mid f_{pd}^{\mu}(X, S) \leq \alpha,\ X \succ 0,\ S \succ 0\right\}.$$

We will prove that L_α is bounded and closed, *i.e.* compact. Let $(X, S) \in L_\alpha$ and let

$$t_i = \frac{\lambda_i(XS)}{\mu}, \quad i = 1, \ldots, n.$$

By the definition of L_α one has:

$$\alpha \;\geq\; f_{pd}^{\mu}(X, S)$$

$$
\begin{aligned}
&\equiv \sum_{i=1}^{n} \psi(t_i - 1) \\
&\geq \psi(t_i - 1) \text{ for any given } i \in \{1, \ldots, n\} \\
&\equiv t_i - 1 - \log(t_i) \\
&= \frac{1}{2}t_i + \left(\frac{1}{2}t_i - \log\left(\frac{1}{2}t_i\right)\right) - \log 2 - 1 \\
&\geq \frac{1}{2}t_i - \log 2 - 1,
\end{aligned}
$$

where the first inequality follows from the nonnegativity of ψ, and the second inequality follows from $\log(t) \leq t$. This shows that

$$\lambda_i(XS) \leq 2\mu\left(\alpha + \log 2 + 1\right), \quad i = 1, \ldots, n. \tag{3.6}$$

On the other hand, the above argument also shows that

$$\alpha \geq t_i - 1 - \log(t_i) \geq -1 - \log(t_i), \quad i = 1, \ldots, n,$$

which in turn implies that

$$\lambda_i(XS) \geq \mu e^{-\alpha - 1}, \quad i = 1, \ldots, n. \tag{3.7}$$

In other words, the eigenvalues of XS are bounded away from zero as well as from above. In particular, (3.6) implies

$$\mathbf{Tr}\,(XS) \equiv \sum_{i=1}^{n} \lambda_i(XS) \leq 2n\mu(\alpha + \log(2) + 1),$$

which implies that L_α is bounded, by Lemma 3.1. It remains to show that L_α is closed. From (3.7) follows that

$$\det(XS) \equiv \prod_{i=1}^{n} \lambda_i(XS) \geq \mu^n e^{-n(\alpha+1)}. \tag{3.8}$$

Now let $(\hat{X}, \hat{S}) \in \mathrm{cl}\,(L_\alpha)$ be given. Note that there must hold $\hat{X} \in \mathcal{P}$ and $\hat{S} \in \mathcal{D}$. By the continuity of the determinant function and (3.8) there must hold

$$\det\left(\hat{X}\hat{S}\right) > 0,$$

which implies that $\hat{X} \succ 0$ and $\hat{S} \succ 0$. The continuity of f^{μ}_{pd} further implies that $f^{\mu}_{pd}(\hat{X}, \hat{S}) \leq \alpha$ and consequently $\left(\hat{X}, \hat{S}\right) \in L_\alpha$. □

Remark 3.1 *In order to prove the existence of the central path, we have not made full use of the fact that (P) and (D) are in the standard form. Rather, we have only used the orthogonality property (Lemma 2.1). This observation will be important in the next chapter where we will study self-dual problems that are not in the standard form.*

3.2 ANALYTICITY OF THE CENTRAL PATH

Our geometric view of the central path is that of an analytic[3] curve through the relative interior of $\mathcal{P} \times \mathcal{D}$ which leads to the optimal set. The analyticity follows from a straightforward application of the *implicit function theorem.*[4]

Theorem 3.2 (Implicit function theorem) *Let $f : \mathbf{R}^{n+m} \mapsto \mathbf{R}^m$ be an analytic function of $w \in \mathbf{R}^n$ and $z \in \mathbf{R}^m$ such that:*

1. *There exist $\bar{w} \in \mathbf{R}^n$ and $\bar{z} \in \mathbf{R}^m$ such that $f(\bar{w}, \bar{z}) = 0$.*
2. *The Jacobian of f with respect to z is nonsingular at $(\bar{w}, \bar{z})$.*

Then there exist open sets $S_{\bar{w}} \subset \mathbf{R}^n$ and $S_{\bar{z}} \subset \mathbf{R}^m$ containing $\bar{w}$ and $\bar{z}$ respectively, and an analytic function $\phi : S_{\bar{w}} \mapsto S_{\bar{z}}$ such that $\bar{z} = \phi(\bar{w})$ and $f(w, \phi(w)) = 0$ for all $w \in S_{\bar{w}}$. Moreover

$$\nabla\phi(w) = -\nabla_z f(w, \phi(w))^{-1} \nabla_w f(w, \phi(w)). \tag{3.9}$$

□

Theorem 3.3 (Analyticity of the central path) *The function*

$$f_{cp} : \mu \mapsto (X(\mu), y(\mu), S(\mu))$$

is an analytic function for $\mu > 0$.

Proof:
Let $\mu_0 > 0$ be given. The proof follows directly by applying the implicit function theorem to the function

$$f : \mathbb{R}^{n \times n} \times \mathbb{R}^m \times \mathbb{R}^{n \times n} \times \mathbf{R} \mapsto \mathbb{R}^m \times \mathbf{R}^{n \times n} \times \mathbf{R}^{n \times n}$$

defined by:

$$f(X, y, S, \mu) := \begin{bmatrix} \mathbf{Tr}\,(A_1 X) - b_1 \\ \vdots \\ \mathbf{Tr}\,(A_m X) - b_m \\ \sum_{i=1}^{m} y_i A_i + S - C \\ XS - \mu I \end{bmatrix}.$$

[3]For a short review on analytic functions see Appendix D.

[4]There are many variations of the implicit function theorem. We will use the version that deals with analytic functions, as given in *e.g.* Dieudonné [52], Theorem 10.2.4.

Note we have dropped the symmetry requirement for X and S, since it is redundant on the central path. Also note that f is zero at $(X(\mu), y(\mu), S(\mu), \mu)$ for any $\mu > 0$. We now associate $[\mathbf{vec}(X)^T \; y^T \; \mathbf{vec}(S)^T]^T$ with z in Theorem 3.2 and μ with w. The minor of the Jacobian matrix of f with respect to (X, y, S) is given by:

$$\nabla_{(X,y,S)} f(X, y, S, \mu) = \begin{bmatrix} \mathcal{A} & 0 & 0 \\ 0 & \mathcal{A}^T & I_{n^2} \\ S \otimes I_n & 0 & I_n \otimes X \end{bmatrix}, \tag{3.10}$$

where $\mathcal{A} = [\mathbf{vec}(A_1), \ldots, \mathbf{vec}(A_m)]^T$, I_n is the identity matrix of size n, and $\otimes$ denotes the Kronecker product. We proceed to show that the matrix in (3.10) is non-singular at $(X(\mu_0), y(\mu_0), S(\mu_0))$.

To this end, assume that

$$\begin{bmatrix} \mathcal{A} & 0 & 0 \\ 0 & \mathcal{A}^T & I_{n^2} \\ S(\mu_0) \otimes I_n & 0 & I_n \otimes X(\mu_0) \end{bmatrix} \begin{bmatrix} x \\ y \\ z \end{bmatrix} = 0,$$

for some $x \in \mathbb{R}^{n^2}$, $y \in \mathbb{R}^m$ and $z \in \mathbb{R}^{n^2}$. This system can be simplified to:

$$\begin{aligned} \mathcal{A}x &= 0 \\ \mathcal{A}^T y + z &= 0 \\ (S(\mu_0) \otimes I_n)x + (I_n \otimes X(\mu_0))z &= 0. \end{aligned}$$

Note that the first two equations imply $x^T z = 0$, and the third equation yields:

$$x + (S(\mu_0) \otimes I_n)^{-1}(I_n \otimes X(\mu_0))z = 0.$$

Taking the inner product with z on both sides and using $x^T z = 0$ yields

$$z^T (S(\mu_0) \otimes I_n)^{-1}(I_n \otimes X(\mu_0))z = 0$$

which is the same as

$$z^T \left[S(\mu_0)^{-1} \otimes X(\mu_0) \right] z = 0, \tag{3.11}$$

if we use the identities $(A \otimes B)^{-1} = A^{-1} \otimes B^{-1}$ and $(AB) \otimes (CD) = (AC) \otimes (BD)$.[5]

It is clear that — since $S^{-1}(\mu_0) \otimes X(\mu_0)$ is positive definite[6] — it follows from (3.11) that $z = 0$, which in turn implies $x = 0$ and $y = 0$.

We can now apply Theorem 3.2 to prove that there exists an analytic function, say

$$\phi : \mu \mapsto \mathbb{R}^{n \times n} \times \mathbb{R}^m \times \mathbb{R}^{n \times n},$$

[5] For proofs of these identities, see Horn and Johnson [86].

[6] If $A \succ 0$ and $B \succ 0$, then $A \otimes B \succ 0$; see Horn and Johnson [86] for a proof.

defined in some interval containing μ_0, and $f(\phi(\mu), \mu) = 0$ if μ belongs to this interval.

Also, $\phi(\mu_0) = (X(\mu_0), y(\mu_0), S(\mu_0))$ and therefore $\phi(\mu) \in \mathcal{S}_n^{++} \times \mathbb{R}^m \times \mathcal{S}_n^{++}$ for μ sufficiently close to μ_0, because ϕ is continuous. We can therefore find an interval containing μ_0 such that the condition $f(\phi(\mu), \mu) = 0$ is equivalent to the centrality conditions (3.1) for values of μ in this interval. Since the centrality conditions have a unique solution, it follows that $\phi(\mu) = (X(\mu), y(\mu), S(\mu))$ in this interval. □

The implicit function theorem also supplies an expression for the tangential direction to the central path. This direction is the solution of a linear system with coefficient matrix (3.10), by (3.9). The tangential direction is the direction used by all primal-dual path-following interior point methods if the current iterate $(X, S) \in \mathcal{P} \times \mathcal{D}$ is on the central path. (The methods differ if the current iterate is *not* on the central path.)

Halická [77] has recently shown that the central path can be analytically extended (see the definition in Appendix D) to $\mu = 0$ in the case of strict complementarity. As a consequence all derivatives of the central path (with respect to μ) have finite limits as $\mu \downarrow 0$, in case of strict complementarity.

3.3 LIMIT POINTS OF THE CENTRAL PATH

In this section we will show that any sequence along the central path has accumulation points in the optimal set, and that these accumulation points are maximally complementary. Then we will prove that as $\mu \downarrow 0$, the central path converges to a maximally complementary solution pair. Under the assumption of strict complementarity, the limit is the so-called *analytic center* of the optimal set which will be defined later on.

In what follows we consider a fixed sequence $\{\mu_t\} \downarrow 0$ with $\mu_t > 0$, $t = 1, \cdots$, and prove that there exists a subsequence of $\{X(\mu_t), S(\mu_t)\}$ which converges to a maximally complementary solution. The existence of limit points of the sequence is an easy consequence of the following lemma.

Lemma 3.2 *Given $\bar{\mu} > 0$, the set*

$$\{(X(\mu), S(\mu)) \ : \ 0 < \mu \le \bar{\mu}\}$$

is contained in a compact subset of $\mathcal{P} \times \mathcal{D}$.

Proof:
The proof follows directly from Lemma 3.1, by noting that $\mathbf{Tr}\,(X(\mu)S(\mu)) \le n\bar{\mu}$ if $\mu \le \bar{\mu}$. □

Now let

$$X(\mu_t) := Q(\mu_t)\Lambda(\mu_t)Q(\mu_t)^T, \quad S(\mu_t) := Q(\mu_t)\Sigma(\mu_t)Q(\mu_t)^T$$

denote the spectral (eigenvector-eigenvalue) decompositions of $X(\mu_t)$ and $S(\mu_t)$. Lemma 3.2 implies that the eigenvalues of $X(\mu_t)$ and $S(\mu_t)$ are bounded. The matrices $Q(\mu_t)$ are orthonormal for all t, and are therefore likewise restricted to a compact set. It follows that the sequence of triples

$$(Q(\mu_t), \Lambda(\mu_t), \Sigma(\mu_t))$$

has an accumulation point, $(Q^*, \Lambda^*, \Sigma^*)$ say. Thus there exists a subsequence of $\{\mu_t\}$ (still denoted by $\{\mu_t\}$ for the sake of simplicity) such that

$$\lim_{t\to\infty} Q(\mu_t) = Q^*, \quad \lim_{t\to\infty} \Lambda(\mu_t) = \Lambda^*, \quad \lim_{t\to\infty} \Sigma(\mu_t) = \Sigma^*.$$

Note that $\Lambda(\mu_t)\Sigma(\mu_t) = \mu I$. Thus, defining

$$X^a := Q^*\Lambda^*{Q^*}^T = \lim_{t\to\infty} X(\mu_t), \quad S^a := Q^*\Sigma^*{Q^*}^T = \lim_{t\to\infty} S(\mu_t), \tag{3.12}$$

we have $\Lambda^*\Sigma^* = 0$ and the pair (X^a, S^a) is optimal.

We are now in the position to show that all limit points of the central path are maximally complementary optimal solutions. The proof given here is based on the proof by De Klerk *et al.* [44].

Theorem 3.4 *The pair* (X^a, S^a) *as defined in (3.12) is a maximally complementary solution pair.*

Proof:

Let (X^*, S^*) be an arbitrary optimal pair. Applying the orthogonality property (Lemma 2.1) and $\mathbf{Tr}\,(X^*S^*) = 0$, $\mathbf{Tr}\,(X(\mu_t)S(\mu_t)) = n\mu_t$ we obtain

$$\mathbf{Tr}\,(X(\mu_t)S^*) + \mathbf{Tr}\,(X^*S(\mu_t)) = n\mu_t.$$

Since $X(\mu_t)S(\mu_t) = \mu_t I$, dividing both sides by μ_t yields

$$\mathbf{Tr}\,(S(\mu_t)^{-1}S^*) + \mathbf{Tr}\,(X^*X(\mu_t)^{-1}) = n \tag{3.13}$$

for all t. This implies

$$\mathbf{Tr}\,(X^*X(\mu_t)^{-1}) \le n, \quad \mathbf{Tr}\,(S(\mu_t)^{-1}S^*) \le n, \tag{3.14}$$

since both terms in the left-hand side of (3.13) are nonnegative. We derive from this that X^a and S^a are maximally complementary. Below we give the derivation for X^a; the derivation for S^a is similar and is therefore omitted.

Denoting the i-th column of the orthonormal (eigenvector) matrix $Q(\mu_t)$ by $q_i(\mu_t)$ and the i-th diagonal element of the (eigenvalue) matrix $\Lambda(\mu_t)$ by $\lambda_i(\mu_t)$ we have

$$X(\mu_t)^{-1} = Q(\mu_t)\Lambda(\mu_t)^{-1}Q(\mu_t)^T = \sum_{i=1}^{n} \frac{1}{\lambda_i(\mu_t)} q_i(\mu_t)q_i(\mu_t)^T. \tag{3.15}$$

Substituting (3.15) into the first inequality in (3.14) yields

$$\begin{aligned}\mathbf{Tr}\,\left(X^*X(\mu_t)^{-1}\right) &= \sum_{i=1}^{n} \mathbf{Tr}\left(\frac{1}{\lambda_i(\mu_t)} X^* q_i(\mu_t)q_i(\mu_t)^T\right) \\ &= \sum_{i=1}^{n} \frac{q_i(\mu_t)^T X^* q_i(\mu_t)}{\lambda_i(\mu_t)} \le n. \end{aligned}\tag{3.16}$$

The last inequality implies

$$q_i(\mu_t)^T X^* q_i(\mu_t) \le n\lambda_i(\mu_t), \quad i = 1, 2, \cdots, n.$$

Letting t go to infinity we obtain

$$q_i^{*T} X^* q_i^* \le n\lambda_i^*, \quad i = 1, 2, \cdots, n,$$

where q_i^* denotes the i-th column of Q^* and λ_i^* the i-th diagonal element of Λ^*. Thus we have $q_i^{*T} X q_i^* = 0$ whenever $\lambda_i^* = 0$. This implies

$$X^* q_i^* = 0 \text{ if } \lambda_i^* = 0, \tag{3.17}$$

since $(q_i^*)^T X^* q_i^* = \left\| X^{*\frac{1}{2}} q_i^* \right\|^2$, where $X^{*\frac{1}{2}}$ is the symmetric square root factor of X^*. In other words, the row space of X is orthogonal to each column q_i^* of Q^* for which $\lambda_i^* = 0$. Hence the row space of X is a subspace of the space generated by the columns q_i^* of Q^* for which $\lambda_i^* > 0$. The latter space is just $\mathcal{R}(Q^*\Lambda^*)$ which equals $\mathcal{R}(X^a)$. Since X^* is symmetric we conclude that $\mathcal{R}(X) \subseteq \mathcal{R}(X^a)$. □

Notation:

In what follows we define

$$\begin{aligned} B &:= \{i : \lambda_i^* > 0\} \\ N &:= \{i : \sigma_i^* > 0\} \\ T &:= \{1, 2, \cdots, n\} \setminus (B \cup N). \end{aligned}$$

Then the sets B, N and T form a partition of the full index set $\{1, 2, \cdots, n\}$. Let Q_J^* denote the submatrix of Q^* consisting of the columns indexed by $J \subseteq \{1, 2, \cdots, n\}$. (The matrices $Q_J(\mu)$ and $\Lambda_J(\mu)$ are defined similarly.) Note that Q_B^*, Q_N^* and Q_T^* are special choices for the matrices Q_B, Q_N and Q_T as defined in Section 2.4. In particular, Q_B^*, Q_N^* and Q_T^* are orthogonal matrices such that

$$\mathcal{R}(Q_B^*) = \mathcal{B}, \ \mathcal{R}(Q_N^*) = \mathcal{N}, \ \mathcal{R}(Q_T^*) = \mathcal{T}.$$

Using this notation, it follows from Theorem 2.7 that any optimal pair (X^*, S^*) can be written as

$$X^* = Q_B^* U_{X^*} Q_B^{*T}, \text{ and } S^* = Q_N^* U_{S^*} Q_N^{*T} \tag{3.18}$$

for suitable matrices $U_{X^*} \succeq 0$ and $U_{S^*} \succeq 0$. In fact, since $Q_B^{*T} Q_B^*$ is equal to the identity matrix $I_{|B|}$ of size $|B|$ and $Q_N^{*T} Q_N^*$ equals the identity matrix $I_{|N|}$ of size $|N|$, U_{X^*} and U_{S^*} follow uniquely from

$$U_{X^*} = Q_B^{*T} X^* Q_B^*, \text{ and } U_{S^*} = Q_N^{*T} S^* Q_N^*. \tag{3.19}$$

Thus, maximally complementary solutions correspond to positive definite U_{X^*} and U_{S^*}.

Definition 3.1 (Analytic center) *The analytic center of $\mathcal{P}^*$ is the (unique) solution of the maximization problem*

$$\max_{U_X \succeq 0} \left\{ \det(U_X) \;\middle|\; Q_B^* U_X {Q_B^*}^T \in \mathcal{P}^* \right\}. \tag{3.20}$$

Similarly, the analytic center of $\mathcal{D}^*$ is the (unique) solution of the maximization problem

$$\max_{U_S \succeq 0} \left\{ \det(U_S) \;\middle|\; Q_N^* U_S {Q_N^*}^T \in \mathcal{D}^* \right\}. \tag{3.21}$$

The uniqueness of the analytic centers follows from the following observations:

1. We know that the optimization problems (3.20) and (3.21) have optimal solutions because the the optimal sets $\mathcal{P}^*$ and $\mathcal{D}^*$ are compact and the determinant is a continuous function. (The compactness of the optimal sets follows by setting $\alpha = 0$ in Lemma 3.1.)

2. We know that $\det(U_X) > 0$ if and only if $X \in \text{ri}(\mathcal{P}^*)$ which implies that the optimal solution in (3.20) will have positive determinant (will be maximally complementary). This means that we can replace $\det(U_X)$ in (3.20) by $\log\det(U_X)$, which is a strictly concave function (by Theorem C.1 in Appendix C).

3. The strict concavity of the $\log\det$ function implies that problem (3.20) has a unique solution.

The analytic center of the optimal set is important in the context of the central path, because it is the (unique) limit point of the central path in the case of strict complementarity.

3.4 CONVERGENCE IN THE CASE OF STRICT COMPLEMENTARITY

We now prove the convergence of the central path to the analytic center of the optimal set under the assumption that a strictly complementary solution exists (*i.e.* $T = \emptyset$, or, equivalently $\mathcal{T} = \{0\}$). The proof is analogous to the proof in the LP case and is taken from De Klerk *et al.* [46].[7]

Theorem 3.5 *If $\mathcal{T} = \{0\}$ (strict complementarity holds) then X^a is the analytic center of $\mathcal{P}^*$ and S^a is the analytic center of $\mathcal{D}^*$.*

Proof:
Just as in the proof of Theorem 3.4, let (X^*, S^*) be an arbitrary optimal pair. We may rewrite (3.13) as

$$\sum_{i=1}^{n} \frac{q_i(\mu_t)^T X^* q_i(\mu_t)}{\lambda_i(\mu_t)} + \sum_{i=1}^{n} \frac{q_i(\mu_t)^T S^* q_i(\mu_t)}{\sigma_i(\mu_t)} = n.$$

[7] An earlier proof of the convergence of the central path in the case of strict complementarity was given by Luo *et al.* [117].

Since all terms in the above sums are nonnegative, this implies

$$\sum_{i\in B}\frac{q_i(\mu_t)^T X^* q_i(\mu_t)}{\lambda_i(\mu_t)}+\sum_{i\in N}\frac{q_i(\mu_t)^T S^* q_i(\mu_t)}{\sigma_i(\mu_t)}\le n.$$

Letting t go to infinity we obtain

$$\sum_{i\in B}\frac{q_i^{*T} X^* q_i^*}{\lambda_i^*}+\sum_{i\in N}\frac{q_i^{*T} S^* q_i^*}{\sigma_i^*}\le n.$$

This can be rewritten as

$$\mathbf{Tr}\left(X^*Q_B^*U_{X^a}^{-1}Q_B^{*T}\right)+\mathbf{Tr}\left(S^*Q_N^*U_{S^a}^{-1}Q_N^{*T}\right)\le n,$$

or

$$\mathbf{Tr}\left(Q_B^{*T}X^*Q_B^*U_{X^a}^{-1}\right)+\mathbf{Tr}\left(Q_N^{*T}S^*Q_N^*U_{S^a}^{-1}\right)\le n.$$

Using the definition of U_X and U_S in (3.19), this implies

$$\mathbf{Tr}\left(U_{X^*}U_{X^a}^{-1}\right)+\mathbf{Tr}\left(U_{S^*}U_{S^a}^{-1}\right)\le n.$$

Since $T=\emptyset$, we have $|B|+|N|=n$. Recall that the matrix $U_{X^*}U_{X^a}^{-1}$ has size $|B|\times|B|$ and $U_{S^*}U_{S^a}^{-1}$ has size $|N|\times|N|$. Applying the arithmetic-geometric mean inequality to the eigenvalues of these matrices we get

$$\det\left(U_{X^*}U_{X^a}^{-1}\right)\det\left(U_{S^*}U_{S^a}^{-1}\right)\le\left(\frac{1}{n}\left(\mathbf{Tr}\left(U_{X^*}U_{X^a}^{-1}\right)+\mathbf{Tr}\left(U_{S^*}U_{S^a}^{-1}\right)\right)\right)^n\le 1,$$

which implies

$$\det\left(U_{X^*}\right)\det\left(U_{S^*}\right)\le\det\left(U_{X^a}\right)\det\left(U_{S^a}\right). \tag{3.22}$$

Substituting $S^*=S^a$ in (3.22) gives $\det(U_{X^*})\le\det(U_{X^a})$ and by setting $X^*=X^a$ we obtain $\det(U_{S^*})\le\det(U_{S^a})$. Thus we have shown that X^a is the analytic center of $\mathcal{P}^*$ and S^a the analytic center of $\mathcal{D}^*$. □

One can also show that the central path passes through the analytic centers of the level sets $\mathbf{Tr}\,(XS)=n\mu$.

Lemma 3.3 *Let $X\in\mathcal{P}$ and $S\in\mathcal{D}$ satisfy $\mathbf{Tr}\,(XS)=n\mu$. One has*

$$\det(XS)\le\det\left(X(\mu)S(\mu)\right),$$

i.e. *the pair $(X(\mu),S(\mu))$ is the analytic center of the level set*

$$\left\{(X,S)\ :\ \mathbf{Tr}\,(XS)=n\mu,\ X\in\mathcal{P},\ S\in\mathcal{D}\right\}.$$

Proof:
Assume that $X\in\mathcal{P}$ and $S\in\mathcal{D}$ satisfy $\mathbf{Tr}\,(XS)=n\mu$. By orthogonality one has

$$\mathbf{Tr}\,\left(X(\mu)-X\right)\left(S(\mu)-S\right)=0,$$

as before. Using $X(\mu)S(\mu) = \mu I$ and $\mathbf{Tr}\,(XS) = n\mu$, simplifies this to

$$\mathbf{Tr}\,\left(X(\mu)^{-1}X\right) + \mathbf{Tr}\,\left(S(\mu)^{-1}S\right) = 2n.$$

Applying the arithmetic-geometric inequality to the eigenvalues of $X(\mu)^{-1}X$ and $S(\mu)^{-1}S$ yields

$$\left[\det\left(X(\mu)^{-1}XS(\mu)^{-1}S\right)\right]^{\frac{1}{2n}} \leq \frac{1}{2n}\left[\mathbf{Tr}\,\left(X(\mu)^{-1}X\right) + \mathbf{Tr}\,\left(S(\mu)^{-1}S\right)\right] = 1,$$

which implies the required result. □

Based on the result of the lemma, it is tempting to conjecture that the central path converges to the analytic center of $\mathcal{P}^* \times \mathcal{D}^*$ even in the absence of strict complementarity, but this is not true.[8]

Example 3.1 (Halická *et al.* [78]) *Let $n = 4$, $m = 4$, $b = [1\ 0\ 0\ 0]^T$ and define (P) and (D) via*

$$C = \begin{bmatrix} 0 & 0 & 0 & 0 \\ 0 & 0 & 0 & 0 \\ 0 & 0 & 0 & 0 \\ 0 & 0 & 0 & 1 \end{bmatrix}, \quad A_1 = \begin{bmatrix} 1 & 0 & 0 & 0 \\ 0 & 1 & 0 & 0 \\ 0 & 0 & 0 & 0 \\ 0 & 0 & 0 & 0 \end{bmatrix},$$

$$A_2 = \begin{bmatrix} 0 & 0 & 0 & 0 \\ 0 & 0 & 0 & 1 \\ 0 & 0 & 1 & 0 \\ 0 & 1 & 0 & 0 \end{bmatrix}, \quad A_3 = \begin{bmatrix} 0 & 0 & 0 & 0 \\ 0 & 0 & 1 & 0 \\ 0 & 1 & 0 & 0 \\ 0 & 0 & 0 & 1 \end{bmatrix}, \quad A_4 = \begin{bmatrix} 0 & 0 & 0 & 0 \\ 0 & 0 & 0 & 0 \\ 0 & 0 & 0 & 1 \\ 0 & 0 & 1 & 0 \end{bmatrix}.$$

The primal problem (P) can be simplified to: minimize x_{44} such that

$$X = \begin{bmatrix} 1 - x_{22} & x_{12} & x_{13} & x_{14} \\ x_{12} & x_{22} & -\frac{1}{2}x_{44} & -\frac{1}{2}x_{33} \\ x_{13} & -\frac{1}{2}x_{44} & x_{33} & 0 \\ x_{14} & -\frac{1}{2}x_{33} & 0 & x_{44} \end{bmatrix} \succeq 0.$$

The optimal set of (P) is given by all the positive semidefinite matrices of the form

$$X^* = \begin{bmatrix} 1 - x_{22} & x_{12} & 0 & 0 \\ x_{12} & x_{22} & 0 & 0 \\ 0 & 0 & 0 & 0 \\ 0 & 0 & 0 & 0 \end{bmatrix}.$$

[8] A proof of this conjecture has been published by Goldfarb and Scheinberg [68], but this proof is incorrect in view of Example 3.1.

Solutions of the form X^* are clearly optimal, since $C \succeq 0$ and therefore $\mathbf{Tr}\,(CX) \geq 0$ $\forall X \in \mathcal{P}$.

The analytic center of $\mathcal{P}^*$ is obviously given by

$$\begin{bmatrix} \frac{1}{2} & 0 & 0 & 0 \\ 0 & \frac{1}{2} & 0 & 0 \\ 0 & 0 & 0 & 0 \\ 0 & 0 & 0 & 0 \end{bmatrix}.$$

However, we will show that the limit point of the primal central path satisfies

$$X(\mu) \to \begin{bmatrix} \frac{2}{5} & 0 & 0 & 0 \\ 0 & \frac{3}{5} & 0 & 0 \\ 0 & 0 & 0 & 0 \\ 0 & 0 & 0 & 0 \end{bmatrix} \quad as\ \mu \downarrow 0.$$

The dual problem is to maximize y_1 such that

$$S = \begin{bmatrix} -y_1 & 0 & 0 & 0 \\ 0 & -y_1 & -y_3 & -y_2 \\ 0 & -y_3 & -y_2 & -y_4 \\ 0 & -y_2 & -y_4 & 1 - y_3 \end{bmatrix} \succeq 0.$$

Thus the dual problem has a unique optimal solution

$$y_i^* = 0\ (i = 1, 2, 3, 4),\ S^* = \begin{bmatrix} 0 & 0 & 0 & 0 \\ 0 & 0 & 0 & 0 \\ 0 & 0 & 0 & 0 \\ 0 & 0 & 0 & 1 \end{bmatrix}.$$

It is also easy to see that strict complementary does not *hold.*

Due to the structure of feasible $S \in \mathcal{D}$, the primal central path has the following structure

$$X(\mu) = \begin{bmatrix} 1 - x_{22}(\mu) & 0 & 0 & 0 \\ 0 & x_{22}(\mu) & -\frac{1}{2}x_{44}(\mu) & -\frac{1}{2}x_{33}(\mu) \\ 0 & -\frac{1}{2}x_{44}(\mu) & x_{33}(\mu) & 0 \\ 0 & -\frac{1}{2}x_{33}(\mu) & 0 & x_{44}(\mu) \end{bmatrix}.$$

By Lemma 3.3, the point on the central path $X(\mu)$ is, for any $\mu > 0$, the analytic center of a level set. The level set is given by the primal feasibility and a level condition which is $x_{44} = x_{44}(\mu) > 0$ in our case. This implies that $X(\mu)$ maximizes

$$\det \begin{bmatrix} 1 - x_{22} & 0 & 0 & 0 \\ 0 & x_{22} & -\frac{1}{2}x_{44}(\mu) & -\frac{1}{2}x_{33} \\ 0 & -\frac{1}{2}x_{44}(\mu) & x_{33} & 0 \\ 0 & -\frac{1}{2}x_{33} & 0 & x_{44}(\mu) \end{bmatrix} \tag{3.23}$$

under the conditions

$$x_{22} \in (0, 1), \quad x_{33} > 0, \quad x_{22}x_{33}x_{44}(\mu) - \frac{x_{33}^3 - x_{44}^3(\mu)}{4} > 0.$$

These constraints are not binding at the optimal solution (the μ-center). We therefore obtain necessary and sufficient conditions for the analytic center by setting the gradient (with respect to x_{22} and x_{33}) of the determinant in (3.23) to zero:

$$x_{33}(\mu)x_{44}(\mu) - 2x_{22}(\mu)x_{33}(\mu)x_{44}(\mu) + \frac{1}{4}x_{44}(\mu)^3 + \frac{1}{4}x_{33}(\mu)^3 = 0 \tag{3.24}$$

$$(1 - x_{22}(\mu))\left(x_{22}(\mu)x_{44}(\mu) - \frac{3}{4}x_{33}(\mu)^2\right) = 0. \tag{3.25}$$

Using $x_{22}(\mu) \in (0, 1)$, we deduce from (3.25) that

$$x_{33}(\mu) = \frac{2}{\sqrt{3}}\sqrt{x_{22}(\mu)x_{44}(\mu)}.$$

Substituting this expression in (3.24) and simplifying, we obtain:

$$\frac{2}{\sqrt{3}}\sqrt{x_{22}(\mu)} - \frac{10}{3\sqrt{3}}x_{22}(\mu)^{3/2} + \frac{1}{4}x_{44}(\mu)^{3/2} = 0.$$

We assume that the central path converges for this example — a proof that the central path always converges will be given in Theorem 3.6. In the limit where $\mu \downarrow 0$, we have $x_{44}(\mu) \to 0$, since the limit point must be an optimal solution. Moreover, we can assume that $x_{22}(\mu)$ is positive in the limit, since the limit point of the central path is maximally complementary (Theorem 3.4). Denoting $\lim_{\mu \downarrow 0} x_{22}(\mu) := x_{22}(0) > 0$, we have:

$$\frac{2}{\sqrt{3}}\sqrt{x_{22}(0)} - \frac{10}{3\sqrt{3}}x_{22}(0)^{3/2} = 0,$$

that implies $x_{22}(0) = \frac{3}{5}$. □

The following example shows that the central path may already fail to converge to the analytic center of the optimal set in the special case of second order cone optimization.

Example 3.2 (Halická *et al.* [78]) *Consider the problem of minimizing x_{12} subject to*

$$\begin{bmatrix} x_{11} & x_{12} & 0 & 0 & 0 \\ x_{12} & x_{22} & 0 & 0 & 0 \\ 0 & 0 & x_{33} & x_{22} & 0 \\ 0 & 0 & x_{22} & x_{12} & 0 \\ 0 & 0 & 0 & 0 & 1-(x_{11}+x_{33}) \end{bmatrix} \succeq 0.$$

Note that this problem is equivalent to a second order cone programming problem (the semidefiniteness constraint can be written as linear and convex quadratic inequality constraints). The optimal set is given by all matrices of the above form where $x_{12} = x_{22} = 0$, and the analytic center of the optimal set is given by the optimal solution where $x_{11} = x_{33} = \frac{1}{3}$.

Using exactly the same technique as in the previous example, one can show that the limit point for the central path is $x_{11} = 2/7$, $x_{33} = 3/7$. However, the proof is more technical for this example due to the larger number of variables, and is therefore omitted. □

3.5 CONVERGENCE PROOF IN THE ABSENCE OF STRICT COMPLEMENTARITY

Since the central path does not converge to the analytic center in general, we give a proof that it always converges.[9]

We will use a well-known result from algebraic geometry, namely the *Curve selection lemma.*[10]

Definition 3.2 (Algebraic set) *A subset $V \in R^k$ is called an algebraic set if V is the locus of common roots of some collection of polynomial functions on $\mathbf{R}^k$.*

Lemma 3.4 (Curve selection lemma) *Let $V \subset \mathbf{R}^k$ be a real algebraic set, and let $U \subset \mathbf{R}^k$ be an open set defined by finitely many polynomial inequalities:*

$$U = \left\{x \in \mathbf{R}^k : g_1(x) > 0, ..., g_l(x) > 0\right\}.$$

---

[9]The convergence property seems to be a 'folklore' result which is, for example, already mentioned in the review paper by Vandenberghe and Boyd [181] on p. 74 without supplying references or a proof. In Kojima *et al.* [104] the convergence of the central path for the linear complementary problem (LCP) is proven with the help of some results from algebraic geometry. In [108], Kojima *et al.* mention that this proof for LCP can be extended to the monotone semidefinite complementarity problem (which is equivalent to SDP), without giving a formal proof. The proof given here is taken from Halická *et al.* [78], and uses ideas from the theory of algebraic sets, but in a different manner than it was done by Kojima *et al.* [104]. An earlier — but more complicated — proof of convergence of the central path in a more general setting is given by Graña Drummond and Peterzil [71].

[10]A proof of the Curve selection lemma is given by Milnor [122] (Lemma 3.1).

If $U \cap V$ contains points arbitrarily close to the origin, then there exists an $\epsilon > 0$ and a real analytic curve

$$p : [0, \epsilon) \mapsto \mathbb{R}^k$$

with $p(0) = 0$ and with $p(t) \in U \cap V$ for $t > 0$.

We will now prove the convergence of the central path by applying the Curve selection lemma.

Theorem 3.6 (Convergence of the central path) *Even in the absence of strict complementarity, the central path*

$$\{(X, S) \mid \mu > 0,\ XS = \mu I,\ X \in \mathcal{P}, S \in \mathcal{D}\}$$

has a unique limit point in $\mathcal{P}^* \times \mathcal{D}^*$.

Proof:
Let (X^*, y^*, S^*) be any limit point of the central path of (P) and (D).

With reference to Lemma 3.4, let the real algebraic set V be defined by

$$V = \left\{ (\bar{X}, \bar{S}, \bar{y}, \mu) \;\middle|\; \begin{array}{rcl} \mathbf{Tr}\, A_i\bar{X} & = & 0 \quad (i = 1, \ldots, m) \\ \sum_i (\bar{y}_i) A_i + \bar{S} & = & 0 \\ (\bar{X} + X^*)(\bar{S} + S^*) - \mu I & = & 0 \end{array} \right\}$$

and let the open set U be defined by: $U = (\bar{X}, \bar{S}, \bar{y}, \mu)$ such that all principal minors of $(\bar{X} + X^*)$ and $(\bar{S} + S^*)$ are positive, and $\mu > 0$.

Now $V \cap U$ corresponds to the central path excluding its limit points, in the sense that if $(\bar{X}, \bar{S}, \bar{y}, \mu) \in U \cap V$, then $X(\mu) = (\bar{X} + X^*)$ and $S(\mu) = (\bar{S} + S^*)$, where $X(\mu)$ (resp. $S(\mu)$) denotes the μ-center of (P) (resp. (D)) as before.

Moreover, the zero element is in the closure of $V \cap U$, by construction.

The required result now follows from the Curve selection lemma. To see this, note that Lemma 3.4 implies the existence of an $\epsilon > 0$ and an analytic function $p : [0, \epsilon) \mapsto \mathcal{S}^n \times \mathcal{S}^n \times \mathbb{R}^m \times \mathbb{R}$ such that

$$p(t) = \left(\bar{X}(t), \bar{S}(t), \bar{y}(t), \mu(t)\right) \to (0_{n\times n}, 0_{n\times n}, 0_m, 0) \text{ as } t \downarrow 0, \tag{3.26}$$

and if $t > 0$, $\left(\bar{X}(t), \bar{S}(t), \bar{y}(t), \mu(t)\right) \in U \cap V$, *i.e:*

$$\begin{array}{rcl} \mathbf{Tr}\, A_i\bar{X}(t) & = & 0 \quad (i = 1, \ldots, m) \\ \sum_i \bar{y}_i(t) A_i + \bar{S}(t) & = & 0 \\ (\bar{X}(t) + X^*)(\bar{S}(t) + S^*) - \mu(t) I & = & 0, \end{array} \tag{3.27}$$

and $\bar{X}(t) \succ 0$, $\bar{S}(t) \succ 0$, $\mu(t) > 0$.

Since the centrality system (3.1) has a unique solution, the system (3.27) also has a unique solution given by

$$\bar{X}(t) + X^* = X(\mu(t)), \quad \bar{S}(t) + S^* = S(\mu(t))$$

if $t > 0$. By (3.26), we therefore have

$$\lim_{t \downarrow 0} X(\mu(t)) = X^*, \quad \lim_{t \downarrow 0} S(\mu(t)) = S^*, \quad \lim_{t \downarrow 0} \mu(t) = 0.$$

Since $\mu(t) > 0$ on $(0, \epsilon)$, $\mu(0) = 0$, and $\mu(t)$ is analytic on $[0, \epsilon)$, there exists an interval, say $(0, \epsilon')$ where $\mu'(t) > 0$ (see Theorem D.1 in Appendix D). Therefore the inverse function $\mu^{-1} : \mu(t) \mapsto t$ exists on the interval $(0, \mu(\epsilon'))$. Moreover, $\mu^{-1}(t) > 0$ for all $t \in (0, \mu(\epsilon'))$, and $\lim_{t \downarrow 0} \mu^{-1}(t) = 0$.

This implies that

$$\lim_{t \downarrow 0} X(t) = \lim_{t \downarrow 0} X(\mu(\mu^{-1}(t))) = \lim_{t \downarrow 0} \bar{X}(\mu^{-1}(t)) + X^* = X^*.$$

Similarly, $\lim_{t \downarrow 0} S(t) = S^*$. Since the limit point (X^*, S^*) was arbitrary, it must therefore be unique. □

At the time of writing, it is an open problem to correctly characterize the limit point of the central path in the absence of strict complementarity.

3.6 CONVERGENCE RATE OF THE CENTRAL PATH

It is natural to ask what the convergence rate of the central path is. In the case of strict complementarity, one can show the following.

Theorem 3.7 (Luo *et al.* [117]) *Assume that $T = \emptyset$ (strict complementarity) and let (X^a, S^a) denote the analytic center of $\mathcal{P}^* \times \mathcal{D}^*$ as before. It holds that*

$$\|X(\mu) - X^a\| = O(\mu), \quad \|S(\mu) - S^a\| = O(\mu).$$

In the absence of strict complementarity the convergence rate can be much worse, as the following example shows.

Example 3.3 (Adapted from Sturm [168]) *Consider the following instance which is in the standard dual form (D):*

$$\max -y_n$$

subject to

$$S := \begin{bmatrix} 1 & y_1 & y_2 & \cdots & y_{n-1} \\ y_1 & y_2 & 0 & \cdots & 0 \\ y_2 & 0 & y_3 & \ddots & \vdots \\ \vdots & \vdots & \ddots & \ddots & 0 \\ y_{n-1} & 0 & \cdots & 0 & y_n \end{bmatrix} \succeq 0.$$

The unique optimal solution is given by $y_i = 0$ $(i = 1, \ldots, n)$, yet on the central path one has

$$y_2(\mu) = c\mu^{2^{-(n-2)}},$$

for some constant c *that is independent of* μ*. We therefore have*

$$\|S(\mu) - S^*\| = \Omega\left(\mu^{2^{-n}}\right),$$

where S^* *denotes the unique optimal solution.* □

At the time of writing, it remains an open problem to determine the worst-case convergence rate of the central path in the absence of strict complementarity.

4
SELF–DUAL EMBEDDINGS

Preamble

If an SDP problem is also its own dual problem, then the duality relations are much simpler. In particular, if such a self–dual problem is strictly feasible, then it is solvable and the optimal value is zero, by the strong duality theorem (Theorem 2.2). In this chapter we show how to embed a given pair of problems in standard form (P) and (D) in a bigger self–dual problem for which a strictly feasible solution on the central path is known. One can therefore solve the embedding problem, and its solution gives information about the feasibility and solvability of (P) and (D).

4.1 INTRODUCTION

Most semidefinite programming algorithms found in the literature require strictly feasible starting points $(X^0 \succ 0, S^0 \succ 0)$ for the primal and dual problems respectively. So-called 'big-M' methods (see *e.g.* Vandenberghe and Boyd [180]) are often employed in practice to obtain feasible starting points. For example, consider an SDP problem in the standard dual form:

$$(D): \quad d^* := \sup_{y,S} \left\{ b^T y \;\middle|\; \sum_{i=1}^{m} y_i A_i + S = C,\ S \succeq 0,\ y \in \mathbf{R}^m \right\}.$$

Assume that a strictly feasible solution (y^0, S^0) of (D) is known, but no strictly primal feasible point is known for its Lagrangian dual (P) (which is in standard primal form):

$$(P): \;\; p^* := \inf_X \left\{ \mathbf{Tr}\,(CX) \mid \mathbf{Tr}\,(A_i X) = b_i \; (i = 1, \ldots, m), \; X \succeq 0 \right\}.$$

In order to apply primal–dual methods (where feasible solutions to both (P) and (D) are required), one could solve the modified problem

$$\sup_{y,S} \left\{ b^T y \;\middle|\; \sum_{i=1}^m y_i A_i + S = C, \; \mathbf{Tr}\,(S) \le \mathrm{M}, \; S \succeq 0. \right\} \tag{4.1}$$

Roughly speaking, problem (4.1) has the same solution status as (D) if M is 'large enough' — if (D) is infeasible, then so is (4.1), and if (D) is solvable then so is (4.1).[1]

A slightly modified dual of problem (4.1) is

$$\inf_{\kappa \ge 0, X \succeq 0} \left\{ \mathbf{Tr}\;(C(X - \kappa I)) + \kappa \mathrm{M} \mid \mathbf{Tr}\;(A_i(X - \kappa I)) = b_i \; (i = 1, \ldots, m) \right\}. \tag{4.2}$$

One can now construct a strictly feasible starting point for problem (4.2) by choosing κ 'large enough', while no such starting point was available for (P).

An analogous strategy is available if only a strictly feasible primal solution is available. If neither a primal nor a dual feasible point is known, a similar strategy can again be employed by introducing two 'big-M' parameters (see Vandenberghe and Boyd [180] for details). The difficulty with these approaches is that no a priori choice for the 'big-M' parameters is available in general. For example, if κ cannot be driven to zero in problem (4.2), then one can only conclude that 'there is no optimal solution S^* of (D) with $\mathbf{Tr}\,(S^*) \le \mathrm{M}$'. We therefore need an a priori bound on $\mathbf{Tr}\,(S^*)$ in order to give a certificate of the problem status of (D), while such information is not available in general.

In the LP case an elegant solution for the initialization problem is to embed the original problem in a skew–symmetric self–dual problem which has a known interior feasible solution on the central path. The solution of the embedding problem then yields the optimal solution to the original problem, or gives a certificate of either primal or dual infeasibility. In this way detailed information about the solution is obtained.[2]

Despite the desirable theoretical properties of self-dual embeddings, the idea did not receive immediate recognition in implementations, due to the fact that the embedding problem has a dense column in the coefficient matrix. This can lead to fill-in of Choleski factorizations during computation. In spite of this perception, Xu *et al.* [185]

[1] If (D) is feasible but not solvable and d^* is finite, then the big-M approach described here will not work. In particular, the modified problem 4.1 will have an optimal solution even though (D) does not.

[2] The idea of self-dual embeddings for LP dates back to the 1950's and the work of Goldman and Tucker [69]. With the arrival of interior point methods, the embedding idea was revived to be used in infeasible start algorithms by Ye *et al.* [191] (see also Jansen *et al.* [89]).

have made a successful implementation for LP using the embedding, and it has even been implemented as an option in the well-known commercial LP solver CPLEX-barrier and as the default option in the commercial solvers XPRESSMP and MOSEK. The common consensus now is that this strategy is a reliable way of detecting infeasibility, and promises to be competitive in practice (see Andersen *et al.* [10]).

For semidefinite programming the homogeneous embedding idea was first developed by Potra and Sheng [148]. The embedding strategy was extended by De Klerk *et al.* in [44] and independently by Luo *et al.* [118] to obtain self-dual embedding problems with nonempty interiors. The resulting embedding problem has a known centered starting point, unlike the homogeneous embedding; this simplifies the analysis, since the central path is well-defined. The embedding technique is implemented in the SDP solver SeDuMi by Sturm [167].

A *maximally complementary* solution (*e.g.* the limit of the central path) of the embedding problem yields one of the following alternatives about the original problem pair:

(I) a complementary solution pair $(x^*, S^*) \in \mathcal{P}^* \times \mathcal{D}^*$ is obtained for the original problem pair (P) and (D);

(II) an improving ray is obtained for either the primal and/or dual problem (so-called *strong infeasibility* (see Definition 2.4) is detected);

(III) a certificate is obtained that no complementary solution pair exists and that neither (P) nor (D) is strongly infeasible. This can only happen if one or both of the primal and dual SDP problems fail to satisfy the Slater regularity condition.

Loosely speaking, the original primal and dual problems are solved if a complementary solution pair exists, or if one or both of the problems are strongly infeasible.

Unfortunately, some pathological duality effects can occur for SDP[3] which are absent from LP, for example:

- A positive duality gap at an optimal primal-dual solution pair;
- an arbitrarily small duality gap can be attained by feasible primal-dual pairs, but no optimal pair exists;
- an SDP problem may have a finite optimal value even though its (Lagrangian) dual is infeasible.

These situations cannot be identified with the embedding approach, unless an additional assumption is made: if the primal and dual problems are in so-called perfect duality, and one is solvable, then additional information can be obtained from the embedding, as was shown by De Klerk *et al.* [46]. This assumption holds if for example the primal is strictly feasible and the dual is feasible, by the strong duality theorem (Theorem 2.2). Moreover, one can replace the problem to be solved by a larger problem which – together with its Lagrangian dual – satisfy these assumptions. In theory at least, we can therefore make the assumption without loss of generality.

[3]Examples of these effects are given in Examples 2.1, 2.2, and 4.2.

4.2 THE EMBEDDING STRATEGY

In what follows, we make no assumptions about feasibility of (P) and (D).

Consider the following *homogeneous embedding*[4] of (P) and (D):

$$
\left.\begin{array}{llllll}
 & \mathbf{Tr}\,(A_iX) & -\tau b_i & & & = 0 \quad \forall i \\
-\sum_{i=1}^m y_iA_i & & +\tau C & -S & & = 0 \\
b^Ty & -\mathbf{Tr}\,(CX) & & & -\rho & = 0 \\
y \in \mathbf{R}^m, & X \succeq 0, & \tau \geq 0, & S \succeq 0, & \rho \geq 0. &
\end{array}\right\} \tag{4.3}
$$

A feasible solution to this system with $\tau > 0$ yields feasible solutions $\frac{1}{\tau}X$ and $\frac{1}{\tau}S$ to (P) and (D) respectively (by dividing the first two equations by τ). The last equation guarantees optimality by requiring a nonpositive duality gap. For this reason there is no strictly feasible solution to (4.3).

Here we will describe an *extended self-dual embedding* due to De Klerk *et al.* [44], in order to have a strictly feasible, self–dual SDP problem with a known starting point on the central path.

The strictly feasible embedding is obtained by extending the constraint set (4.3) and adding extra variables to obtain:

$$
\min_{y,X,\tau,\theta,S,\rho,\nu} \theta\beta
$$

subject to

$$
\begin{array}{llllllll}
 & \mathbf{Tr}\,(A_iX) & -\tau b_i & +\theta\bar{b}_i & & & & = 0 \\
-\sum_{j=1}^m y_jA_j & & +\tau C & -\theta\bar{C} & -S & & & = 0 \\
b^Ty & -\mathbf{Tr}\,(CX) & & +\theta\alpha & & -\rho & & = 0 \\
-\bar{b}^Ty & +\mathbf{Tr}\,(\bar{C}X) & -\tau\alpha & & & & -\nu & = -\beta \\
y \in \mathbf{R}^m, & X \succeq 0, & \tau \geq 0, & \theta \geq 0, & S \succeq 0, & \rho \geq 0, & \nu \geq 0 &
\end{array} \tag{4.4}
$$

where

$$
\begin{aligned}
\bar{b}_i &:= b_i - \mathbf{Tr}\,(A_i) \\
\bar{C} &:= C - I \\
\alpha &:= 1 + \mathbf{Tr}\,(C) \\
\beta &:= n + 2,
\end{aligned}
$$

and $i = 1, \ldots, n$.

---

[4]The formulation (4.3) was first solved by Potra and Sheng [148, 149] using an infeasible interior point method, and recently it has been incorporated in the SeDuMi software of Sturm [167].

It is straightforward to verify that a feasible interior starting solution is given by $y^0 = 0$, $X^0 = S^0 = I$, and $\theta^0 = \rho^0 = \tau^0 = \nu^0 = 1$.

Note also that the solution with $\nu = \beta$ and all other variables zero is optimal, since the objective function is always nonnegative. In other words, $\theta = 0$ in any optimal solution. It is therefore a trivial matter to find an optimal solution, but we are only interested in *maximally complementary solutions*.

The underlying idea is as follows: we wish to know whether an optimal solution to the embedding exists with $\tau^* > 0$. (Recall that this yields a complementary solution pair to (P) and (D).) A maximally complementary solution will yield such a solution, if it exists. In the next section it will become clear how maximally complementary solutions may be obtained as the limit of the *central path* of the embedding problem.

SELF–DUALITY OF THE EMBEDDING PROBLEM

The dual of the embedding problem is essentially the same as the embedding problem itself, which explains the 'self-dual' terminology. To prove it, we first consider a generic form of self-dual conic optimization problems.

Theorem 4.1 *Let $\mathcal{K} \subset \mathbb{R}^k$ be a closed convex cone with dual cone $\mathcal{K}^*$ and let $A \in \mathbb{R}^{k\times k}$ be skew-symmetric. Then the Lagrangian dual of the optimization problem*

$$p^*_{sf} = \inf_{x,s} \left\{ c^T x \mid Ax - s = -c,\ x \in \mathcal{K},\ s \in \mathcal{K}^* \right\} \tag{4.5}$$

is given by

$$-\inf_{x,s} \left\{ c^T x \mid Ax - s = -c,\ x \in \mathcal{K},\ s \in \mathcal{K}^* \right\}.$$

It follows that if problem (4.5) is strictly feasible, then $p^*_{sf} = 0$.

Proof:
The Lagrangian associated with problem (4.5) is

$$L(x, s, y) = c^T x + y^T (Ax - s + c) = (A^T y + c)^T x - y^T s + y^T c,$$

and the Lagrangian dual of problem (4.5) is defined by

$$\sup_{y\in\mathbb{R}^k} \inf_{x\in\mathcal{K},\ s\in\mathcal{K}^*} L(x, s, y) = \sup_{y\in\mathbb{R}^k} \left\{ c^T y + \inf_{x\in\mathcal{K},\ s\in\mathcal{K}^*} \left\{ (A^T y + c)^T x - y^T s \right\} \right\}.$$

The inner minimization problem will only be bounded from below if $A^T y + c \in \mathcal{K}^*$ and $-y \in \mathcal{K}$, in which case the optimal value of the inner minimization problem is zero. We can therefore simplify the Lagrangian dual to

$$\sup_{-y\in\mathcal{K}} \left\{ c^T y \mid A^T y + c \in \mathcal{K}^* \right\}.$$

We now use the skew symmetry $A^T = -A$ and substitute new variables $u = -y$ and $v = A^T y + c = Au + c$ to obtain the problem

$$\sup_{u,v} \left\{ -c^T u \mid Au - v = -c,\ u \in \mathcal{K},\ v \in \mathcal{K}^* \right\}.$$

Switching from maximization to minimization we obtain the required result. □

We can now show that problem (4.4) is self–dual by casting it in the generic form (4.5).

Corollary 4.1 *The embedding problem (4.4) is self–dual.*

Proof:
With reference to problem (4.4), we construct the skew–symmetric matrix A in (4.5) as follows:

$$A := \begin{bmatrix} 0 & \mathcal{A} & -b & \bar{b} \\ -\mathcal{A}^T & 0 & \mathbf{svec}(C) & -\mathbf{svec}\left(\bar{C}\right) \\ b^T & -\mathbf{svec}(C)^T & 0 & \alpha \\ -\bar{b}^T & \mathbf{svec}(\bar{C})^T & -\alpha & 0 \end{bmatrix},$$

where $\mathcal{A}$ is the $m \times \frac{1}{2}n(n+1)$ matrix with row i given by $\mathbf{svec}(A_i)^T$ $(i = 1, \ldots, m)$.
We further group the variables in (4.4) as

$$x = \left[y^T, \mathbf{svec}(X)^T, \tau, \theta\right]^T, \quad s = \left[z^T, \mathbf{svec}(S)^T, \rho, \nu\right]^T,$$

where $z \in \mathbf{R}^m$ is an auxiliary vector of variables which will be restricted to the zero vector; this is accomplished by choosing the cone $\mathcal{K}$ as[5]

$$\mathcal{K} = \mathbf{R}^m \times \mathbf{svec}\left(\mathcal{S}_n^+\right) \times \mathbf{R}_+ \times \mathbf{R}_+.$$

The dual cone is therefore given by

$$\mathcal{K}^* = 0_m \times \mathbf{svec}\left(\mathcal{S}_n^+\right) \times \mathbf{R}_+ \times \mathbf{R}_+.$$

Note that the condition $s \in \mathcal{K}^*$ in (4.5) implies $z = 0_m$, as required. Finally, we define $c \in \mathbf{R}^{m+\frac{1}{2}n(n+1)+2}$ as having last component β and all other components zero.
It is straightforward to verify that we have now cast the embedding problem (4.4) in the generic form (4.5), so that the self-duality of the embedding follows from Theorem 4.1. □

COMPLEMENTARITY FOR THE EMBEDDING PROBLEM

The strict feasibility and self-duality of the embedding problem (4.4) imply that the duality gap equals $2\theta\beta$. It is readily verified that

$$\theta\beta = \mathbf{Tr}\ (XS) + \tau\rho + \theta\nu. \tag{4.6}$$

[5]We use the notation $\mathbf{svec}\left(\mathcal{S}_n^+\right) := \left\{x \in \mathbf{R}^{\frac{1}{2}n(n+1)} \;\middle|\; \mathbf{smat}(x) \in \mathcal{S}_n^+\right\}$.

This shows that an optimal solution (where $\theta\beta = 0$) must satisfy the complementarity conditions:

$$\left.\begin{array}{rcl} XS &=& 0 \\ \rho\tau &=& 0 \\ \theta\nu &=& 0. \end{array}\right\} \tag{4.7}$$

4.3 SOLVING THE EMBEDDING PROBLEM

The embedding problem can be solved by any interior point method that 'follows' the central path. But first we must formalize the notion of the central path of the embedding, since the embedding problem is not in standard form. To this end, one can relax the complementarity optimality conditions (4.7) of the embedding problem to

$$\begin{array}{rcl} XS &=& \mu I \\ \tau\rho &=& \mu \\ \nu\theta &=& \mu. \end{array}$$

If one defines new 'primal and dual variables' $\tilde{X}, \tilde{S}$ of dimension $n+2$ as follows:

$$\tilde{X} = \begin{bmatrix} X & & \\ & \tau & \\ & & \nu \end{bmatrix}, \quad \tilde{S} = \begin{bmatrix} S & & \\ & \rho & \\ & & \theta \end{bmatrix},$$

then the central path of the embedding problem can be defined as the (unique) solution of

$$\tilde{X}\tilde{S} = \mu I, \quad \mu > 0,$$

subject to (4.4), denoted by $(\tilde{X}(\mu), \tilde{S}(\mu))$ for each $\mu > 0$.

Search directions for the embedding problem

Several interior point algorithms will be analyzed in the following chapters. This analysis, however, will be for problems in the standard form (P) and (D). The only part of the analysis that is different for the embedding problem, is the actual computation of feasible search directions. In other words, once the search directions have been computed the rest of the analysis uses only the orthogonality property, which also holds here. To see this, let $(\Delta\tilde{X}, \Delta\tilde{S})$ denote any feasible direction for the embedding problem at the feasible point $(\tilde{X}, \tilde{S})$. It is straightforward to verify that $\mathbf{Tr}\left(\Delta\tilde{X}\Delta\tilde{S}\right) = 0$, *i.e.* the *orthogonality principle* holds. Note that this is enough to ensure the existence and uniqueness of the central path, by the proof of Theorem 3.1 (see Remark 3.1). Moreover, it is enough to ensure that the limit of the central path is maximally complementary (see the proof of Theorem 3.4).

In other words, all we have to show here is that the search directions used in interior point algorithms are well-defined for the embedding problem.

For continuity of presentation we defer this analysis to Appendix F.

4.4 INTERPRETING THE SOLUTION OF THE EMBEDDING PROBLEM

We can now use a maximally complementary solution of the embedding problem (4.4) to obtain information about the original problem pair (P) and (D). In particular, one can distinguish between the three possibilities as discussed in the introduction, namely

(I) A primal–dual complementary pair $(X^*, S^*) \in \mathcal{P}^* \times \mathcal{D}^*$ is obtained;

(II) A primal and/or dual improving ray is detected;

(III) A certificate is obtained that no complementary pair exists, and that neither (P) nor (D) has an improving ray.

Given a maximally complementary solution of the embedding problem, these cases are distinguished as follows:

Theorem 4.2 *Let $(y^*,\ X^*,\ \tau^*,\ \theta^*,\ S^*,\ \rho^*,\ \nu^*)$ be a maximally complementary solution of the self–dual embedding problem (4.4). Then:*

(i) if $\tau^* > 0$, then case (I) *holds;*

(ii) if $\tau^* = 0$ and $\rho^* > 0$, then case (II) *holds;*

(iii) if $\tau^* = \rho^* = 0$, then case (III) *holds.*

Proof:
Consider the two possibilities $\tau^* = 0$ and $\tau^* > 0$.

Case: $\tau^* > 0$
Here, $\frac{1}{\tau^*}X^*$ and $\frac{1}{\tau^*}S^*$ are maximally complementary and optimal for (P) and (D) respectively, *i.e.* case (I) holds.

Case: $\tau^* = 0$
In this case, one has $\tau = 0$ in any optimal solution of the embedding problem. This implies no complementary solution pair for (P) and (D) exists, because if such a pair exists we can construct an optimal solution of the embedding problem with $\tau > 0$ as follows: Let a complementary pair $(X_P, S_D) \in \mathcal{P}^* \times \mathcal{D}^*$ be given, and set $\theta^* = \rho^* = 0$, $X^* = \tau^* X_P$, and $S^* = \tau^* S_D$, where $\tau^* > 0$ will be specified presently. This choice of variables already satisfies the first three constraints of the embedding problem for any choice of $\tau^* > 0$ (see (4.4)). The fourth equation of the embedding problem can now be simplified to:

$$\begin{aligned} \nu^* &= n + 2 - \mathbf{Tr}\,(X^* + S^*) - \tau^* \\ &= n + 2 - \tau^* \left(\mathbf{Tr}\,(X_P + S_P) + 1\right). \end{aligned}$$

One can therefore choose any $\tau^* > 0$ which satisfies

$$\tau^* < \frac{n+2}{\mathbf{Tr}\,(X_P + S_P) + 1}, \tag{4.8}$$

to obtain a value $\nu^* \geq 0$. We have therefore constructed an optimal solution of the embedding problem with $\tau^* > 0$.

If $\tau^* = 0$ it also follows that $\mathbf{Tr}\,(A_i X^*) = 0$ for all i and $\sum_{i=1}^m y_i^* A_i \preceq 0$. Now we distinguish between two sub-cases: $\rho^* > 0$ and $\rho^* = 0$.

Sub-case: $\tau^* = 0$ and $\rho^* > 0$
Here one has $b^T y^* - \mathbf{Tr}\,(CX^*) > 0$, *i.e.* $b^T y^* > 0$ and/or $\mathbf{Tr}\,(CX^*) < 0$. In other words, there are primal and/or dual improving rays and case (II) applies. If $b^T y^* > 0$ then y^* is a dual improving ray. In this case (P) is infeasible by Lemma 2.2, and if (D) is feasible it is unbounded. If $\mathbf{Tr}\,(CX^*) < 0$ then there exists a primal improving ray. In this case (D) is (strictly) infeasible, and if (P) is feasible it is unbounded. If both $b^T y^* > 0$ and $\mathbf{Tr}\,(CX^*) < 0$, then both a primal and a dual improving ray exist and in this case both (P) and (D) are infeasible.

Conversely, one must show that if there exists a primal and/or dual improving ray, then any maximally complementary solution of the embedding problem must have $\rho^* > 0$ and $\tau^* = 0$. Given a primal improving ray $\bar{X} \succeq 0$, one can construct an optimal solution to the embedding by setting $X^* = \kappa \bar{X}$, where $\kappa > 0$ is a constant to be specified later, and further setting $\tau^* = 0$, $\theta^* = 0$ (which guarantees optimality), and $y^* = 0$, to obtain:

$$\begin{aligned}
\rho^* &= -\kappa \mathbf{Tr}\,(C\bar{X}) > 0 \\
\kappa \mathbf{Tr}\,(A_i \bar{X}) = \mathbf{Tr}\,(A_i X^*) &= 0, \quad i = 1, \ldots, m \\
S^* &= 0 \\
\nu^* &= n + 2 + \kappa \mathbf{Tr}\,\left(C\bar{X} - \bar{X}\right).
\end{aligned}$$

The first three equations show that ρ^*, X^* and S^* are feasible. It remains to prove that ν^* is nonnegative. This is ensured by choosing

$$\kappa = \frac{-1}{\mathbf{Tr}\,\left(C\bar{X} - \bar{X}\right)} > 0,$$

where the inequality follows from the definition of an improving ray. The proof for a dual improving ray proceeds analogously.

Sub-case: $\tau^* = \rho^* = 0$
Finally, if a maximally complementary solution is obtained with $\tau^* = \rho^* = 0$, then we again have that all optimal solutions yield $\rho = \tau = 0$, *i.e.* cases (I) and (II) cannot occur. This completes the proof. □

Two important questions now arise:

- How does one decide if $\tau^* > 0$ and $\rho^* > 0$ in a maximally complementary solution, if only an ϵ-optimal solution of the embedding problem is available?
- What additional information can be obtained if case (III) holds?

These three questions will be addressed in turn in the following sections.

4.5 SEPARATING SMALL AND LARGE VARIABLES

A path following interior point method only yields an ϵ-optimal solution to the embedding problem. This solution may yield small values of ρ and τ, and to distinguish between cases (I) to (III) it is necessary to know if these values are zero in a maximally complementary solution. This is the most problematic aspect of the analysis at this time, and only partial solutions are given here.

In what follows the set of feasible $\tilde{X}$ for the embedding problems is denoted by $\tilde{\mathcal{P}}$ and the optimal set by $\tilde{\mathcal{P}}^*$. The sets $\tilde{\mathcal{D}}$ and $\tilde{\mathcal{D}}^*$ are defined similarly.

To separate 'small' and 'large' variables we need the following definition.[6]

Definition 4.1 *The primal and dual condition numbers of the embedding are defined as*

$$\sigma_P := \sup_{\tilde{X}\in\tilde{\mathcal{P}}^*} f(\tilde{X}), \quad \sigma_D := \sup_{\tilde{S}\in\tilde{\mathcal{D}}^*} f(\tilde{S}),$$

where f is defined by

$$f(\tilde{X}) := \begin{cases} \infty & \text{if } \tilde{X} = 0 \\ \min_{i:\lambda_i(\tilde{X})>0}\lambda_i(\tilde{X}) & \text{otherwise.} \end{cases}$$

The condition number σ of the embedding is defined as $\sigma := \min\{\sigma_P, \sigma_D\}$.

Note that σ is positive and finite because the solution set of the self-dual embedding problem is bounded.

In linear programming a positive lower bound for σ can be given in terms of the problem data (see Roos *et al.* [161], Theorem I.33). It is an open problem to give a similar bound in the semidefinite case.

If we have a centered solution to the embedding problem with centering parameter μ, then we can use any knowledge of σ to decide the following:

Lemma 4.1 *For any positive μ one has:*

$$\begin{array}{llll} \tau(\mu) \geq \sigma/\tilde{n} & \text{and} & \rho(\mu) \leq \tilde{n}\mu/\sigma & \text{if } \tau^* > 0 \text{ and } \rho^* = 0 \\ \tau(\mu) \leq \tilde{n}\mu/\sigma & \text{and} & \rho(\mu) \geq \sigma/\tilde{n} & \text{if } \tau^* = 0 \text{ and } \rho^* > 0, \end{array}$$

where the superscript $*$ indicates a maximally complementary solution.

Proof:
Assume that ρ^* is positive in a maximally complementary solution. Let $\tilde{S}^* \in \tilde{\mathcal{D}}^*$ be such that ρ^* is as large as possible. By definition one therefore has $\rho^* \geq \sigma$. Recall that by the orthogonality property one has

$$\mathbf{Tr}\left(\tilde{X}(\mu)\tilde{S}^*\right) \leq \tilde{n}\mu,$$

[6]This definition is due to Ye [187] in the special case of LP.

which implies that the eigenvalues of $\tilde{X}(\mu)\tilde{S}^*$ satisfy

$$\lambda_i\left(\tilde{X}(\mu)\tilde{S}^*\right) \le \tilde{n}\mu, \quad \forall\, i.$$

In particular

$$\tau(\mu)\rho^* \le \tilde{n}\mu.$$

This shows that

$$\tau(\mu) \le \frac{\tilde{n}\mu}{\rho^*} \le \frac{\tilde{n}\mu}{\sigma}.$$

Since $\tau(\mu)\rho(\mu) = \mu$ one also has

$$\rho(\mu) \ge \frac{\sigma}{\tilde{n}}.$$

The case where $\tau^* > 0$ and $\rho^* = 0$ is proved in the same way. □

The lemma shows that once the barrier parameter μ has been reduced to the point where $\mu \le \left(\frac{\sigma}{\tilde{n}}\right)^2$, then it is known which of τ or ρ is positive in a maximally complementary solution, provided that one is indeed positive. The smaller the condition number, the more work will be needed in general to solve the embedding to sufficient accuracy.

The proof of Lemma 4.1 can easily be extended to the case where the ϵ-optimal solution is only approximately centered, where approximate centrality is defined by

$$\kappa(\tilde{X}\tilde{S}) := \frac{\lambda_{\max}(\tilde{X}\tilde{S})}{\lambda_{\min}(\tilde{X}\tilde{S})} \le \bar{\kappa},$$

for some parameter $\bar{\kappa} > 1$. Formally one has the following result.

Lemma 4.2 *Let $(\tilde{X}, \tilde{S})$ be a feasible solution of the embedding problem such that $\kappa(\tilde{X}\tilde{S}) \le \bar{\kappa}$ for some $\bar{\kappa} > 1$. One has the relations:*

$$\begin{array}{llll} \tau \ge \frac{\sigma}{\bar{\kappa}n} & \text{and} & \rho \le \frac{\mathbf{Tr}\,(\tilde{X}\tilde{S})}{\sigma} & \text{if } \tau^* > 0 \text{ and } \rho^* = 0 \\ \tau \le \frac{\mathbf{Tr}\,(\tilde{X}\tilde{S})}{\sigma} & \text{and} & \rho \ge \frac{\sigma}{\bar{\kappa}n} & \text{if } \tau^* = 0 \text{ and } \rho^* > 0 \end{array} \tag{4.9}$$

where the superscript $*$ indicates a maximally complementary solution.

The condition number is related to the concept of *ill-posedness* (see *e.g.* Renegar [157]) of the SDP problem to be solved. Consider the following example.

Example 4.1 *The problem in Example 2.2 is weakly infeasible, but it can be considered as ill-posed in the following sense: if we perturb the data slightly to obtain the problem*

$$\sup\left\{ y_1 \;\middle|\; y_1 \begin{bmatrix} 1 & 0 \\ 0 & \epsilon^2 \end{bmatrix} \preceq \begin{bmatrix} 0 & 1 \\ 1 & 0 \end{bmatrix} \right\} \tag{4.10}$$

for some small $\epsilon > 0$, then the problem (4.10) and its dual are strictly feasible and therefore solvable. Problem (4.10) has the unique solution

$$S^* = \begin{bmatrix} 1/\epsilon & 1 \\ 1 & \epsilon \end{bmatrix}, \; y_1^* = -\frac{1}{\epsilon},$$

and its dual has solution

$$X^* = \begin{bmatrix} \frac{1}{2} & -\frac{\epsilon}{2} \\ -\frac{\epsilon}{2} & \frac{1}{2\epsilon^2} \end{bmatrix}.$$

Assume now that we solve problem (4.10) via the embedding approach. It follows from inequality (4.8) that the optimal value of the embedding variable τ will satisfy

$$\tau^* = \frac{4}{1 + \frac{1}{\epsilon} + \epsilon + \frac{1}{2} + \frac{1}{2\epsilon^2}} < 8\epsilon^2.$$

This shows that the condition number σ will be $O(\epsilon^2)$. Thus, we will have to reduce μ to the point where $\mu < O\left(\epsilon^4\right)$ in order to correctly classify the problem status via the embedding approach. The required value of μ will typically be smaller than machine precision, if say $\epsilon \leq 10^{-4}$.

This illustrates the inherent numerical difficulty with problems like the weakly infeasible problem in Example 2.2. □

Note that Lemma 4.1 has very limited predictive powers: if $\rho(\mu) \leq \frac{\tilde{n}\mu}{\sigma}$ and $\tau(\mu) \leq \frac{\tilde{n}\mu}{\sigma}$, then one can conclude that $\tau^* = \rho^* = 0$ in a maximally complementary solution. In all other cases no conclusion can be drawn unless some a priori information is available about (P) and (D).

In order to improve the results of Lemma 4.1 one must establish the rate at which $\tau(\mu)$ and $\rho(\mu)$ converges to zero in the case where both are zero in a maximally complementary solution. This remains an open problem. Example 3.3 shows that the convergence rate will certainly not be better than $O\left(\mu^{2^{-(n-2)}}\right)$ in general, but this may not be the worst case behaviour.

4.6 REMAINING DUALITY AND FEASIBILITY ISSUES

If $\rho^* = \tau^* = 0$ in a maximally complementary solution of the embedding problem (i.e. case (III) holds), then one of the following situations has occurred:

1) The problems (P) and (D) are solvable but have a positive duality gap;

2) either (P) or (D) (or both) are weakly infeasible;

3) both (P) and (D) are feasible, but one or both are unsolvable, e.g. p^* is finite but is not attained.

Case 1) was illustrated in Example 2.1, and case 2) by Example 2.2. The remaining case occurs in the following example.

Example 4.2 *The following problem (adapted from Vandenberghe and Boyd [180]) is in the form (D): find*

$$\sup y_2$$

subject to

$$y_1 \begin{bmatrix} 0 & 0 & 0 \\ 0 & 1 & 0 \\ 0 & 0 & 0 \end{bmatrix} + y_2 \begin{bmatrix} 1 & 0 & 0 \\ 0 & 0 & 1 \\ 0 & 1 & 0 \end{bmatrix} \preceq \begin{bmatrix} 1 & 0 & 0 \\ 0 & 0 & 0 \\ 0 & 0 & 0 \end{bmatrix}.$$

This problem is solvable with optimal value $y_2^* = 0$ and the corresponding primal problem is solvable with optimal value 1. □

Note that in all cases the problems under consideration are ill-posed, in the sense discussed in Example 4.1. In other words, an arbitrary small perturbation can change the feasibility status. We therefore have little hope of solving such problems on a computer where rounding errors in effect introduce small perturbations.

From a purely theoretical point of view one can get more information on the solution status of such problems from the embedding if we make an additional assumption. To fix our ideas, we wish to determine the status of a given problem in the standard dual form (D). We now make an additional assumption.

Assumption 4.1 *Problems (P) and (D) are in perfect duality, and $\mathcal{P}^*$ is non-empty if d^* is finite.*

We can make this assumption without loss of generality by replacing (D) and (P) by the regularized problems (D_{cor}) and (P_{gf}) (see Appendix G).

Theorem 4.3 (De Klerk *et al.* [46]) *Consider a problem (D) in standard form such that (D) and its dual (P) satisfy Assumption 4.1. Assume that a maximally complementary solution*

$$(y^*,\ X^*,\ \tau^*,\ \theta^*,\ S^*,\ \rho^*,\ \nu^*)$$

of the embedding problem for (P) and (D) is known. The following now holds:

1. *If $\tau^* > 0$, then $\frac{1}{\tau^*}S^*$ is an optimal solution of (D); STOP*
2. *If $\rho^* > 0$, then strong infeasibility of either (P) or (D) or both is detected; STOP*
3. *If $\tau^* = \rho^* = 0$, then*

$$\lim_{\mu \downarrow 0} \mathbf{Tr}\left(\frac{CX(\mu)}{\tau(\mu)}\right) = \begin{cases} \infty & \text{if } (D) \text{ is unbounded;} \\ -\infty & \text{if } (D) \text{ is weakly infeasible;} \\ d^* & \text{if } d^* \text{ is finite,} \end{cases}$$

where $X(\mu)$ and $\tau(\mu)$ refer to the values of the embedding variables on the central path.

5

THE PRIMAL LOGARITHMIC BARRIER METHOD

Preamble

Primal logarithmic barrier methods for SDP are analysed in this chapter. These algorithms are also known as primal path-following methods, since they follow the primal central path approximately to reach the optimal set. In particular, a simple analysis of the so-called method of approximate centers for SDP is presented here, based on He *et al.* [82]. This method is an extension of a method by Roos and Vial [162] for LP. Purely primal (or purely dual) methods only use information regarding the primal (or dual) feasible iterate, but not both. Primal-dual methods use both primal and dual information in forming a search direction. For many semidefinite programming problems arising in combinatorial optimization, either the primal feasible X or the dual feasible S will always be sparse, but not both. The best way to exploit sparsity is often to work with a purely primal, or purely dual method.

5.1 INTRODUCTION

In the rest of this monograph we assume strict feasibility of (P) and (D) (Assumption 2.2 on page 23). As before, let $(X(\mu), y(\mu), S(\mu))$ denote the unique solution of the system of centrality conditions

$$\begin{aligned} \mathbf{Tr}\,(A_iX) &= b_i, \quad X \succ 0, \quad i = 1, \ldots, m \\ \sum_{i=1}^{m} y_iA_i + S &= C, \quad S \succ 0 \end{aligned}$$

$$XS = \mu I.$$

Recall from Section 3.1 that the existence and uniqueness of the solution follows from the fact that $(X(\mu), y(\mu), S(\mu))$ is the unique minimum of the strictly convex primal–dual barrier function

$$f^{\mu}_{pd}(X,S) = \frac{1}{\mu}\mathbf{Tr}\,(XS) - \log\det(XS) - n,$$

that is defined on ri $(\mathcal{P} \times \mathcal{D})$. The primal–dual barrier is easily shown to be the difference (up to the constant n) between the primal and dual barrier functions, defined respectively on ri $(\mathcal{P})$ and ri $(\mathcal{D})$ by

$$f^{\mu}_{p}(X) = \frac{1}{\mu}\mathbf{Tr}\,(CX) - \log\det X$$

and

$$f^{\mu}_{d}(y,S) = -\frac{1}{\mu}b^T y - \log\det S.$$

The primal central path corresponds to the minimizers $X(\mu)$ of $f^{\mu}_{p}(X)$. For this reason μ is referred to as either the *centering parameter* or the *barrier parameter*.

The short step algorithm to be presented follows the primal central path closely, and the search direction ΔX is simply the projected Newton direction of the primal barrier; the projected Newton direction is obtained by minimizing the quadratic Taylor approximation of f^{μ}_{p} subject to the condition $\Delta X \in \mathcal{L}^{\perp}$ for a feasible primal direction. Formally the definition is (see *e.g.* Gill *et al.* [63]):

$$\Delta X = \arg\min_{\Delta X} \left\langle \nabla f^{\mu}_{p}(X), \Delta X\right\rangle + \frac{1}{2}\left\langle \nabla^2 f^{\mu}_{p}(X)\Delta X, \Delta X\right\rangle$$

subject to the feasibility conditions

$$\mathbf{Tr}\,(A_i \Delta X) = 0, \quad i = 1, \ldots, m.$$

We can now give a formal description of the algorithm for (P).

Notation

In order to keep notation simple, we do not indicate iterates generated by algorithms by using subscripts or superscripts. This necessitates 'programming language'-type expressions such as $X := X + \Delta X$ in the statement of algorithms, although notation like $X^{(k+1)} := X^{(k)} + (\Delta X)^{(k)}$ is certainly more pleasing from a mathematical point of view. We will also use statements like: 'Let $X \in \mathcal{P}$ denote the current iterate', if we consider the situation at the start of an arbitrary iteration.

Short step primal logarithmic barrier algorithm

Input
A pair (X^0, μ_0) such that X^0 is strictly feasible and 'sufficiently centered';

Parameters
An accuracy parameter $\epsilon > 0$.
An updating parameter $\theta := \frac{1}{4\sqrt{n}+2}$;
begin
$X := X^0$; $\mu := \mu_0$;
while $n\mu > \epsilon$ **do**
$X := X + \Delta X$;
$\mu := (1 - \theta)\mu$;
end
end

In Section 5.2 we will quantify the requirement 'sufficiently centered' that appears in the statement of the algorithm. Loosely speaking, we require that the starting point $X^0 \in \mathrm{ri}\,(\mathcal{P})$ is 'sufficiently close' to the primal μ^0-center $X(\mu^0)$. Moreover, we will show in the following sections that the algorithm converges to an ϵ-optimal solution in $O\left(\sqrt{n}\log(n\mu^0/\epsilon)\right)$ iterations.

Outline of the chapter

The chapter is structured as follows: In Section 5.2 the centrality function is introduced, which is then related to the primal search direction in Section 5.3. The behaviour of the primal step near the central path is analysed in Section 5.5. The analysis of a centering parameter update in Section 5.6 allows the complexity analysis of Section 5.7. Different versions of the dual algorithm are discussed in Section 5.8 and Section 5.9. The chapter ends with a section on the application of the dual method to SDP's arising from relaxations of combinatorial optimization problems.

5.2 A CENTRALITY FUNCTION

For $X \in \mathrm{ri}\,(\mathcal{P})$ and a given parameter $\mu > 0$, we define

$$(S(X,\mu), y(X,\mu)) := \arg \min_{S\in\mathcal{S}_n, y\in\mathbb{R}^m} \left\{ \left\| \frac{1}{\mu} X^{\frac{1}{2}} S X^{\frac{1}{2}} - I \right\| \;:\; \sum_{i=1}^m y_i A_i + S = C \right\}. \tag{5.1}$$

In other words, $S(X,\mu)$ satisfies the dual feasibility constraints without the semi–definiteness condition. We define the 'centrality function'[1]

$$\delta_p(X,\mu) := \left\| \frac{X^{\frac{1}{2}}S(X,\mu)X^{\frac{1}{2}}}{\mu} - I \right\|. \tag{5.2}$$

Note that one has

$$\delta_p(X,\mu) = 0 \iff X = X(\mu).$$

We will refer to X as being *sufficiently centered* with respect to μ if $\delta_p(X,\mu)$ is smaller than some prescribed parameter.

The matrix $S(X,\mu)$ plays an important role in the analysis of the algorithm. In particular, the search direction can be expressed in terms of it, as is shown in the next section.

5.3 THE PROJECTED NEWTON DIRECTION FOR THE PRIMAL BARRIER FUNCTION

Recall that the projected Newton direction for the primal barrier

$$f_p^\mu(X) = \frac{1}{\mu}\mathbf{Tr}\,(CX) - \log\det X$$

at a given pair (X,μ) is defined as:

$$\begin{aligned} \Delta X &= \arg\min_{\Delta X} \langle \nabla f_p^\mu(X,\mu), \Delta X\rangle + \frac{1}{2}\langle \nabla^2 f_p^\mu(X,\mu)\Delta X, \Delta X\rangle & (5.3)\\ &\equiv \arg\min_{\Delta X} \mathbf{Tr}\,\left(\nabla f_p^\mu(X,\mu)\Delta X\right) + \frac{1}{2}\mathbf{Tr}\,\left(\nabla^2 f_p^\mu(X,\mu)\Delta X^2\right) & (5.4)\end{aligned}$$

subject to the feasibility conditions

$$\mathbf{Tr}\,(A_i\Delta X) = 0, \quad i = 1,\ldots,m,$$

where ∇f_p^μ and $\nabla^2 f_p^\mu$ denote the gradient and Hessian of f_p^μ respectively. In other words, the projected Newton direction minimizes the quadratic Taylor approximation to f_p^μ subject to the condition for a feasible direction. We will denote the projected Newton direction at X by ΔX to keep notation simple.

As in the LP case, we can derive an explicit expression for ΔX. To this end, it is shown in Appendix C that

$$\nabla f_p^\mu(X) = \frac{1}{\mu}C - X^{-1},$$

[1]By centrality function we mean a function $f : \mathrm{ri}\,(\mathcal{P}\times\mathcal{D}) \mapsto \mathbf{R}_+$ such that if (X,S) is a global minimizer of f, then $(X,S) = (X(\mu),S(\mu))$ for some $\mu > 0$.

and $\nabla^2 f_p^\mu(X,\mu) : \mathcal{S}_n \mapsto \mathcal{S}_n$ is the linear operator that satisfies

$$\nabla^2 f_p^\mu(X)\Delta X = X^{-1}\Delta X X^{-1} \quad \forall\, \Delta X \in \mathcal{S}_n.$$

Substitution of the gradient and Hessian in (5.3) yields

$$\Delta X \quad = \quad \text{argmin}_{\Delta X}\left\{\frac{1}{\mu}\mathbf{Tr}\,(C\Delta X) - \mathbf{Tr}\,\left(X^{-1}\Delta X\right) + \frac{1}{2}\mathbf{Tr}\,\left(\left(X^{-1}\Delta X\right)^2\right)\right\},$$

subject to

$$\mathbf{Tr}\,(A_i\Delta X) = 0, \quad i = 1,\ldots,m.$$

The KKT optimality conditions for this problem are

$$\begin{aligned} \frac{1}{\mu}C - X^{-1} + X^{-1}\Delta X X^{-1} + \sum_{i=1}^{m} y_i A_i &= 0 \\ \mathbf{Tr}\,(A_i\Delta X) &= 0, \quad i = 1,\ldots,m. \end{aligned}$$

Straightforward manipulation of the optimality conditions yields

$$\mathbf{vec}\left(X^{-\frac{1}{2}}\Delta X X^{-\frac{1}{2}}\right) = -\left[I - \mathcal{A}_X{}^T(\mathcal{A}_X\mathcal{A}_X{}^T)^{-1}\mathcal{A}_X\right]\mathbf{vec}\left(\frac{1}{\mu}X^{\frac{1}{2}}CX^{\frac{1}{2}} - I\right), \tag{5.5}$$

where $\mathcal{A}_X$ is the $m \times n^2$ matrix with rows $\mathbf{vec}\left(X^{\frac{1}{2}}A_jX^{\frac{1}{2}}\right)^T$, for $j = 1,\ldots,m$. Expression (5.5) is simply the orthogonal projection of the vector

$$\mathbf{vec}\left(\frac{1}{\mu}X^{\frac{1}{2}}CX^{\frac{1}{2}} - I\right)$$

onto the null-space of $\mathcal{A}_X$.[2] Note that the row space of $\mathcal{A}_X$ is given by

$$\text{span}\left\{\mathbf{vec}\left(X^{\frac{1}{2}}A_1X^{\frac{1}{2}}\right),\ldots,\mathbf{vec}\left(X^{\frac{1}{2}}A_mX^{\frac{1}{2}}\right)\right\},$$

and the null-space is the orthogonal complement of this space.

Reverting to the space of symmetric matrices $\mathcal{S}_n$, it is clear that the search direction ΔX is obtained via a projection of the matrix $\left(\frac{1}{\mu}X^{\frac{1}{2}}CX^{\frac{1}{2}} - I\right)$ onto the orthogonal complement of

$$\text{span}\,\{X^{\frac{1}{2}}A_1X^{\frac{1}{2}},\ldots,X^{\frac{1}{2}}A_mX^{\frac{1}{2}}\}.$$

The relevant projection operator $P_{\mathcal{A}_X} : \mathcal{S}_n \mapsto \mathcal{S}_n$ is given by

$$P_{\mathcal{A}_X}(M) := \arg\min_{W\in\mathcal{S}_n}\{\|W - M\| \ : \ \mathbf{Tr}\,(X^{\frac{1}{2}}A_iX^{\frac{1}{2}}W) = 0, \ \ i = 1,\ldots,m\}. \tag{5.6}$$

[2]A different but equivalent approach may be found in a paper by Nemirovski and Gahinet [133], where they consider the projection onto span $\{A_1,\ldots,A_m\}^\perp$ relative to the metric induced by the inner product $\langle A,B\rangle_{X^{\frac{1}{2}}} := \mathbf{Tr}\,(AX^{\frac{1}{2}}BX^{\frac{1}{2}})$ for symmetric matrices A and B.

We can now write the search direction ΔX in terms of $S(X,\mu)$.

Lemma 5.1 *The projected Newton direction at $X \in \operatorname{ri}(\mathcal{P})$ has the following two representations:*

$$\Delta X = -X^{\frac{1}{2}}\left(P_{\mathcal{A}_X}\left(\frac{X^{\frac{1}{2}}CX^{\frac{1}{2}}}{\mu} - I\right)\right)X^{\frac{1}{2}} = -\left(\frac{XS(X,\mu)X}{\mu} - X\right).$$

Proof:

We will compare the two representations by looking at the optimality conditions for the respective underlying optimization problems, namely (5.6) and (5.1).

Note that the optimization problem that yields $P_{\mathcal{A}_X}\left(\frac{X^{\frac{1}{2}}CX^{\frac{1}{2}}}{\mu} - I\right)$ via (5.6) is

$$\min_{W\in\mathcal{S}_n}\left\{\left\|W - \left(\frac{X^{\frac{1}{2}}CX^{\frac{1}{2}}}{\mu} - I\right)\right\| \;\middle|\; \mathbf{Tr}\left(X^{\frac{1}{2}}A_iX^{\frac{1}{2}}W\right) = 0,\ i = 1,\ldots,m\right\}. \tag{5.7}$$

The KKT (first-order) optimality conditions for this problem are

$$\left.\begin{aligned} W - \left(\tfrac{X^{\frac{1}{2}}CX^{\frac{1}{2}}}{\mu} - I\right) + \textstyle\sum_{i=1}^m \xi_i X^{\frac{1}{2}}A_iX^{\frac{1}{2}} &= 0, \\ \mathbf{Tr}\,(A_iX^{\frac{1}{2}}WX^{\frac{1}{2}}) &= 0,\quad i = 1,\ldots,m. \end{aligned}\right\} \tag{5.8}$$

Similarly, the optimization problem that yields $S(X,\mu), y(X,\mu)$ is

$$\min_{y\in\mathbb{R}^m, S\in\mathcal{S}_n}\left\{\left\|\frac{X^{\frac{1}{2}}SX^{\frac{1}{2}}}{\mu} - I\right\| \;\middle|\; \sum_{i=1}^m y_iA_i + S = C\right\}, \tag{5.9}$$

and the associated KKT optimality conditions take the form

$$\left.\begin{aligned} \tfrac{XSX}{\mu^2} - Q &= \tfrac{X}{\mu} \\ \mathbf{Tr}\,(A_iQ) &= 0,\quad i = 1,\ldots,m \\ \textstyle\sum_{i=1}^m y_iA_i + S &= C, \end{aligned}\right\} \tag{5.10}$$

where $Q \in \mathcal{S}_n$. If we denote the solution of System (5.10) by

$$\big(y(X,\mu), S(X,\mu), Q(X,\mu)\big),$$

then it follows that

$$\xi(X,\mu) := \frac{1}{\mu}y(X,\mu) \text{ and } W(X,\mu) := \mu X^{-\frac{1}{2}}Q(X,\mu)X^{-\frac{1}{2}}$$

satisfy the first equation of System (5.8). The second equation of System (5.10) shows that

$$\mathbf{Tr}\,(A_iX^{\frac{1}{2}}W(X,\mu)X^{\frac{1}{2}}) = 0,\quad i = 1,\ldots,m.$$

Thus an optimal solution to problem (5.7) can be constructed from an optimal solution to (5.9) and *vice versa*. Since problem (5.9) is simply a linear least squares problem and consequently has a unique solution, the equivalence of the two definitions of ΔX follows. □

Computation of the search direction in practice

The optimality conditions (5.10) may be solved by rewriting them as

$$\sum_{i=1}^{m} y_i \mathbf{Tr}\,(XA_iXA_j) = \mathbf{Tr}\,(XA_jXC) - \mu\mathbf{Tr}\,(A_jX), \quad j = 1, \ldots, m. \tag{5.11}$$

The solution of this $m \times m$ linear system yields $y(X, \mu)$. The coefficient matrix $[\mathbf{Tr}\,(XA_iXA_j)]$ of the linear system (5.11) is symmetric positive definite because the matrices A_i $(i = 1, \ldots, m)$ are linearly independent (see Appendix A, Lemma A.3). Letting $S(X, \mu) = \sum_{i=1}^{m} y_i(X, \mu)A_i - C$, the search direction is calculated from

$$\Delta X = -\frac{1}{\mu}XS(X, \mu)X + X.$$

5.4 THE AFFINE–SCALING DIRECTION

Lemma 5.1 shows that the search direction ΔX may be split into two terms, say

$$\Delta X := \frac{1}{\mu}\Delta X^a + \Delta X^c,$$

where

$$\Delta X^a := -X^{\frac{1}{2}}\left(P_{\mathcal{A}_X}\left(X^{\frac{1}{2}}CX^{\frac{1}{2}}\right)\right)X^{\frac{1}{2}} \tag{5.12}$$

and

$$\Delta X^c := X^{\frac{1}{2}}\left(P_{\mathcal{A}_X}(I)\right)X^{\frac{1}{2}}. \tag{5.13}$$

The terms ΔX^a and ΔX^c are respectively called the *affine–scaling* and *centering* components of the search direction. Note that the affine–scaling component of the search direction ΔX becomes dominant for small values of μ. Remember that we are trying to compute the 'target' point $X(\mu)$ on the primal central path. We can therefore interpret the affine–scaling direction as a 'greedy' approach where we 'target' the limit point of the primal central path. In other words, the affine–scaling direction aims at maximal decrease of the objective in a single iteration without attempting to stay near the central path.

This geometric interpretation will now be further motivated by a different formulation of ΔX^a.

A SECOND FORMULATION FOR THE AFFINE–SCALING DIRECTION

By the definition of the projection $P_{\mathcal{A}_X}$ in (5.6) we can write the definition of ΔX^a as

$$X^{-\frac{1}{2}}\Delta X^a X^{-\frac{1}{2}} = \arg\min_{W\in\mathcal{S}_n}\{\|W - X^{\frac{1}{2}}CX^{\frac{1}{2}}\| \ : \ \mathbf{Tr}\,(X^{\frac{1}{2}}A_iX^{\frac{1}{2}}W) = 0\}, \quad (5.14)$$

for $i = 1, \ldots, m$.

The affine–scaling direction can also be defined in a different way, namely

$$\Delta X^a := \arg\min_{\Delta X}\left\{\mathbf{Tr}\,(C\Delta X) \ : \ \left\|X^{-\frac{1}{2}}\Delta X X^{-\frac{1}{2}}\right\|^2 \le 1, \ \mathbf{Tr}\,(A_i\Delta X) = 0\right\} \quad (5.15)$$

for $i = 1, \ldots, m$. It is easily verified that two definitions are equivalent by comparing the optimality conditions of the two minimization problems (5.14) and (5.15).

One has $X + \Delta X^a \in \mathcal{P}$ (the full step is always feasible), as the next lemma shows.

Lemma 5.2 *Let $X \in \mathcal{P}$ be given. If ΔX is a feasible direction* (i.e. $\mathbf{Tr}\,(A_i\Delta X) = 0$ $(i = 1, \ldots, m)$) *and also belongs to the ellipsoid*

$$\left\{\Delta X \ : \ \left\|X^{-\frac{1}{2}}\Delta X X^{-\frac{1}{2}}\right\|^2 \le 1\right\},$$

then $X + \Delta X \in \mathcal{P}$.

Proof:

The condition

$$\left\|X^{-\frac{1}{2}}\Delta X X^{-\frac{1}{2}}\right\|^2 = \sum_{i=1}^n \lambda_i\left(X^{-\frac{1}{2}}\Delta X X^{-\frac{1}{2}}\right)^2 \le 1 \quad (5.16)$$

implies that

$$\left|\lambda_i\left(X^{-\frac{1}{2}}\Delta X X^{-\frac{1}{2}}\right)\right| \le 1, \quad (i = 1, \ldots, n),$$

which in turn shows that

$$I + X^{-\frac{1}{2}}\Delta X X^{-\frac{1}{2}} \succeq 0.$$

Pre and post-multiplying by $X^{\frac{1}{2}}$ gives $X + \Delta X \succeq 0$, which is the required result. □

The ellipsoid defined by (5.16) is called the *Dikin ellipsoid* at X.

Muramatsu [130] proved that an algorithm using the affine–scaling direction can converge to a non-optimal point, regardless of which (fixed) step length is used. In order to formulate a convergent algorithm, it is therefore important to add the centering component ΔX^c to the search direction. (For LP it is possible to choose the step length such that the affine–scaling algorithm is convergent.)

5.5 BEHAVIOUR NEAR THE CENTRAL PATH

We now consider the situation where, for a given $\mu > 0$, we know an $X \in \operatorname{ri}(\mathcal{P})$ such that $\delta_p(X,\mu) < 1$. We wish to know what the effect of a full projected Newton step

$$X^+ := X + \Delta X = 2X - \frac{1}{\mu} X S(X,\mu) X.$$

is in this situation.

The pair $(X^+, S(X,\mu))$ now satisfies the primal and dual equality constraints but not necessarily the semidefiniteness requirements. The next two lemmas show that the semidefiniteness requirements are also satisfied if X is sufficiently centered with respect to μ.

Lemma 5.3 *If* $X \succ 0$ *and* $\delta_p(X,\mu) < 1$, *then* $S(X,\mu) \succ 0$.

Proof:

By the definition of $\delta_p(X,\mu)$ in (5.2) we have

$$\begin{aligned} \delta_p(X,\mu)^2 &= \left\| \frac{X^{\frac{1}{2}} S(X,\mu) X^{\frac{1}{2}}}{\mu} - I \right\|^2 \\ &= \mathbf{Tr}\left(\left(\frac{1}{\mu} X^{\frac{1}{2}} S(X,\mu) X^{\frac{1}{2}} - I \right)^2 \right) \\ &= \sum_{i=1}^{n} \left(\frac{1}{\mu} \lambda_i \left(X^{\frac{1}{2}} S(X,\mu) X^{\frac{1}{2}} \right) - 1 \right)^2. \end{aligned}$$

Using $\delta_p(X,\mu) < 1$, we have

$$\sum_{i=1}^{n} \left(\frac{1}{\mu} \lambda_i \left(X^{\frac{1}{2}} S(X,\mu) X^{\frac{1}{2}} \right) - 1 \right)^2 < 1,$$

which shows that $\lambda_i \left(X^{\frac{1}{2}} S(X,\mu) X^{\frac{1}{2}} \right) > 0$ $(i = 1\ldots,n)$, and thus $S(X,\mu) \succ 0$. $\square$

The last lemma shows a very interesting feature of the method. Although we only use primal feasible $X \in \operatorname{ri}(\mathcal{P})$ in the execution of the algorithm, we obtain feasible dual points $S(X,\mu) \in \operatorname{ri}(\mathcal{D})$ as a bonus whenever $\delta_p(X,\mu) < 1$. This is important, because it gives us the upper bound

$$\mathbf{Tr}(CX) - p^* \leq \mathbf{Tr}(XS(X,\mu))$$

on the difference between the objective value at the current iterate and the optimal value, by the weak duality theorem.

The next step is to show that $X^+ := X + \Delta X$ is feasible if X is sufficiently centered.

Lemma 5.4 *Let $X^+ = X + \Delta X = 2X - \frac{1}{\mu}XS(X,\mu)X$. If $X \succ 0$ and $\delta_p(X,\mu) < 1$, then $X^+ \succ 0$.*

Proof:

Note that X^+ may be written as

$$X^+ = X^{\frac{1}{2}}\left(2I - X^{\frac{1}{2}}\frac{S(X,\mu)}{\mu}X^{\frac{1}{2}}\right)X^{\frac{1}{2}}. \tag{5.17}$$

Because $\delta_p(X,\mu) < 1$, *i.e.* $\left\|\frac{1}{\mu}X^{\frac{1}{2}}S(X,\mu)X^{\frac{1}{2}} - I\right\| < 1$, it follows that

$$\sum_{i=1}^{n}\lambda_i^2\left(\frac{X^{\frac{1}{2}}S(X,\mu)X^{\frac{1}{2}}}{\mu} - I\right) < 1.$$

Thus we have

$$\lambda_i\left(\frac{1}{\mu}X^{\frac{1}{2}}S(X,\mu)X^{\frac{1}{2}}\right) \in (0,2), \quad i = 1,\ldots,n$$

which implies

$$\lambda_i\left(2I - X^{\frac{1}{2}}\frac{S(X,\mu)}{\mu}X^{\frac{1}{2}}\right) \in (0,2), \quad i = 1,\ldots,n$$

and consequently $X^+ \succ 0$, by (5.17). □

One also has quadratic convergence of the primal iterate to the central path.[3]

Lemma 5.5 *If $X \in \mathrm{ri}\,(\mathcal{P})$ and $\delta_p(X,\mu) < 1$ then*

$$X^+ := X + \Delta X = 2X - \frac{1}{\mu}XS(X,\mu)X$$

satisfies $\delta_p(X^+,\mu) \le \delta_p^2(X,\mu)$.

---

[3]The quadratic convergence result was first established by Faybusovich [55], and later by He *et al.* [82] and Anstreicher and Fampa [12]. It was also obtained in the general setting of convex programming problems in conic form for self-scaled cones by Nesterov and Todd [138].

Proof:

By the definition of $\delta_p(X,\mu)$ in (5.2) we have

$$\begin{aligned}\delta_p(X^+,\mu)^2 &= \left\|\frac{X^{+\frac{1}{2}}S(X^+,\mu)X^{+\frac{1}{2}}}{\mu} - I\right\|^2 \\ &\leq \left\|\frac{X^{+\frac{1}{2}}S(X,\mu)X^{+\frac{1}{2}}}{\mu} - I\right\|^2 \\ &= \mathbf{Tr}\left(\left(\frac{1}{\mu}S(X,\mu)X^+ - I\right)^2\right).\end{aligned}$$

Substituting $X^+ = 2X - \frac{1}{\mu}XS(X,\mu)X$ yields

$$\begin{aligned}\delta_p(X^+,\mu)^2 &\leq \mathbf{Tr}\left(\left(\frac{1}{\mu}S(X,\mu)\left[2X - \frac{1}{\mu}XS(X,\mu)X\right] - I\right)^2\right) \\ &= \mathbf{Tr}\left(\left(\frac{1}{\mu}S(X,\mu)X - I\right)^4\right) \\ &= \mathbf{Tr}\left(\left(\frac{1}{\mu}X^{\frac{1}{2}}S(X,\mu)X^{\frac{1}{2}} - I\right)^4\right) \\ &= \left\|\left(\frac{1}{\mu}X^{\frac{1}{2}}S(X,\mu)X^{\frac{1}{2}} - I\right)^2\right\|^2 \\ &\leq \left\|\frac{1}{\mu}X^{\frac{1}{2}}S(X,\mu)X^{\frac{1}{2}} - I\right\|^4 = \delta_p^4(X,\mu),\end{aligned}$$

where the second inequality follows from the sub-multiplicativity of the Frobenius norm (see Appendix A). □

5.6 UPDATING THE CENTERING PARAMETER

Once the primal iterate X is sufficiently centered, *i.e.* $\delta_p(X,\mu) \leq \tau$ for some tolerance τ, the parameter μ can be reduced. To fix our ideas, we update the barrier parameter in such a way that we will still have $\delta_p(X,\mu^+) \leq \frac{1}{2}$ after an update $\mu \rightarrow \mu^+$. The following step $X + \Delta X$ will then yield a feasible X^+ satisfying $\delta_p(X^+,\mu^+) \leq \frac{1}{4}$, by Lemma 5.5.

In order to realize these ideas, we must analyse the effect of a μ-update on the centrality function.

Lemma 5.6 *Define a μ–update by $\mu^+ := (1-\theta)\mu$, where $0 < \theta < 1$ is a given parameter. It then follows that*

$$\delta_p(X,\mu^+) \le \frac{1}{1-\theta}\left(\delta_p(X,\mu) + \theta\sqrt{n}\right).$$

Proof:

Using the definition of $S(X,\mu^+)$ we may write

$$\begin{aligned}
\delta_p(X,\mu^+) &= \left\| \frac{X^{\frac{1}{2}} S(X,\mu^+) X^{\frac{1}{2}}}{(1-\theta)\mu} - I \right\| \\
&\le \left\| \frac{X^{\frac{1}{2}} S(X,\mu) X^{\frac{1}{2}}}{(1-\theta)\mu} - I \right\| \\
&= \left\| \frac{X^{\frac{1}{2}} S(X,\mu) X^{\frac{1}{2}}}{(1-\theta)\mu} - \frac{1}{1-\theta} I + \frac{\theta}{1-\theta} I \right\| \\
&\le \frac{1}{1-\theta}\left(\left\| \frac{X^{\frac{1}{2}} S(X,\mu) X^{\frac{1}{2}}}{\mu} - I \right\| + \theta \|I\| \right) \\
&= \frac{1}{1-\theta}\left(\delta_p(X,\mu) + \theta\sqrt{n}\right),
\end{aligned}$$

where the second inequality follows from the triangle inequality. □

The above result enables us to choose an updating parameter θ which guarantees that the primal iterate remains sufficiently centered with respect to the new parameter $\mu^+ := (1-\theta)\mu$.

Lemma 5.7 *Let $\delta_p(X,\mu) \le \frac{1}{2}$ and $\theta = 1/(4\sqrt{n}+2)$. After a step $X^+ = X + \Delta X$ and a subsequent update $\mu^+ = (1-\theta)\mu$, one has $\delta_p(X^+,\mu^+) \le \frac{1}{2}$.*

Proof:

Using Lemma 5.6 and Lemma 5.5 successively we get

$$\begin{aligned}
\delta_p(X^+,\mu^+) &\le \frac{1}{1-\theta}\left(\delta_p(X^+,\mu) + \theta\sqrt{n}\right) \\
&\le \frac{1}{1-\theta}\left(\delta_p^2(X,\mu) + \theta\sqrt{n}\right).
\end{aligned}$$

Substitution of $\theta = 1/(4\sqrt{n}+2)$ gives

$$\delta_p(X^+,\mu^+) \le \frac{4\sqrt{n}+2}{4\sqrt{n}+1}\left(\frac{1}{4} + \frac{\sqrt{n}}{4\sqrt{n}+2}\right) = \frac{1}{2},$$

which completes the proof. □

Dynamic μ-updates

It is easily verified that if $\delta_p(X,\mu) \leq \frac{1}{2}$, the dynamic update

$$\theta = \frac{\frac{1}{2} - \delta_p(X,\mu)}{\sqrt{n}+\frac{1}{2}} \geq \frac{1}{4\sqrt{n}+2}$$

ensures that $\delta_p(X,\mu^+) \leq \frac{1}{2}$, if $\mu^+ = (1-\theta)\mu$. A natural question is whether it is possible to find the smallest value of μ^+ such that the proximity condition $\delta_p(X,\mu^+) \leq \frac{1}{2}$ still holds. This is indeed possible; the key observation is that $\delta_p(X,\mu)$ can be rewritten as

$$\delta_p(X,\mu) = \left\| X^{-\frac{1}{2}} \left(\Delta X^c + \frac{1}{\mu}\Delta X^a \right) X^{-\frac{1}{2}} \right\|,$$

by the definition of δ_p and Lemma 5.1.

Denoting $D^a := X^{-\frac{1}{2}}\Delta X^a X^{-\frac{1}{2}}$ and $D^c := X^{-\frac{1}{2}}\Delta X^c X^{-\frac{1}{2}}$, we see that the smallest value of μ^+ which still satisfies $\delta_p(X,\mu^+) \leq \frac{1}{2}$ is the smallest positive root of the equation

$$\delta_p(X,\mu) = \left\| D^c + \frac{1}{\mu} D^a \right\| = \frac{1}{2}.$$

Squaring both sides of the last equation yields the following quadratic equation in $\frac{1}{\mu}$:

$$\frac{1}{\mu^2}\left\|D^a\right\|^2 + \frac{2}{\mu}\mathbf{Tr}\left(D^a D^c\right) + \left\|D^c\right\|^2 - \frac{1}{4} = 0,$$

which can be solved to obtain the desired value μ^+.[4]

5.7 COMPLEXITY ANALYSIS

To prove the polynomial complexity of the algorithm, we need the following lemma that bounds the duality gap in terms of the centrality function δ_p.

Lemma 5.8 *If $X \in \mathrm{ri}(\mathcal{P})$ and $\delta_p(X,\mu) \leq 1$ for a given $\mu > 0$, then*

$$\mu\left(n - \delta_p(X,\mu)\sqrt{n}\right) \leq \mathbf{Tr}(CX) - b^T y(X,\mu) \leq \mu\left(n + \delta_p(X,\mu)\sqrt{n}\right).$$

Proof:

Note that

$$\mathbf{Tr}(CX) - b^T y(X,\mu) = \mathbf{Tr}(XS(X,\mu)) = \mathbf{Tr}(X^{\frac{1}{2}}S(X,\mu)X^{\frac{1}{2}}).$$

[4]This dynamic updating strategy is the extension of the strategy for LP described by Roos *et al.* [161], §6.8.3.

Using the Cauchy–Schwartz inequality yields

$$\delta_p(X,\mu)\sqrt{n} = \left\| \frac{X^{\frac{1}{2}}S(X,\mu)X^{\frac{1}{2}}}{\mu} - I \right\| \|I\| \geq \left| \frac{\mathbf{Tr}\,(XS(X,\mu))}{\mu} - n \right|,$$

which implies that

$$n - \delta_p(X,\mu)\sqrt{n} \leq \frac{\mathbf{Tr}\,(XS(X,\mu))}{\mu} \leq n + \delta_p(X,\mu)\sqrt{n},$$

which in turn gives the required result. □

We can now derive the worst case complexity bound of the algorithm.

Theorem 5.1 *Let $\epsilon > 0$ be an accuracy parameter, $\theta = \frac{1}{4\sqrt{n}+2}$ and $\mu^0 > 0$. Let $X^0 \succ 0$ be a strictly feasible starting point such that $\delta_p(X^0,\mu^0) \leq \frac{1}{2}$. The algorithm terminates after at most $\left\lceil 6\sqrt{n}\log\frac{n\mu^0}{\epsilon}\right\rceil$ steps, the last generated points X and $S(X,\mu)$ are strictly feasible, and the duality gap is bounded by* $\mathbf{Tr}\,(XS(X,\mu)) \leq \frac{3}{2}\epsilon$.

Proof:

After each iteration of the algorithm, X will be strictly feasible, and $\delta_p(X,\mu) \leq 1/2$, due to Lemma 5.7. After the k-th iteration one has $\mu = (1-\theta)^k\mu_0$. The algorithm stops if k is such that

$$n\mu_0(1-\theta)^k < \epsilon.$$

Taking logarithms on both sides, this inequality can be rewritten as

$$-k\ln(1-\theta) > \log\frac{n\mu^0}{\epsilon}.$$

Since $-\ln(1-\theta) > \theta$, the last inequality will certainly hold if

$$k\theta > \log\frac{n\mu^0}{\epsilon},$$

which implies

$$k > 6\sqrt{n}\log\frac{n\mu^0}{\epsilon}$$

for the default setting $\theta := \frac{1}{4\sqrt{n}+2}$. This proves the first statement in the theorem. Now let X be the last generated point; then it follows from Lemma 5.3 that $S(X,\mu) \succ 0$. Moreover, the duality gap is then bounded by

$$\begin{aligned}\mathbf{Tr}\,(XS(X,\mu)) &\leq n\mu\left(1+\frac{\delta_p(X,\mu)}{\sqrt{n}}\right)\\ &\leq \epsilon\left(1+\frac{\delta_p(X,\mu)}{\sqrt{n}}\right) \leq \frac{3}{2}\epsilon,\end{aligned}$$

where the first inequality follows from Lemma 5.8. This completes the proof. □

5.8 THE DUAL ALGORITHM

The algorithm for the dual problem is perfectly analogous to that of the primal problem. If one defines

$$X(S,\mu) := \arg\min_{X \in \mathcal{S}} \left\{ \left\| \frac{S^{\frac{1}{2}} X S^{\frac{1}{2}}}{\mu} - I \right\| \;\middle|\; \mathbf{Tr}\, A_i X = b_i, \quad i = 1, \ldots, m \right\},$$

for a strictly feasible dual variable $S \succ 0$, then the first–order optimality conditions which yield $X(S, \mu)$ are

$$S \left[\frac{XS}{\mu} - I \right] - \sum_{i=1}^{m} \Delta y_i A_i \;=\; 0 \tag{5.18}$$

$$\mathbf{Tr}\,(A_i X) \;=\; b_i, \quad i = 1, \ldots, m. \tag{5.19}$$

Pre- and post-multiplying the first equation with S^{-1} and subsequently using the second equation yields:

$$\sum_{i=1}^{m} \Delta y_i \mathbf{Tr}\, \left(A_i S^{-1} A_j S^{-1} \right) = -\frac{1}{\mu} b_j + \mathbf{Tr}\, \left(A_j S^{-1} \right), \quad j = 1, \ldots, m. \tag{5.20}$$

If we now define

$$\delta_d(S,\mu) := \left\| \frac{S^{\frac{1}{2}} X(S,\mu) S^{\frac{1}{2}}}{\mu} - I \right\|,$$

then we can repeat the analysis for the primal algorithm, but with the roles of X and S interchanged. The search direction of the algorithm, *i.e.* the projected Newton direction of the dual barrier f_d^{μ}, becomes

$$\Delta S = S^{\frac{1}{2}} \left(I - \frac{1}{\mu} S^{\frac{1}{2}} X(S,\mu) S^{\frac{1}{2}} \right) S^{\frac{1}{2}} = -\sum_{i=1}^{m} \Delta y_i A_i, \tag{5.21}$$

where Δy is obtained by solving (5.20), and ΔS subsequently follows from

$$\Delta S = -\sum_{i=1}^{m} \Delta y_i A_i.$$

We now give a formal statement of the short step dual logarithmic barrier method.

Short step dual logarithmic barrier algorithm

Input
  A strictly dual feasible pair (S^0, y^0);
  A parameter $\mu_0 > 0$ such that $\delta_d(S^0, \mu_0) \leq \frac{1}{2}$.
Parameters
  An accuracy parameter $\epsilon > 0$.
  An updating parameter $\theta := \frac{1}{4\sqrt{n}+2}$;
begin
  $S := S^0$; $\mu := \mu_0$;
  while $n\mu > \epsilon$ **do**
    $S := 2S - \frac{1}{\mu} SX(S, \mu)S$;
    $\mu := (1 - \theta)\mu$;
  end
end

Due to the symmetry in the analysis, the dual algorithm has the same complexity bound as the primal algorithm.

Remark 5.1 *An important observation is that it is* not *necessary to form $X(S, \mu)$ explicitly in order to decide whether or not it is positive definite, or to subsequently calculate the duality gap if it is indeed positive definite: by (5.18) we know that*

$$X(S, \mu) \succeq 0 \Longleftrightarrow S + \sum_{i=1}^{m} \Delta y_i A_i \succeq 0.$$

Moreover, note that if $X(S, \mu) \succeq 0$, then the duality gap at $(X(S, \mu), S)$ is given by

$$\mathbf{Tr}\left(X(S, \mu)S\right) = \mu \mathbf{Tr}\left(S - \Delta S\right),$$

by (5.21). These observations are important when exploiting sparsity in the problem data, as we will see in Section 5.10.

5.9 LARGE UPDATE METHODS

The updating strategy for μ described thus far is too conservative for practical use. In this section we describe two updating strategies that allow much greater reductions of μ.

THE DUAL SCALING METHOD

This method uses a dynamic updating strategy for μ, namely

$$\mu := \frac{\mathbf{Tr}\,(XS)}{n + \nu\sqrt{n}},$$

where $S \in \mathrm{ri}\,(\mathcal{D})$ is the current dual iterate, $X \in \mathcal{P}$ is the best-known primal solution,[5] and $\nu \geq 1$ is a given parameter.

Dual scaling algorithm

Input
  A pair $(X^0, S^0) \in \mathrm{ri}\,(\mathcal{P} \times \mathcal{D})$;
  A parameter $\mu_0 > 0$ such that $\delta_d(S^0, \mu_0) \leq \frac{1}{2}$.
Parameters
  An accuracy parameter $\epsilon > 0$;
  A parameter $\nu \geq 1$;

begin
  $X = X^0, S := S^0; \mu := \mu_0;$
  while $\mathbf{Tr}\,(XS) > \epsilon$ **do**
    $S := 2S - \frac{1}{\mu} SX(S, \mu)S;$
    if $X(S, \mu) \succ 0$ **then** $X := X(S, \mu);$
    $\mu := \frac{\mathbf{Tr}\,(XS)}{n+\nu\sqrt{n}};$
  end
end

Again, the steps in the algorithm that involve $X(S, \mu)$ should be interpreted in light of Remark 5.1: we do not have to form $X(S, \mu)$ explicitly in order to do the dual step $S + \Delta S$, or to decide whether $X(S, \mu) \succ 0$. Moreover, the role of the matrix X in the statement of the algorithm is also symbolic — we do not need to store this matrix in an implementation of the algorithm.

We have the following complexity result for the dual scaling method.

Theorem 5.2 (Ye [188]) *The dual scaling method stops after at most*

$$O\left(\nu\sqrt{n}\log\left(\frac{\mathbf{Tr}\,\left(X^0 S^0\right)}{\epsilon}\right)\right)$$

---

[5] We assume that a primal feasible solution $X \in \mathcal{P}$ is known a priori. One can use $X = X(S, \mu)$ if $X(S, \mu) \succeq 0$.

iterations. The output is a pair $(X(S,\mu),S) \in \mathcal{P} \times \mathcal{D}$ *where* $\mathbf{Tr}\,(X(S,\mu)S) \le \epsilon$.

THE LONG STEP METHOD

The short step primal method presented in this chapter has been extended by Faybusovich [55] and later by Anstreicher and Fampa [12] to use larger μ-updates.

Given the results in this chapter, this 'large update' (or 'long step') algorithm can be derived as a mechanical extension of the corresponding LP analysis by Roos *et al.* [161], §6.9.

The algorithm in question performs damped projected Newton steps with respect to a given value of μ until $\delta_d(S,\mu) \le \frac{1}{\sqrt{2}}$ holds for the current iterate S (*inner iterations*). Only then is μ reduced by a fixed fraction via $\mu \leftarrow (1-\theta)\mu$ (*outer iteration*). The algorithm can be stated as follows.

Long step dual logarithmic barrier method

Input
  A pair $(S^0, y^0) \in \mathrm{ri}\,(\mathcal{D})$;
  A centering parameter $\tau > 0$ (default $\tau = \frac{1}{\sqrt{2}}$);
  A value μ^0 such that $\delta_d(S^0,\mu^0) \le \tau$;
  An accuracy parameter $\epsilon > 0$;
  An updating parameter $0 < \theta < 1$.
begin
  $S := S^0$; $y := y^0$; $\mu = \mu^0$;
  while $\mathbf{Tr}\,(XS) > \epsilon$ **do**
    if $\delta_d(S,\mu) \le \tau$ **do** (*outer iteration*)
      $\mu := (1-\theta)\mu$;
    else if $\delta_d(S,\mu) > \tau$ **do** (*inner iteration*)
      Compute $(\Delta S, \Delta y)$;
      Find $\alpha = \mathrm{argmin}\; f_d^{\mu}(y + \alpha\Delta y, S + \alpha\Delta S, \mu)$;
      $S := S + \alpha\Delta S$;
      $y := y + \alpha\Delta y$;
    end
  end
end

This algorithm has the same worst-case iteration bound as in the LP case: if $\theta = O(1)$ the iteration bound is $O(nL)$ and if $\theta = O\left(\frac{1}{\sqrt{n}}\right)$ the iteration bound becomes $O(\sqrt{n}L)$.

5.10 THE DUAL METHOD AND COMBINATORIAL RELAXATIONS

Many SDP problems arising in combinatorial optimization involve rank one coefficient matrices, say

$$A_i = a_i a_i^T, \quad a_i \in \mathbb{R}^n, \quad i = 1, \ldots, m.$$

In this case the linear system in (5.20) can be formed very efficiently. Indeed, note that the coefficient matrix (say M) in (5.20) now reduces to:

$$\begin{aligned} m_{ij} &:= \mathbf{Tr}\left(A_i S^{-1} A_j S^{-1}\right) \\ &= \mathbf{Tr}\left(a_i a_i^T S^{-1} a_j a_j^T S^{-1}\right) \\ &= \left(a_i^T S^{-1} a_j\right)^2. \end{aligned}$$

For SDP relaxations of Boolean quadratic problems (like the MAX-CUT relaxation (1.3) on page 6) the expression simplifies even more, as a_i $(i = 1, \ldots, n)$ is then simply the ith standard unit vector. In this case we have:

$$m_{ij} = \left(S^{-1}\right)_{ij}^2, \quad i, j = 1, \ldots, n.$$

The coefficient matrix M can therefore be formed efficiently if S is sparse. For many applications this is indeed the case, since

$$S = -\sum_{i=1}^{m} y_i A_i + C,$$

and C often has the same sparsity structure as the Laplacian of a sparse graph. (Recall that this is the case for the MAX-CUT relaxation on page 6.) If S is indeed sparse, one can compute S^{-1} by doing a sparse Choleski factorization of S with pre-ordering to reduce fill in. Although S^{-1} is not necessarily sparse if S is, this is often the case in practice, and then the coefficient matrix of the system (5.20) is also sparse and can likewise be solved using sparse Choleski techniques.

The importance of Remark 5.1 on page 90 is now clear: we only wish to work with the sparse matrix S, and would like to avoid computation and storage of $X(S, \mu)$, since this matrix will be dense in general.

Software The dual scaling algorithm has been implemented by Benson *et al.* [23, 22] in the DSDP software package. This software out-performs primal–dual solvers on the MAX-CUT relaxation and similar problems where sparsity can be exploited when forming and solving the linear system (5.20). Choi and Ye [34] have improved the performance on large problems by solving the linear system (5.20) using a preconditioned conjugate gradient method (instead of a Choleski factorization). The authors report results for the MAX-CUT relaxation of sparse graphs with up to 14000 vertices.

6

PRIMAL–DUAL AFFINE–SCALING METHODS

Preamble

Perhaps the most obvious solution approach for the dual pair of SDP problems (P) and (D) is to solve the nonlinear system of optimality conditions (2.15) on page 33 via some iterative scheme. There are different ways to do this, and the resulting algorithms are called primal–dual affine–scaling methods. We will analyse in this chapter the algorithm due to De Klerk *et al.* [48]. Primal–dual affine–scaling directions are also known as the predictor directions in predictor–corrector algorithms (see Chapter 7), and are therefore of significant theoretical interest.

6.1 INTRODUCTION

The LP case

The introduction of Karmarkar's polynomial–time projective method for LP in 1984 [100] was accompanied by claims of some superior computational results. Later it seemed that the computation was done with a variant of the *affine–scaling* method, proposed by Dikin nearly two decades earlier in 1967 [53]. The two algorithms are closely related, and modifications of Karmarkar's algorithm by Vanderbei *et al.* [182] and Barnes [15] proved to be a rediscovery of the affine–scaling method. Dikin's affine–scaling method is a purely primal method, and the underlying idea is to minimize the objective function over an ellipsoid which is inscribed in the primal feasible region. Interestingly enough, polynomial complexity of Dikin's affine–scaling method in its original form has still not been proved. Even more interesting is that the extension of this method to SDP may fail to converge to an optimal solution.[1]

In the primal–dual setting, the natural extension of the notion of affine–scaling is to minimize the duality gap over some inscribed ellipsoid in the primal–dual space. A primal–dual affine–scaling method for LP is studied by Monteiro *et al.* in [125], where the primal–dual search direction minimizes the duality gap over a sphere in the primal–dual space. This algorithm may be viewed as a 'greedy' primal–dual algorithm, that aims to reach optimality in a single iteration, without attempting to stay close to the central path. The worst–case iteration complexity for this method is $O(nL^2)$, where[2]

$$L := \log\left(\frac{\text{initial duality gap}}{\epsilon}\right).$$

As such, it is an algorithm of significant theoretical interest.

Jansen *et al.* [91] proposed a related primal–dual direction for LP — called the *Dikin-type affine–scaling direction* — where the resulting algorithm has an improved $O(nL)$ worst-case complexity bound. This search direction minimizes the duality gap over the so-called *primal–dual Dikin ellipsoid*. The interesting feature of this method is that each step involves both centering and reduction of the duality gap.

It was shown by Jansen *et al.* [92] that the Dikin-type direction and the primal-dual affine–scaling direction by Monteiro *et al.* [125] both belong to a generalized family of 'affine–scaling' directions. This motivates the name Dikin-type *affine–scaling* direction.

Extensions to SDP

There is no obvious unique way to extend the method by Monteiro *et al.* [125] to SDP. We can define a family of directions at a strictly feasible pair (X, S) as the solutions of the system:

$$\begin{aligned} \mathbf{Tr}\,(A_i \Delta X) &= 0, \quad i = 1, \ldots, m \\ \sum_{i=1}^{m} \Delta y_i A_i + \Delta S &= 0 \\ H_P\,(\Delta X S + X \Delta S) &= -H_P(XS), \end{aligned}$$

---

[1] See Muramatsu [130] and the remarks in Section 5.4.

[2] In LP, if ϵ is chosen such that L equals the bit-length of the input, then it is possible to round an ϵ-optimal solution pair (X^*, S^*) (*i.e.* $\mathbf{Tr}\,(X^*S^*) \leq \epsilon$) to exact primal and dual solutions (see *e.g.* Roos *et al.* [161]). In SDP, this association between L and the bit-length of the input data is meaningless — there is no choice for ϵ which would yield exact optimal solutions. In fact, it is easy to construct instances of SDP problems with integer data but unique irrational solutions (see Section 1.9).

where H_P is the linear transformation given by

$$H_P(M) := \frac{1}{2}\left[PMP^{-1} + P^{-T}M^TP^T\right],$$

for any $M \in \mathbb{R}^{n\times n}$, and P is one of the scaling matrices from Table 1.1. Note that this system corresponds to $\mu = 0$ in (1.12). In other words, it can be seen as a strategy to solve the optimality conditions (2.15) for (P) and (D).

Recall from Section 5.4 that the primal affine–scaling direction is associated with $\mu = 0$. Loosely speaking, we can view affine–scaling directions as a 'greedy' attempt to reach the limit point of the central path in a single step.

In the special case of LP, each choice of scaling matrix P from Table 1.1 yields a search direction which coincides with the primal–dual affine–scaling direction of Monteiro *et al.*

We will only consider the direction using the Nesterov-Todd scaling (see Table 1.1); this is important, since the primal–dual affine-scaling method can fail for SDP for the scaling I (the AHO direction) in Table 1.1. This interesting negative result was proven by Muramatsu and Vanderbei [131]. At the time of writing, the primal–dual affine–scaling method has not been extended for the other choices of scaling matrices in the table.

Outline of this chapter

In this chapter *both* the primal–dual affine–scaling method of Monteiro *et al.* [125] and the method of Jansen *et al.* [91] are generalized to SDP, by using the NT scaling. The former will be referred to as the *primal–dual affine–scaling method*, and the latter as the *primal–dual Dikin-type method*, or Dikin-type method, for short.

The Dikin-type method will be presented first, and its simple analysis will then be extended to the primal–dual affine–scaling method. Some preliminaries are discussed first. In, particular, symmetric primal–dual scaling is discussed in Section 6.2, and the algorithms are introduced in Section 6.3. It is shown how the two directions correspond to the minimization of the duality gap over two different ellipsoids in the scaled space. In Section 6.5 conditions to ensure a feasible step–length are derived, and the polynomial complexity result for the Dikin-type method is proven in Section 6.6. In Section 6.7 the analysis is extended to include the primal-dual affine–scaling method.

6.2 THE NESTEROV–TODD (NT) SCALING

For strictly feasible solutions $X \succ 0$ and $S \succ 0$ to (P) and (D) respectively, the scaling–matrix

$$D := S^{-\frac{1}{2}}\left(S^{\frac{1}{2}}XS^{\frac{1}{2}}\right)^{\frac{1}{2}}S^{-\frac{1}{2}}, \tag{6.1}$$

introduced in Section 1.7, satisfies $D^{-1}X = SD$, or

$$D^{-\frac{1}{2}}XD^{-\frac{1}{2}} = D^{\frac{1}{2}}SD^{\frac{1}{2}} := V. \tag{6.2}$$

In other words, the matrix D may be used to scale the variables X and S to the same symmetric positive definite matrix V.[3]

Note that

$$V^2 = D^{-\frac{1}{2}}XSD^{\frac{1}{2}} \sim XS, \tag{6.3}$$

i.e. V^2 has the same eigenvalues as XS and is symmetric positive definite.

We will sometimes refer to this way of 'symmetrizing' XS as NT scaling.[4]

As a consequence of (6.3), the duality gap at $(X, S) \in \mathrm{ri}\,(\mathcal{P} \times \mathcal{D})$ is given by

$$\mathbf{Tr}\,(XS) = \mathbf{Tr}\,(V^2) = \|V\|^2 = \sum_{i=1}^{n} \lambda_i^2(V).$$

We can scale any pair primal–dual search directions in a similar way; feasible search directions $(\Delta X, \Delta y, \Delta S)$ must satisfy

$$\left.\begin{array}{rcl} \mathbf{Tr}\,(A_i \Delta X) & = & 0, \quad i = 1, \ldots, m \\ \sum_{i=1}^{m} \Delta y_i A_i + \Delta S & = & 0. \end{array}\right\} \tag{6.4}$$

Recall that $\mathbf{Tr}\,(\Delta X \Delta S) = 0$ (see Lemma 2.1).

The *scaled search directions* are defined by

$$D_X := D^{-\frac{1}{2}} \Delta X D^{-\frac{1}{2}}$$

and

$$D_S := D^{\frac{1}{2}} \Delta S D^{\frac{1}{2}}.$$

For the scaled directions D_X and D_S we therefore also have the orthogonality property $\mathbf{Tr}\,(D_X D_S) = 0$. Using (6.4), we obtain

$$D_S := -\sum_{i=1}^{m} \Delta y_i D^{\frac{1}{2}} A_i D^{\frac{1}{2}},$$

i.e. D_S must be in the span of matrices $D^{\frac{1}{2}} A_i D^{\frac{1}{2}}$ and D_X in its orthogonal complement, *i.e.*

$$\mathbf{Tr}\left(D^{\frac{1}{2}} A_i D^{\frac{1}{2}} D_X\right) = 0, \quad i = 1, \ldots, m.$$

The *scaled primal–dual direction* is defined by

$$D_V := D_X + D_S.$$

[3]In the matrix analysis literature, the matrix D is sometimes referred to as the *geometric mean* of S^{-1} and X; see *e.g.* Aarts [1], Definition C.0.2, and the references therein. As mentioned in Chapter 1, the matrix D was introduced by Nesterov and Todd in [135] and later by Sturm and Zhang [170] from a different perspective.

[4]In practice the scaling matrix D may be computed from the Choleski factorizations of X and S, and one additional singular value decomposition (SVD) (see Todd *et al.* [173]).

After a feasible primal–dual step $(\Delta X, \Delta S)$ the duality gap becomes

$$\mathbf{Tr}\ ((X+\Delta X)(S+\Delta S)) = \mathbf{Tr}\ ((V+D_X)(V+D_S)) = \mathbf{Tr}\,(V^2+VD_V), \tag{6.5}$$

where we have used the linearity of the trace as well as the property $\mathbf{Tr}\ (AB) = \mathbf{Tr}\ (BA)$ (see Appendix A, Section A.3).

Notation

Note that there is a one-to-one correspondence between $(X,S) \in \mathrm{ri}\ (\mathcal{P}\times\mathcal{D})$ and V^2 via (6.3). Similarly, there are one-to-one correspondences between ΔX and D_X, and ΔS and D_S. In what follows we will sometimes use the original variables (X,S) and original search directions $(\Delta X, \Delta S)$, but we will also use the scaled expressions V, D_X, D_S and D_V when this is convenient.

6.3 THE PRIMAL–DUAL AFFINE–SCALING AND DIKIN–TYPE METHODS

We will now introduce the Dikin-type and primal–dual affine–scaling directions from a geometric perspective, namely by minimizing the duality gap over two different ellipsoids.

THE DIKIN-TYPE DIRECTION

The search direction of the primal–dual Dikin-type affine–scaling algorithm is derived by minimizing the duality gap over the so called *primal–dual Dikin ellipsoid* as follows:

$$D_V^* := \arg\min_{D_V}\left\{\mathbf{Tr}\,(V^2+VD_V)\ :\ \|V^{-\frac{1}{2}}D_VV^{-\frac{1}{2}}\| \leq 1\right\}. \tag{6.6}$$

Note that $V + D_V \succeq 0$ if D_V is feasible in (6.6). It is easily verified that the optimal solution is given by

$$D_V^* = D_X^* + D_S^* = -\frac{V^3}{\|V^2\|}. \tag{6.7}$$

The transformation back to the unscaled space is done by pre- and post-multiplying (6.7) by $D^{\frac{1}{2}}$ to obtain

$$\Delta X + D\Delta SD = \frac{-XSX}{(\mathbf{Tr}\,(XS)^2)^{\frac{1}{2}}}. \tag{6.8}$$

The Dikin-type direction is obtained by solving (6.8) subject to the conditions (6.4).

THE PRIMAL–DUAL AFFINE–SCALING DIRECTION

The primal–dual affine–scaling direction can be defined by simply changing the norm in (6.6) to the spectral norm:

$$D_V^* := \arg\min_{D_V} \left\{ \mathbf{Tr}\,(V^2 + VD_V) \;:\; \|V^{-\frac{1}{2}} D_V V^{-\frac{1}{2}}\|_2 \leq 1 \right\}. \tag{6.9}$$

Note that $\|V^{-\frac{1}{2}} D_V V^{-\frac{1}{2}}\|_2 \leq 1$ is equivalent to the condition

$$I \succeq V^{-\frac{1}{2}} D_V V^{-\frac{1}{2}} \succeq -I.$$

This is the same as $V + D_V \succeq 0$ and $V - D_V \succeq 0$. Since $V + D_V \succeq 0$ one has $\mathbf{Tr}\,(V(V + D_V)) \geq 0$, which implies that the optimal solution in (6.9) is given by $D_V^* = -V$. Pre- and post-multiplying $D_V^* = -V$ by $D^{\frac{1}{2}}$ as before, one obtains

$$\Delta X + D \Delta S D = -X. \tag{6.10}$$

The solution of this equation subject to conditions (6.4) yields the primal-dual affine–scaling direction.

We get the same direction if we minimize the duality gap over the sphere $\|D_V\|^2 \leq \|V\|^2$. This shows the analogy with the LP situation: the primal–dual affine–scaling direction is obtained by minimizing the duality gap over a sphere in the scaled primal–dual space.

A NOTE ON THE DIKIN ELLIPSOID

There is some inconsistency in the literature concerning the definition of the primal–dual Dikin ellipsoid. In the paper by Nemirovski and Gahinet [133] it is defined as

$$\|X^{-\frac{1}{2}} \Delta X X^{-\frac{1}{2}}\|^2 + \|S^{-\frac{1}{2}} \Delta S S^{-\frac{1}{2}}\|^2 \leq 1, \tag{6.11}$$

which is the same as

$$\|V^{-\frac{1}{2}} D_X V^{-\frac{1}{2}}\|^2 + \|V^{-\frac{1}{2}} D_S V^{-\frac{1}{2}}\|^2 \leq 1.$$

A primal–dual step $(X + \Delta X, S + \Delta S)$ which satisfies (6.11) is always feasible: From (6.11) one has $-1 \leq \lambda_i \left(X^{-\frac{1}{2}} \Delta X X^{-\frac{1}{2}} \right) \leq 1$ $(i = 1, \ldots, n)$, or

$$X^{-\frac{1}{2}} \Delta X X^{-\frac{1}{2}} \succeq -I,$$

which in turn implies $X + \Delta X \succeq 0$. Similarly one has $S + \Delta S \succeq 0$.

COMPUTATION OF THE TWO SEARCH DIRECTIONS

It is easily shown that (6.8) and (6.4) imply

$$\sum_{j=1}^{n} \Delta y_j \mathbf{Tr}\,(A_i D A_j D) = \frac{-\mathbf{Tr}\,(A_i X S X)}{(\mathbf{Tr}\,(XS)^2)^{\frac{1}{2}}}, \quad i = 1, \ldots, m, \tag{6.12}$$

for the Dikin-type direction, and that (6.10) and (6.4) imply

$$\sum_{j=1}^{n} \Delta y_j \mathbf{Tr}\,(A_i D A_j D) = -\mathbf{Tr}\,(A_i X)\,, \quad i = 1, \ldots, m, \tag{6.13}$$

for the primal-dual affine–scaling direction. The solution of these $m \times m$ linear systems yield Δy for the respective search directions. The coefficient matrices of the systems (6.12) and (6.13) are positive definite; a simple proof of this is given in Appendix A, Lemma A.3.

Once Δy is known, ΔS follows from $\sum_{i=1}^{m} \Delta y_i A_i = -\Delta S$, and ΔX is subsequently obtained from (6.8) (Dikin-type steps) or (6.10) (primal-dual affine–scaling steps).

The linear systems (6.12) and (6.13) — or linear systems with a different right-hand side vector — can be formed and solved in

$$\frac{2}{3}mn^3 + \frac{1}{2}m^2n^2 + O(\max\{m,n\}^3)$$

flops. The reader is referred to Monteiro and Zanjácomo [127] for details. All the primal–dual search directions in this monograph are defined via a linear system with the same coefficient matrix as (6.13) and can therefore be solved in the same number of operations. The problem is that forming the coefficient matrix in (6.13) is an expensive operation. The coefficient matrix is also always dense, even if the data matrices $C, A_1, \ldots, A_m$ or X or S is sparse. For this reason one cannot exploit sparsity as efficiently as in the LP case when doing the numerical linear algebra. Specialized techniques for forming and solving systems like (6.13) are described by Fujisawa *et al.* [60].

THE ALGORITHMS

The two primal–dual algorithms can both be described in the following framework.

Short step primal-dual affine-scaling algorithms

Input
A strictly feasible pair (X^0, S^0);

Parameters
$\tau_0 > 1$ such that $\kappa\left(X^0 S^0\right) \le \tau_0$;
accuracy $\epsilon > 0$;
$L := \log \frac{\mathbf{Tr}\,(X^0 S^0)}{\epsilon}$;
$\alpha := \frac{1}{\sqrt{n\tau_0}}$ (Dikin-type steps), or
$\alpha := \frac{1}{nL\tau_0}$ (Primal–dual affine–scaling steps);

begin
$X := X^0;\ S := S^0$;
while $\mathbf{Tr}\,(XS) > \epsilon$ **do**
Compute ΔX, ΔS from (6.8) and (6.4) (Dikin-type steps)
or from (6.10) and (6.4) (primal–dual affine–scaling steps);
$X := X + \alpha\Delta X$;
$S := S + \alpha\Delta S$;
end
end

We will prove that the Dikin-type algorithm computes a strictly feasible ϵ-optimal solution (X^*, S^*) in $O(\tau_0 nL)$ steps, and this solution satisfies $\kappa(X^* S^*) \le \tau_0$, where $\kappa(XS)$ is a 'function of centrality' to be described shortly (in (6.14)). The primal–dual affine–scaling algorithm converges in $O(\tau_0 nL^2)$ steps, and the solution satisfies $\kappa(X^* S^*) \le 3\tau_0$.

6.4 FUNCTIONS OF CENTRALITY

The Dikin-type steps have the feature that the proximity to the central path is maintained, where this proximity is quantified by

$$\kappa(XS) := \frac{\lambda_{\max}(XS)}{\lambda_{\min}(XS)} \tag{6.14}$$

with $\lambda_{\max}(XS)$ the largest eigenvalue of XS and $\lambda_{\min}(XS)$ the smallest.

This 'centering effect' of the Dikin-type steps is illustrated by the following example.

Example 6.1 *The centering effect is clearly visible in Figure 6.1, for the small example with data*

$$A_1 = \begin{bmatrix} 2 & -1 & 3 \\ -1 & 1 & 1 \\ 3 & 1 & -2 \end{bmatrix}, \; A_2 = \begin{bmatrix} 5 & 4 & 2 \\ 4 & 2 & -1 \\ 2 & -1 & 1 \end{bmatrix}, \; A_3 = \begin{bmatrix} 1 & 1 & 3 \\ 1 & 6 & 4 \\ 3 & 4 & -2 \end{bmatrix},$$

and with feasible starting solution

$$X^0 = \begin{bmatrix} 2.2 & 0.1 & 0.1 \\ 0.1 & 1.5 & 0.08 \\ 0.1 & 0.08 & 1.5 \end{bmatrix}, \; S^0 = \begin{bmatrix} 1.5 & 0.005 & 0 \\ 0.005 & 0.95 & 0 \\ 0 & 0 & 1.5 \end{bmatrix}.$$

The minimum and maximum eigenvalues of XS are plotted at successive iterations

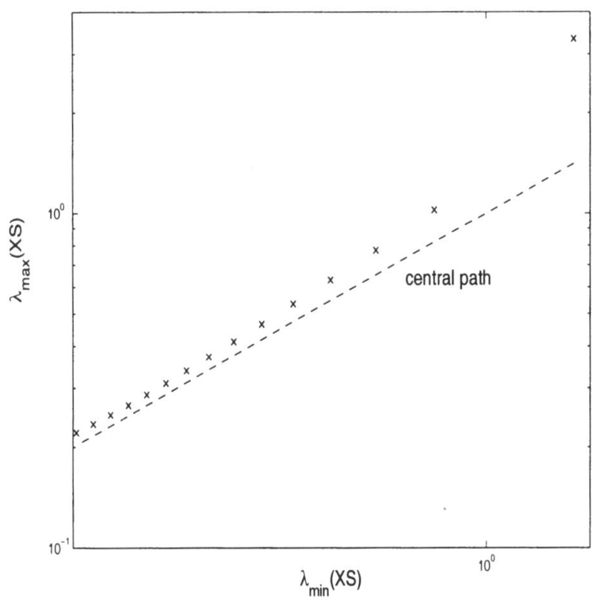

Figure 6.1. The centering effect of primal–dual Dikin steps, seen from iterates of the short step algorithm. The dashed line corresponds to the central path.

for the short step primal–dual Dikin-type method. In this figure the central path corresponds to the diagonal where largest and smallest eigenvalues are equal. □

The primal–dual affine–scaling steps may become increasingly less centered with respect to the function $\kappa(XS)$, which complicates the analysis somewhat.

Note that one always has $\kappa(XS) \geq 1$ with equality if and only if $XS = \mu I$ for some $\mu > 0$, *i.e.* if the pair (X, S) is centered with parameter μ.

6.5 FEASIBILITY OF THE DIKIN-TYPE STEP

We proceed with the analysis of the Dikin-type affine–scaling method, after which we will extend the analysis to cover the primal–dual affine–scaling method as well.

Having computed the Dikin-type step direction $(\Delta X, \Delta S)$ from (6.8) and (6.4), a feasible step length must be established. Denoting

$$X_\alpha := X + \alpha\Delta X, \quad S_\alpha := S + \alpha\Delta S,$$

we establish a value $\bar{\alpha} > 0$ such that $X_{\bar{\alpha}} \succ 0$ and $S_{\bar{\alpha}} \succ 0$. The following lemma gives a sufficient condition for a feasible step length $\bar{\alpha}$.

Lemma 6.1 *Let $X \succ 0$ and $S \succ 0$. If one has*

$$\det\left(X_\alpha S_\alpha\right) > 0 \quad \forall\, 0 \leq \alpha \leq \bar{\alpha},$$

then $X_{\bar{\alpha}} \succ 0$ and $S_{\bar{\alpha}} \succ 0$.

Proof:

Since $\det(X_\alpha S_\alpha) = \det(X_\alpha)\det(S_\alpha)$, one has

$$\det(X_\alpha S_\alpha) = \prod_i \lambda_i(X_\alpha) \prod_i \lambda_i(S_\alpha). \tag{6.15}$$

The left-hand side of equation (6.15) is strictly positive on $[0, \bar{\alpha}]$. This shows that the eigenvalues of X_α and S_α remain positive on $[0, \bar{\alpha}]$ (the eigenvalues $\lambda_i(X_\alpha$ and $\lambda_i(S_\alpha$ $(i = 1, \ldots, n)$ are smooth functions of α in the sense of Theorem A.5 in Appendix A). □

In order to derive bounds on α which are sufficient to guarantee a feasible step length, we need the following three technical results.

Lemma 6.2 *Let $D_X \in \mathcal{S}_n$ and $D_S \in \mathcal{S}_n$ be such that* $\mathbf{Tr}\left(D_X D_S\right) = 0$. *The spectral radius of $D_X D_S + D_S D_X$ is bounded by*

$$\rho(D_X D_S + D_S D_X) \leq \frac{1}{2}\|D_X + D_S\|^2.$$

Proof:

It is trivial to verify that

$$D_X D_S + D_S D_X = \frac{1}{2}\left[(D_X + D_S)^2 - (D_X - D_S)^2\right]$$

which implies

$$-\frac{1}{2}(D_X - D_S)^2 \preceq D_X D_S + D_S D_X \preceq \frac{1}{2}(D_X + D_S)^2.$$

It follows that

$$-\frac{1}{2}\|D_X - D_S\|^2 I \preceq D_X D_S + D_S D_X \preceq \frac{1}{2}\|D_X + D_S\|^2 I.$$

Since $\mathbf{Tr}\,(D_X D_S) = 0$, the matrices $(D_X + D_S)$ and $(D_X - D_S)$ have the same norm. Consequently

$$-\frac{1}{2}\|D_X + D_S\|^2 I \preceq D_X D_S + D_S D_X \preceq \frac{1}{2}\|D_X + D_S\|^2 I$$

from which the required result follows. □

Corollary 6.1 *For the Dikin-type step* $D_X + D_S = -V^3/\|V^2\|$*, one has*

$$\rho(D_X D_S + D_S D_X) \leq \frac{1}{2}\rho(V^2).$$

Proof:

By Lemma 6.2 one has

$$\begin{aligned} 2\rho\,(D_X D_S + D_S D_X) &\leq \|D_X + D_S\|^2 \\ &= \left(\frac{\|V^3\|}{\|V^2\|}\right)^2 = \frac{\mathbf{Tr}\,(V^6)}{(\|V^2\|)^2} \\ &\leq \rho(V^2)\frac{\mathbf{Tr}\,(V^4)}{(\|V^2\|)^2} = \rho(V^2), \end{aligned}$$

which is the required result. □

The following lemma contains two useful results from linear algebra concerning semidefinite matrices. It is proven in Appendix A (Lemma A.1).

Lemma 6.3 *Let* $Q \in \mathcal{S}_n^{++}$*, and let* $M \in \mathbb{R}^{n\times n}$ *be skew–symmetric,* i.e. $M = -M^T$. *One has* $\det(Q + M) > 0$. *Moreover, if* $\lambda_i(Q + M) \in \mathbb{R}$ $(i = 1, \ldots, n)$*, then*

$$0 < \lambda_{\min}(Q) \leq \lambda_{\min}(Q + M) \leq \lambda_{\max}(Q + M) \leq \lambda_{\max}(Q),$$

which implies $\kappa(Q + M) \leq \kappa(Q)$.

We are now in a position to find a step size α which guarantees that the Dikin-type step will be feasible. To simplify the analysis we introduce a parameter $\tau > 1$ such that $\kappa(XS) = \kappa(V^2) \leq \tau$. This implies the existence of numbers τ_1 and τ_2 such that

$$\tau_1 I \preceq V^2 \preceq \tau_2 I, \quad \tau_2 = \tau_1 \tau. \tag{6.16}$$

Lemma 6.4 *The steps $X_\alpha = X + \alpha\Delta X$ and $S_\alpha = S + \alpha\Delta S$ are feasible if the step size α satisfies $\alpha \leq \bar{\alpha}$ where*

$$\bar{\alpha} = \min\left\{\frac{\|V^2\|}{2\tau_2}, \frac{4\tau_1}{\|V^2\|}\right\}.$$

Furthermore

$$\kappa\left(X_{\bar{\alpha}}S_{\bar{\alpha}}\right) \leq \tau.$$

Proof:

We show that the determinant of $X_\alpha S_\alpha$ remains positive for all $\alpha \leq \bar{\alpha}$. One then has $X_{\bar{\alpha}}, S_{\bar{\alpha}} \succ 0$ by Lemma 6.1.

To this end note that

$$\begin{aligned}
X_\alpha S_\alpha &\sim (V + \alpha D_X)(V + \alpha D_S) \\
&= V^2 + \alpha D_X V + \alpha V D_S + \alpha^2 D_X D_S \\
&= V^2 - \frac{\alpha V^4}{\|V^2\|} + \frac{1}{2}\alpha^2(D_X D_S + D_S D_X) \\
&+ \left[\frac{1}{2}\alpha^2(D_X D_S - D_S D_X) + \frac{1}{2}\alpha(D_X V + V D_S - V D_X - D_S V)\right],
\end{aligned}$$

since $D_X + D_S = -V^3/\|V^2\|$. The matrix in square brackets is skew–symmetric. Lemma 6.3 therefore implies that the determinant of $X_\alpha S_\alpha$ will be positive if the matrix

$$Q(\alpha) := V^2 - \frac{\alpha V^4}{\|V^2\|} + \frac{1}{2}\alpha^2(D_X D_S + D_S D_X)$$

is positive definite. Note that $Q(0) = V^2 \succ 0$ and $\kappa\left(Q(0)\right) \leq \tau$. We proceed to prove that $\kappa\left(Q(\alpha)\right)$ remains bounded by $\kappa(Q(\alpha)) \leq \tau$ for $0 \leq \alpha \leq \bar{\alpha}$. This is sufficient to prove that $Q(\alpha) \succ 0,\ 0 \leq \alpha \leq \bar{\alpha}$, and therefore that a step of length $\bar{\alpha}$ is feasible.

Moreover, after such a feasible step we will have $X_{\bar{\alpha}} \succ 0,\ S_{\bar{\alpha}} \succ 0$. The matrix $X_{\bar{\alpha}}S_{\bar{\alpha}}$ therefore has positive eigenvalues and we can apply the second part of Lemma 6.3 to obtain

$$\kappa\left(X_{\bar{\alpha}}S_{\bar{\alpha}}\right) \leq \kappa\left(Q(\bar{\alpha})\right) \leq \tau.$$

We start the proof by noting that if λ is an eigenvalue of V^2, then $(\lambda - \alpha\lambda^2/\|V^2\|)$ is an eigenvalue of $[V^2 - \alpha V^4/\|V^2\|]$. The function

$$\phi(t) := t - \alpha\frac{t^2}{\|V^2\|}$$

is monotonically increasing on $t \in [0, \tau_2]$ if $\alpha \leq \bar{\alpha}$, since $\bar{\alpha} \leq \|V^2\|/(2\tau_2)$. Thus

$$\phi(\tau_1)I \preceq V^2 - \frac{\alpha V^4}{\|V^2\|} \preceq \phi(\tau_2)I \quad \forall\, 0 \leq \alpha \leq \bar{\alpha}$$

or, equivalently,

$$\phi(\tau_1)I + \frac{1}{2}\alpha^2(D_X D_S + D_S D_X) \preceq Q(\alpha) \preceq \phi(\tau_2)I + \frac{1}{2}\alpha^2(D_X D_S + D_S D_X),$$

for all $0 \le \alpha \le \bar{\alpha}$. We will therefore certainly have $\kappa(Q(\alpha)) \le \tau$ if

$$\tau\left[\phi(\tau_1)I + \frac{1}{2}\alpha^2(D_X D_S + D_S D_X)\right] \succeq \phi(\tau_2)I + \frac{1}{2}\alpha^2(D_X D_S + D_S D_X).$$

This matrix inequality can be simplified using $\tau_2 = \tau\tau_1$ and subsequently dividing by α. This yields

$$\left(\frac{\tau_2^2 - \tau\tau_1^2}{\|V^2\|}\right) I + \alpha(\tau - 1)\left(\frac{1}{2}(D_X D_S + D_S D_X)\right) \succeq 0. \tag{6.17}$$

Expression (6.17) may be further simplified using

$$\tau_2^2 - \tau\tau_1^2 = (\tau - 1)\tau_1\tau_2$$

to obtain

$$\left(\frac{\tau_1\tau_2}{\|V^2\|}\right) I + \frac{1}{2}\alpha(D_X D_S + D_S D_X) \succeq 0$$

which will surely hold if

$$\left(\frac{\tau_1\tau_2}{\|V^2\|}\right) I - \alpha\rho\left(\frac{1}{2}(D_X D_S + D_S D_X)\right) I \succeq 0.$$

Substituting the bound

$$\rho\left(\frac{1}{2}(D_X D_S + D_S D_X)\right) \le \frac{1}{4}\rho(V^2) \le \frac{1}{4}\tau_2$$

from Corollary 6.1 yields

$$\frac{\tau_1\tau_2}{\|V^2\|} - \frac{1}{4}\alpha\tau_2 \ge 0,$$

or

$$\alpha \le \frac{4\tau_1}{\|V^2\|},$$

which is the second bound in the lemma. □

6.6 COMPLEXITY ANALYSIS FOR THE DIKIN-TYPE METHOD

A feasible Dikin-type step of length α reduces the duality gap by at least a factor $(1 - \frac{\alpha}{\sqrt{n}})$. Formally we have the following result.

Lemma 6.5 *Given a feasible primal–dual pair* (X, S) *and a step length* α *such that the Dikin-type step is feasible,* i.e. $X_\alpha := X + \alpha\Delta X \succ 0$, *and* $S_\alpha := S + \alpha\Delta S \succ 0$, *it holds that*

$$\mathbf{Tr}\,(X_\alpha S_\alpha) \leq \left(1 - \frac{\alpha}{\sqrt{n}}\right) \mathbf{Tr}\,(XS).$$

Proof:

The duality gap after the Dikin-type step is given by

$$\begin{aligned}
\mathbf{Tr}\,(X_\alpha S_\alpha) &= \mathbf{Tr}\,((V + \alpha D_X)(V + \alpha D_S)) \\
&= \mathbf{Tr}\,\left(V^2 + \alpha V(D_X + D_S)\right) \\
&= \mathbf{Tr}\,\left(V^2 - \alpha\frac{V^4}{\|V^2\|}\right) \\
&= \|V\|^2 - \alpha\|V^2\| \\
&= \left(1 - \alpha\frac{\|V^2\|}{\|V\|^2}\right) \mathbf{Tr}\,(XS).
\end{aligned}$$

By the Cauchy–Schwartz inequality one has

$$\|V\|^2 = \mathbf{Tr}\,\left(IV^2\right) \leq \|I\|\,\left\|V^2\right\| = \sqrt{n}\left\|V^2\right\|,$$

which gives the required result. □

We are now ready to prove a worst-case iteration complexity bound.

Theorem 6.1 *Let* $\epsilon > 0$ *be an accuracy parameter, and let* $\tau_0 > 1$ *be such that* $\kappa(X^0S^0) \leq \tau_0$. *Further let* $L = \log\left(\mathbf{Tr}\,(X^0S^0)/\epsilon\right)$, *and* $\alpha = \frac{1}{\tau_0\sqrt{n}}$. *The Dikin-type step algorithm requires at most* $\lceil \tau_0 nL \rceil$ *iterations to compute a feasible primal–dual pair* (X^*, S^*) *satisfying* $\kappa(X^*S^*) \leq \tau_0$ *and* $\mathbf{Tr}\,(X^*S^*) \leq \epsilon$.

Proof:

We first prove that the default choice of α always allows a feasible step. To this end, note that

$$\alpha = \frac{1}{\tau_0\sqrt{n}} = \frac{\tau_1}{\tau_2\sqrt{n}} \leq \frac{\tau_1\sqrt{n}}{2\tau_2} = \frac{\|\tau_1 I\|}{2\tau_2} \leq \frac{\|V^2\|}{2\tau_2},$$

since $0 \preceq \tau_1 I \preceq V^2$. This shows that α meets the first condition of Lemma 6.4. Moreover, it holds that $\|V^2\| \leq \tau_2\sqrt{n}$, which implies

$$\frac{4\tau_1}{\|V^2\|} \geq \frac{4\tau_1}{\tau_2\sqrt{n}} = \frac{4}{\tau_0\sqrt{n}} > \alpha.$$

The default choice of α therefore meets the conditions of Lemma 6.4 and ensures a feasible Dikin-type step.

We know by Lemma 6.5 that the duality gap is reduced at each iteration by at least a factor $(1 - \frac{1}{n\tau_0})$. As the initial duality gap equals $\mathbf{Tr}\,(X^0 S^0)$, the duality gap will be smaller than ϵ after k iterations if

$$\left(1 - \frac{1}{n\tau_0}\right)^k \mathbf{Tr}\,(X^0 S^0) \leq \epsilon.$$

Taking logarithms yields

$$k \log\left(1 - \frac{1}{n\tau_0}\right) + \log\left(\mathbf{Tr}\,(X^0 S^0)\right) \leq \log(\epsilon). \tag{6.18}$$

Since

$$-\log\left(1 - \frac{1}{n\tau_0}\right) \geq \left(\frac{1}{n\tau_0}\right),$$

condition (6.18) will certainly be satisfied if

$$\frac{k}{n\tau_0} \geq \log\left(\mathbf{Tr}\,(X^0 S^0)\right) - \log \epsilon = L,$$

which implies the required result. □

The $O(\tau_0 n)$ complexity bound is a factor $\sqrt{n}$ worse than the best known bound for primal–dual algorithms.

6.7 ANALYSIS OF THE PRIMAL–DUAL AFFINE–SCALING METHOD

We return to the primal–dual affine–scaling algorithm. This analysis is analogous to that of the Dikin-type method, but there is one significant difference: whereas the Dikin-type iterates stay in the same neighbourhood of the central path, the same is not true of the affine–scaling steps. The deviation from centrality may increase at each step, but this can be bounded, and polynomial complexity can be retained at a price: The step length has to be shortened to

$$\alpha = \frac{1}{nL\tau_0}, \tag{6.19}$$

and the worst case iteration complexity bound becomes $O(\tau_0 n L^2)$.

We need to modify the analysis of the Dikin-type step algorithm with regard to the following:

- We allow for an increase in the distance $\kappa(XS)$ from the central path by a constant factor $t > 1$ at each step;
- The step length α in (6.19) is shown to be feasible for $\tau_0 n L^2$ iterations, provided that we choose the factor t in such a way that the distance from the central path

stays within the bound $\kappa(XS) < 3\tau_0$ for $O(\tau n L^2)$ iterations – the convergence criterion is met before the deviation from centrality becomes worse than $3\tau_0$.

Recall that the primal-dual affine–scaling direction is obtained by solving

$$\Delta X + D\Delta S D = -X,$$

subject to (6.4). A feasible step in this direction gives the following reduction in the duality gap.

Lemma 6.6 *Given a pair $(X,S) \in \mathrm{ri}\,(\mathcal{P} \times \mathcal{D})$, assume that the primal–dual affine–scaling step with step length α is feasible,* i.e. *$X_\alpha := X + \alpha\Delta X \in \mathcal{P}$, and $S_\alpha := S + \alpha\Delta S \in \mathcal{D}$. It holds that*

$$\mathbf{Tr}\,(X_\alpha S_\alpha) = (1-\alpha)\mathbf{Tr}\,(XS).$$

Proof:

Analogous to the proof of Lemma 6.5. □

As with the Dikin-type step analysis, we will also need the following bound.

Lemma 6.7 *For the primal–dual affine–scaling step $D_V = D_X + D_S = -V$, one has*

$$\rho(D_X D_S + D_S D_X) \le \frac{1}{2}\|V\|^2.$$

Proof:

Follows from Lemma 6.2. □

Now let $\tau = \kappa(XS)$ and $\tau_0 = \kappa(X^0 S^0)$ for the current pair of iterates (X,S) and starting solution (X^0, S^0) respectively, and let τ_1, τ_2 satisfy (6.16).[5]

We also define the amplification factor

$$t := 1 + \frac{1}{nL^2\tau_0},$$

which is used to bound the loss of centrality in a given iteration.

[5]The value τ_0 had to be strictly greater than one, *i.e.* $\tau_0 > 1$, for the Dikin-type algorithm. Here it is sufficient to require $\tau_0 \ge 1$. The value $\tau_0 = \kappa(X^0 S^0)$ can therefore be used for the primal–dual affine–scaling method.

Lemma 6.8 *If $\tau \leq \frac{3\tau_0}{t}$, then the steps $X_\alpha = X + \alpha\Delta X$ and $S_\alpha = S + \alpha\Delta S$ are feasible for the step size*

$$\bar{\alpha} = \frac{1}{nL\tau_0},$$

and the deviation from centrality is bounded by

$$\kappa\left(X_{\bar{\alpha}} S_{\bar{\alpha}}\right) \leq t\tau,$$

where $\tau = \kappa(XS)$.

Proof:

As in the proof of Lemma 6.4, we show that the determinant of $X_\alpha S_\alpha$ remains positive for all $\alpha \leq \bar{\alpha}$, which ensures $X_{\bar{\alpha}}, S_{\bar{\alpha}} \succ 0$ by Lemma 6.1.

As before, note that

$$\begin{aligned}
X_\alpha S_\alpha &\sim (V + \alpha D_X)(V + \alpha D_S) \\
&= V^2 + \alpha D_X V + \alpha V D_S + \alpha^2 D_X D_S \\
&= (1-\alpha)V^2 + \frac{1}{2}\alpha^2(D_X D_S + D_S D_X) \\
&+ \left[\frac{1}{2}\alpha^2(D_X D_S - D_S D_X) + \frac{1}{2}\alpha(D_X V + V D_S - V D_X - D_S V)\right],
\end{aligned}$$

since $D_X + D_S = -V$. The matrix in square brackets is skew–symmetric. Lemma 6.3 therefore implies that the determinant of $[X_\alpha S_\alpha]$ will be positive if the matrix

$$Q(\alpha) := (1-\alpha)V^2 + \frac{1}{2}\alpha^2(D_X D_S + D_S D_X)$$

is positive definite. Note that $Q(0) = V^2 \succ 0$ and $\kappa\left(Q(0)\right) = \tau$. We proceed to prove that $\kappa\left(Q(\alpha)\right)$ remains bounded by $\kappa(Q(\alpha)) \leq t\tau$ for $0 \leq \alpha \leq \bar{\alpha}$, for the fixed amplification factor t. This is sufficient to prove that $Q(\alpha) \succ 0,\ 0 \leq \alpha \leq \bar{\alpha}$, and therefore that a step of length $\bar{\alpha}$ is feasible.

Moreover, after such a feasible step we will have $X_{\bar{\alpha}} \succ 0,\ S_{\bar{\alpha}} \succ 0$. The matrix $X_{\bar{\alpha}} S_{\bar{\alpha}}$ therefore has positive eigenvalues and we can apply the second part of Lemma 6.3 to obtain

$$\kappa\left(X_{\bar{\alpha}} S_{\bar{\alpha}}\right) \leq \kappa\left(Q(\bar{\alpha})\right) \leq t\tau.$$

To start the proof, note that for all $0 \leq \alpha \leq \bar{\alpha}$ one has

$$\tau_1(1-\alpha)I + \frac{1}{2}\alpha^2(D_X D_S + D_S D_X) \preceq Q(\alpha) \preceq \tau_2(1-\alpha)I + \frac{1}{2}\alpha^2(D_X D_S + D_S D_X).$$

We will therefore certainly have $\kappa(Q(\alpha)) \leq t\tau$ if

$$t\tau[\tau_1(1-\alpha)I + \frac{1}{2}\alpha^2(D_X D_S + D_S D_X)] \succeq \tau_2(1-\alpha)I + \frac{1}{2}\alpha^2(D_X D_S + D_S D_X).$$

Using $\tau_2 = \tau\tau_1$ the last relation becomes

$$\tau_2(1-\alpha)(t-1)I + \frac{1}{2}\alpha^2(t\tau - 1)(D_X D_S + D_S D_X) \succeq 0. \tag{6.20}$$

Since one has $\rho(D_X D_S + D_S D_X) \le \frac{1}{2}\|V\|^2 \le \frac{1}{2}\tau_2 n$ by Lemma 6.7, inequality (6.20) will hold if

$$(1-\alpha)(t-1) - \frac{1}{4}\alpha^2(t\tau - 1)n \ge 0. \tag{6.21}$$

Using the assumption $t\tau \le 3\tau_0$, it follows that (6.21) will surely hold if

$$(1-\alpha)\left(\frac{1}{nL^2\tau_0}\right) - \frac{1}{4}\alpha^2(3\tau_0 - 1)n \ge 0,$$

which is satisfied by $\bar{\alpha} = \frac{1}{nL\tau_0}$. □

We now investigate how many iterations can be performed while still satisfying the assumption $\kappa(XS) \le 3\tau_0/t$ of Lemma 6.8.

Lemma 6.9 *One has*

$$\kappa(XS) \le 3\tau_0$$

for the first $\lceil nL^2\tau_0 \rceil$ iterations of the primal–dual affine–scaling algorithm.

Proof:

By Lemma 6.8 one has

$$\kappa(XS) \le \tau_0 t^k \text{ after } k \text{ iterations},$$

provided that k is sufficiently small to guarantee $\tau_0 t^k \le 3\tau_0$. Using $t = 1 + \frac{1}{nL^2\tau_0}$, we obtain

$$t^k = \left(1 + \frac{1}{nL^2\tau_0}\right)^k < 3 \text{ if } k \le \lceil nL^2\tau_0 \rceil,$$

which gives the required result. □

It only remains to prove that $\lceil nL^2\tau_0 \rceil$ iterations are sufficient to guarantee convergence. The proof is analogous to that of Theorem 6.1. Formally we have the following result.

Theorem 6.2 *Let $\epsilon > 0$ be an accuracy parameter, and let τ_0 be such that $\kappa(X^0S^0) \le \tau_0$. Further let $L = \log\left(\mathbf{Tr}\,(X^0S^0)/\epsilon\right)$ and $\alpha = \frac{1}{nL\tau_0}$. The primal–dual affine–scaling algorithm requires at most $\lceil nL^2\tau_0 \rceil$ iterations to compute a feasible primal–dual pair (X^*, S^*) satisfying $\kappa(X^*S^*) \le 3\tau_0$ and $\mathbf{Tr}\,(X^*S^*) \le \epsilon$.*

Proof:

By Lemma 6.8 a step of size $\alpha = \frac{1}{nL\tau_0}$ is feasible as long as the iterates (X, S) satisfy $\kappa(XS) \leq 3\tau_0$. Such a step reduces the duality gap by a factor $(1 - \frac{1}{nL\tau_0})$ (Lemma 6.6).

By the proof of Theorem 6.1, this reduction of the duality gap ensures that the convergence criterion is met after k steps, if

$$k\alpha \geq \log \frac{\mathbf{Tr}\,(X^0 S^0)}{\epsilon} = L,$$

i.e. if $k \geq nL^2\tau_0$. Lemma 6.9 guarantees that the first $nL^2\tau_0$ steps will be feasible, which completes the proof. □

7

PRIMAL–DUAL PATH–FOLLOWING METHODS

Preamble

Primal–dual path–following methods are aptly named: the underlying idea for these algorithms is to 'follow' the primal–dual central path approximately in order to reach the optimal set. More precisely, the centrality conditions (3.1) on page 41 are solved approximately for a given value of $\mu > 0$, after which μ is reduced and the process is repeated. Primal-dual path–following methods have emerged as the most successful interior point algorithms for linear programming (LP). Predictor-corrector methods are particularly in favour, following successful implementations (see *e.g.* Andersen and Andersen [9]). The extension of algorithms from linear to semidefinite programming (SDP) has followed the same trends. We will consider methods here that use the Nesterov–Todd (NT) scaling, namely a short step method as analysed by De Klerk *et al.* [47], and a long step (large update) method due to Jiang [96]. We will also review some predictor–corrector methods that use the NT direction.

7.1 THE PATH–FOLLOWING APPROACH

For a given value $\mu > 0$, the μ-center $(X(\mu), S(\mu))$ can be regarded as a *target point* on the central path with associated target duality gap $\mathbf{Tr}\,(X(\mu)S(\mu)) = n\mu$. In other words, if the μ-center $(X(\mu), S(\mu))$ can be computed exactly, then the duality gap will be equal to $n\mu$.

Path–following algorithms iteratively compute an approximation to $(X(\mu), S(\mu))$; this is then followed by a decrease in the value of μ.

Assume that a strictly feasible pair $(X, S) \in \mathrm{ri}\,(\mathcal{P} \times \mathcal{D})$ is given as well as a value $\mu > 0$. Ideally, we wish to compute $(\Delta X, \Delta S)$ such that $X + \Delta X \in \mathcal{P}$, $S + \Delta S \in \mathcal{D}$, and

$$(X + \Delta X)(S + \Delta S) = \mu I. \tag{7.1}$$

As discussed on page 14, there are different ways to approximate the solution of the resulting overdetermined, nonlinear system of equations. The different solution approaches lead to different search directions.[1] One of the popular primal-dual directions is the so-called Nesterov-Todd (NT) direction, introduced in [138] (see page 14).[2] We will only study the NT direction in this chapter.

To derive the NT search directions, the notation for the primal–dual (NT) scaling as introduced in Section 6.2 on page 97 is used. Using the scaling matrix D defined in (6.1) on page 97, we can rewrite (7.1) as

$$(V + D_X)(V + D_S) = \mu I, \tag{7.2}$$

where $V = D^{-\frac{1}{2}} X D^{-\frac{1}{2}} = D^{\frac{1}{2}} S D^{\frac{1}{2}}$, $D_X = D^{-\frac{1}{2}} \Delta X D^{-\frac{1}{2}}$, $D_S = D^{\frac{1}{2}} \Delta S D^{\frac{1}{2}}$, and $D_V = D_X + D_S$, as before.

We now weaken condition (7.2) by replacing the left-hand side with its symmetric part, to obtain

$$\frac{1}{2}\left[(V + D_X)(V + D_S) + ((V + D_X)(V + D_S))^T\right] = \mu I.$$

The next step is to linearize this system by neglecting the cross terms $D_X D_S$ and $D_S D_X$, to obtain

$$\frac{1}{2}\left((D_X + D_S)V + V(D_X + D_S)\right) = \mu I - V^2. \tag{7.3}$$

Equation (7.3) (called a Lyapunov equation) has a unique symmetric solution (see Theorem E.2 on page 250), given by

$$D_V \equiv D_X + D_S = \mu V^{-1} - V.$$

Pre- and post-multiplying with $D^{\frac{1}{2}}$ yields the *Nesterov-Todd* (or NT) equations:

$$\Delta X + D \Delta S D = \mu S^{-1} - X \tag{7.4}$$

subject to

$$\left.\begin{array}{rcl} \mathbf{Tr}\,(A_i \Delta X) & = & 0, \quad i = 1, \ldots, m \\ \Delta S & = & \sum_{i=1}^m \Delta y_i A_i. \end{array}\right\} \tag{7.5}$$

---

[1]A comparison of the best known search directions was done by Todd [172] and by Aarts [1].

[2]Kojima *et al.* [106] showed the NT direction to be a special case of the class of primal–dual directions for monotone semidefinite complementarity problems introduced in Kojima *et al.* [108].

As before, we can easily deduce

$$\begin{aligned}\sum_{j=1}^{n} \Delta y_j \mathbf{Tr}\,(A_i D A_j D) &= \mu \mathbf{Tr}\,\left(A_i S^{-1}\right) - \mathbf{Tr}\,(A_i X) \\ &= \mu \mathbf{Tr}\,\left(A_i S^{-1}\right) - b_i, \quad i = 1, \ldots, m.\end{aligned}$$

This linear system has the same coefficient matrix as the system (6.13) on page 101; recall that this matrix is positive definite. We can therefore solve for Δy and consequently obtain ΔS from the second equation in (7.5), after which we obtain ΔX from (7.4). The resulting direction $(\Delta X, \Delta S)$ is the *NT direction*.

Note that the primal–dual affine–scaling direction in Chapter 6 is obtained by setting $\mu = 0$. In other words, the primal–dual affine–scaling direction 'targets' the limit point of the central path.

A CENTRALITY FUNCTION

Assume we are given $(X, S) \in \mathrm{ri}\,(\mathcal{P} \times \mathcal{D})$ and a value $\mu > 0$. We will use the centrality function

$$\delta(X, S, \mu) := \frac{1}{2}\frac{1}{\sqrt{\mu}}\|D_V\| = \frac{1}{2}\left\|\sqrt{\mu}V^{-1} - \frac{1}{\sqrt{\mu}}V\right\|,$$

that was introduced by Jiang [96] (without the constant $\frac{1}{2}$). Note that $\delta(X, S, \mu) \geq 0$ and

$$\delta(X, S, \mu) = 0 \iff V^2 = \mu I \iff XS = \mu I.$$

This function generalizes the LP function of Jansen *et al.* [93] to semidefinite programming, and will be used extensively in this chapter. It was shown by Jiang [96] that $\delta(X, S, \mu)$ is related to the directional derivative of the primal-dual barrier along the NT direction. To derive this relation, let $(\Delta X, \Delta S)$ denote the NT direction at (X, S) and let f_{pd}^{μ} denote the primal–dual log barrier

$$f_{pd}^{\mu}(X, S, \mu) = \frac{1}{\mu}\mathbf{Tr}\,(XS) - \log\det(XS).$$

The directional derivative of f_{pd} at (X, S) along the NT direction is given by:

$$\begin{aligned}&\langle \nabla_X f(X, S, \mu), \Delta X\rangle + \langle \nabla_S f(X, S, \mu), \Delta S\rangle \\ =\ & \mathbf{Tr}\left(\left(\frac{1}{\mu}S - X^{-1}\right)\Delta X\right) + \mathbf{Tr}\left(\left(\frac{1}{\mu}X - S^{-1}\right)\Delta S\right) \\ =\ & \mathbf{Tr}\left(\left(\frac{1}{\mu}V - V^{-1}\right)D_X\right) + \mathbf{Tr}\left(\left(\frac{1}{\mu}V - V^{-1}\right)D_S\right) \\ =\ & \mathbf{Tr}\left(\left(\frac{1}{\mu}V - V^{-1}\right)D_V\right) \\ =\ & -\frac{1}{\mu}\mathbf{Tr}\,\left(D_V^2\right) = -4\delta^2.\end{aligned}$$

This equality shows that δ is a natural centrality function associated with the NT direction.

THE GENERIC ALGORITHM

The algorithms presented in this chapter all fit in the following framework.

Generic primal-dual path–following algorithm

Input
A pair $(X^0, S^0) \in \mathcal{P} \times \mathcal{D}$;

Parameters
Parameters $\tau < 1$ and $\mu_0 > 0$ such that $\delta\left(X^0, S^0, \mu_0\right) \le \tau$;
An accuracy parameter $\epsilon > 0$;

begin
$X := X^0$; $S := S^0$; $\mu = \mu_0$
while **Tr** $(XS) > \epsilon$ **do**
Compute ΔX, ΔS from (7.4) and (7.5);
Choose a step length $\alpha \in (0, 1]$;
$X := X + \alpha \Delta X$;
$S := S + \alpha \Delta S$;
Choose an updating parameter $0 < \theta < 1$;
$\mu := (1-\theta)\mu$;
end
end

We will refer to

$$(X^+, S^+) := (X + \Delta X, S + \Delta S)$$

as a *full NT step*, and to

$$(X_\alpha, S_\alpha) := (X + \alpha \Delta X, S + \alpha \Delta S)$$

as a *damped NT step* if $0 < \alpha < 1$.

7.2 FEASIBILITY OF THE FULL NT STEP

As before, let $(X, S) \in \text{ri}\,(\mathcal{P} \times \mathcal{D})$ and a value $\mu > 0$ be given. One can now prove the following two results which are analogous to the LP case: If $\delta(X, S, \mu) < 1$ then the full NT step is feasible, and the duality gap after the step attains its target value, namely $n\mu$.

The feasibility of the full NT step is proved in the following lemma.

Lemma 7.1 (Condition for a feasible full NT step) *Let* $(X,S) \in \mathrm{ri}(\mathcal{P} \times \mathcal{D})$ *and* $\mu > 0$. *If* $\delta := \delta(X,S,\mu) < 1$, *then the full NT step is strictly feasible.*

Proof:

We show that the determinant of $X_\alpha S_\alpha$ remains positive for all $\alpha \leq 1$. One then has $X(1), S(1) \succ 0$ by Lemma 6.1.

To this end note that

$$\begin{aligned}
X_\alpha S_\alpha &\sim (V + \alpha D_X)(V + \alpha D_S) \\
&= V^2 + \alpha D_X V + \alpha V D_S + \alpha^2 D_X D_S \\
&= V^2 + \alpha(\mu I - V^2) + \frac{1}{2}\alpha^2 \left(D_X D_S + D_S D_X\right) \\
&+ \left[\frac{1}{2}\alpha^2(D_X D_S - D_S D_X) + \frac{1}{2}\alpha(D_X V + V D_S - V D_X - D_S V)\right],
\end{aligned}$$

where we have used equation (7.3).

The matrix in square brackets in the last equation is skew–symmetric. Lemma 6.3 therefore implies that the determinant of $[X_\alpha S_\alpha]$ will be positive if the matrix

$$M(\alpha) := V^2 + \alpha(\mu I - V^2) + \frac{1}{2}\alpha^2 \left(D_X D_S + D_S D_X\right)$$

is positive definite. Since we can rewrite the expression for $M(\alpha)$ as

$$M(\alpha) = (1-\alpha)V^2 + \alpha\mu\left[I + \frac{\alpha}{\mu}\frac{1}{2}\left(D_X D_S + D_S D_X\right)\right],$$

one will have $M(\alpha) \succ 0$ if $\alpha \leq 1$ and $\left\|\frac{1}{2}\left(D_X D_S + D_S D_X\right)/\mu\right\|_2 < 1$. The last condition is easily shown to hold by using Lemma 6.2 and $\delta < 1$ successively:

$$\left\|\frac{1}{2}\left(D_X D_S + D_S D_X\right)/\mu\right\|_2 = \frac{1}{\mu}\left\|\frac{1}{2}\left(D_X D_S + D_S D_X\right)\right\|_2 \leq \frac{1}{4\mu}\|D_V\|^2 = \delta^2 < 1.$$

This completes the proof. □

The next lemma shows that the target duality gap is attained after a full NT step.

Corollary 7.1 *Let* $(X,S) \in \mathrm{ri}(\mathcal{P} \times \mathcal{D})$ *and* $\mu > 0$ *such that* $\delta(X,S,\mu) < 1$. *Then*

$$\mathbf{Tr}\left(X^+ S^+\right) = n\mu,$$

i.e. *the target duality gap is attained after one full NT step.*

Proof:

By the proof of Lemma 7.1 we have

$$\begin{aligned} & X^+S^+ & (7.6)\\ \sim\ & \mu I + \frac{1}{2}\left(D_X D_S + D_S D_X\right) \\ +\ & \left[\frac{1}{2}(D_X D_S - D_S D_X) + \frac{1}{2}(D_X V + V D_S - V D_X - D_S V)\right]. & (7.7)\end{aligned}$$

Because $A \sim B$ implies $\mathbf{Tr}\,(A) = \mathbf{Tr}\,(B)$ $\left(A, B \in \mathbf{R}^{n\times n}\right)$, we deduce

$$\mathbf{Tr}\,\left(X^+S^+\right) = \mathbf{Tr}\,(\mu I) = n\mu,$$

by using $\mathbf{Tr}\,(D_X D_S) = 0$ and the skew symmetry of the matrix in square brackets.[3]

□

7.3 QUADRATIC CONVERGENCE TO THE CENTRAL PATH

Notation

In what follows we denote the skew-symmetric matrix in (7.7) by M. We can also simplify the notation by defining

$$D_{XS} := \frac{1}{2}\left(D_X D_S + D_S D_X\right),$$

i.e. D_{XS} is the symmetric part of $D_X D_S$.

We proceed to prove quadratic convergence of full NT steps to the target point $(X(\mu), S(\mu))$. To this end, we need three technical results that give information about the spectrum of X^+S^+. We denote the NT scaling of X^+S^+ by $(V^+)^2$.

Lemma 7.2 *One has*

$$\lambda_{\min}\left(\left(V^+\right)^2\right) \geq \mu(1-\delta^2),$$

where $\lambda_{\min}$ *denotes the smallest eigenvalue.*

Proof:

From (7.7) it follows that

$$\lambda_{\min}\left(\left(V^+\right)^2\right) = \lambda_{\min}\left(\mu I + D_{XS} + M\right).$$

[3] Recall that the trace of a skew symmetric matrix is zero.

The skew-symmetry of M implies

$$\begin{aligned}\lambda_{\min}\left(\left(V^{+}\right)^{2}\right) &\geq \lambda_{\min}\left(\mu I + D_{XS}\right)\\ &\geq \mu - \|D_{XS}\|_{2}.\end{aligned}$$

Substitution of the bound for $\|D_{XS}\|_2$ from Lemma 6.2 now yields:

$$\lambda_{\min}\left(\left(V^{+}\right)^{2}\right) \geq \mu - \frac{1}{4}\|D_V\|^2 = \mu\left(1-\delta^2\right),$$

which completes the proof. □

Lemma 6.2 gave a bound on the spectral norm of D_{XS}. We now derive a similar bound on its Frobenius norm.

Lemma 7.3 *One has*

$$\|D_{XS}\|^2 \leq \frac{1}{8}\|D_V\|^4.$$

Proof:

It is trivial to verify that

$$D_X D_S + D_S D_X = \frac{1}{2}\left[(D_X + D_S)^2 - (D_X - D_S)^2\right].$$

Since $\mathbf{Tr}\,(D_X D_S) = 0$, the matrices $D_V = D_X + D_S$ and $Q_V := D_X - D_S$ have the same norm. Consequently

$$\begin{aligned}\|D_{XS}\|^2 &= \left\|\frac{1}{4}\left(D_V^2 - Q_V^2\right)\right\|^2\\ &= \frac{1}{16}\mathbf{Tr}\,\left(D_V^4 + Q_V^4 - 2D_V^2 Q_V^2\right)\\ &\leq \frac{1}{16}\left(\left\|D_V^2\right\|^2 + \left\|Q_V^2\right\|^2\right)\\ &\leq \frac{1}{16}\left(\|D_V\|^4 + \|Q_V\|^4\right) = \frac{1}{8}\|D_V\|^4.\end{aligned}$$

□

The quadratic convergence result will now be proved.

Lemma 7.4 (Quadratic convergence) *After a feasible full NT step the centrality function satisfies*

$$\delta^{+} := \delta(X^{+}, S^{+}, \mu) \leq \frac{\delta^2}{\sqrt{2(1-\delta^2)}}.$$

Proof:

The centrality function after the full NT step is given by

$$\begin{aligned}
\left(\delta^{+}\right)^{2} &= \frac{1}{4\mu}\left\|\mu\left(V^{+}\right)^{-1}-V^{+}\right\|^{2} \\
&= \frac{1}{4\mu}\left\|\left(V^{+}\right)^{-1}\left(\mu I-\left(V^{+}\right)^{2}\right)\right\|^{2} \\
&\leq \frac{1}{4\mu}\lambda_{\max}^{2}\left(\left(V^{+}\right)^{-1}\right)\left\|\mu I-\left(V^{+}\right)^{2}\right\|^{2}.
\end{aligned}$$

We now substitute the bound from Lemma 7.2 to obtain

$$\left(\delta^{+}\right)^{2} \leq \frac{1}{4\mu^{2}(1-\delta^{2})}\left\|\mu I-\left(V^{+}\right)^{2}\right\|^{2}.$$

To complete the proof we show that:

$$\left\|\mu I-\left(V^{+}\right)^{2}\right\|^{2} \leq \left\|D_{XS}\right\|^{2}. \tag{7.8}$$

In order to prove (7.8), note that

$$\begin{aligned}
\left\|\mu I-\left(V^{+}\right)^{2}\right\|^{2} &= \sum_{i=1}^{n}\left[\lambda_{i}\left(\mu I+D_{XS}+M\right)-\lambda_{i}\left(\mu I\right)\right]^{2} \\
&= \sum_{i=1}^{n}\left[\lambda_{i}\left(D_{XS}+M\right)\right]^{2} \\
&= \mathbf{Tr}\left(\left(D_{XS}+M\right)^{2}\right).
\end{aligned}$$

Using the skew-symmetry of M one obtains

$$\begin{aligned}
\left\|\mu I-\left(V^{+}\right)^{2}\right\|^{2} &= \mathbf{Tr}\left(\left(D_{XS}\right)^{2}-MM^{T}\right) \\
&\leq \mathbf{Tr}\left(D_{XS}\right)^{2} = \|D_{XS}\|^{2}.
\end{aligned}$$

The final result now follows from Lemma 7.3. □

The local convergence result has the following implications.

Corollary 7.2 *If $\delta(X,S,\mu) < \frac{1}{\sqrt{2}}$, then $\delta(X^{+},S^{+},\mu) < \delta^{2}(X,S,\mu)$,* i.e. *quadratic convergence to the μ-center is obtained. The weaker condition $\delta(X,S,\mu) < \sqrt{\frac{2}{3}}$ implies $\delta(X^{+},S^{+},\mu) < \delta(X,S,\mu)$ and is therefore sufficient for convergence.*

7.4 UPDATING THE BARRIER PARAMETER μ

If the current iterates $(X, S) \in \text{ri}\,(\mathcal{P} \times \mathcal{D})$ are sufficiently close to the target point $(X(\mu), S(\mu))$, say $\delta(X, S, \mu) \le \frac{1}{2}$, then the parameter μ is updated via

$$\mu^+ = (1 - \theta)\mu$$

where $0 < \theta < 1$ is a given parameter. We show that a default value of $\theta = \frac{1}{2\sqrt{n}}$ ensures that $\delta(X, S, \mu^+) \le \frac{1}{\sqrt{2}}$. The next full NT step then again yields a feasible pair $(X^+, S^+) \in \text{ri}\,(\mathcal{P} \times \mathcal{D})$ with $\delta(X^+, S^+, \mu^+) \le \frac{1}{2}$, due to the quadratic convergence property (Corollary 7.2).

We first prove a lemma that relates the centrality function after the μ-update to the centrality function before the μ-update.

Lemma 7.5 *Let $(X, S) \in \text{ri}\,(\mathcal{P} \times \mathcal{D})$, $\mu := \mathbf{Tr}\,(XS)/n$, and $\delta := \delta(X, S, \mu)$. If $\mu^+ = (1 - \theta)\mu$ for some $0 < \theta < 1$, one has*

$$(\delta(X, S, \mu^+))^2 = \frac{n\theta^2}{4(1-\theta)} + (1-\theta)\delta^2.$$

Proof:

To simplify notation we introduce $U := \frac{1}{\sqrt{\mu}}V$. In terms of this notation one has

$$\begin{aligned} 4\left(\delta(X, S, \mu^+)\right)^2 &= \left\| \sqrt{1-\theta}\,U^{-1} - \frac{1}{\sqrt{1-\theta}}U \right\|^2 \\ &= \left\| \frac{\theta U}{\sqrt{1-\theta}} - \sqrt{1-\theta}\left(U^{-1} - U\right) \right\|^2. \end{aligned}$$

Note that $\|U\|^2 = \mathbf{Tr}\,(U^2) = \frac{1}{\mu}\mathbf{Tr}\,(V^2) = n$. This implies that U is orthogonal to $U^{-1} - U$:

$$\mathbf{Tr}\,\left(U\left(U^{-1} - U\right)\right) = n - \|U\|^2 = 0.$$

Consequently

$$4\left(\delta(X, S, \mu^+)\right)^2 = \frac{\theta^2\|U\|^2}{1-\theta} + (1-\theta)\left\|U^{-1} - U\right\|^2.$$

The required result now follows from the observation $\left\|U^{-1} - U\right\| = 2\delta$ together with $\|U\|^2 = n$. □

An immediate corollary of the lemma is the following: If one has a primal-dual pair $(X, S) \in \text{ri}\,(\mathcal{P} \times \mathcal{D})$ and parameter μ such that $\delta(X, S, \mu) \le \frac{1}{2}$, and μ is updated via

$\mu^+ := (1 - \frac{1}{2\sqrt{n}})\mu$, then one has $\delta(X, S, \mu^+) \leq \frac{1}{\sqrt{2}}$. As discussed above, the next NT step now yields a pair $(X^+, S^+)\mathrm{ri}\,(\mathcal{P} \times \mathcal{D})$ that satisfies

$$\delta\left(X^+, S^+, \mu^+\right) \leq \frac{1}{2}.$$

The algorithm therefore generates a sequence of iterates that always satisfy $\delta \leq \frac{1}{2}$. Moreover, the duality gap is reduced by a factor $(1 - \frac{1}{2\sqrt{n}})$ at each iteration, since the duality gap after the full NT step equals the target duality gap.

These observations imply the following result which establishes the polynomial iteration complexity of the algorithm. The proof is similar to that of Theorem 6.2 on page 112, and is omitted here.

Theorem 7.1 *If $\tau = \frac{1}{\sqrt{2}}$ and $\theta = \frac{1}{2\sqrt{n}}$, then the primal–dual path–following algorithm with full NT steps on page 118 stops after at most*

$$\left\lceil 2\sqrt{n}\log\frac{n\mu^0}{\epsilon}\right\rceil$$

iterations. The output is a primal-dual pair $(X, S) \in \mathrm{ri}\,(\mathcal{P}\times\mathcal{D})$ satisfying $\mathbf{Tr}\,(XS) \leq \epsilon$.

7.5 A LONG STEP PATH–FOLLOWING METHOD

This algorithm performs damped NT steps with respect to a given μ until the condition $\delta(X, S, \mu) \leq \frac{1}{\sqrt{2}}$ is met. These steps are termed *inner iterations*. Only then is the parameter μ updated via $\mu^+ = (1 - \theta)\mu$ (*outer iteration*). The step lengths are determined by line searches of the primal-dual logarithmic barrier function

$$f^{\mu}_{pd}(X, S, \mu) = \frac{1}{\mu}\mathbf{Tr}\,(XS) - \log\det(XS) - n.$$

Formally, the algorithm is as follows.

Long step primal-dual path–following method

Input
  A centering parameter $\tau > 0$;
  A pair $(X^0, S^0) \in \mathrm{ri}(\mathcal{P} \times \mathcal{D})$ and a $\mu^0 > 0$ such that $\delta\left(X^0, S^0, \mu^0\right) \leq \tau$;
  An accuracy parameter $\epsilon > 0$;
  An updating parameter $\theta < 1$;
begin
  $X := X^0$; $S := S^0$;
  while $\mathbf{Tr}(XS) > \epsilon$ **do**
    if $\delta(X, S, \mu) \leq \tau$ **do** (*outer iteration*)
      $\mu := (1 - \theta)\mu$;
    else if $\delta(X, S, \mu) > \tau$ **do** (*inner iteration*)
      Compute $\Delta X, \Delta S$ from (7.4) and (7.5) ;
      Find $\alpha := \arg\min f_{pd}^{\mu}(X + \alpha\Delta X, S + \alpha\Delta S, \mu)$;
      $X := X + \alpha\Delta X$, $S := S + \alpha\Delta S$;

    end
  end
end

The complexity analysis for the long step method is very similar to that of potential reduction algorithms (see Chapter 8), and we only state the worst-case iteration complexity result here. A detailed complexity analysis of the long step method is given by Jiang [96] (see also Sturm and Zhang [169]).

Theorem 7.2 *The long step primal–dual path–following algorithm requires at most*

$$O\left(n \log\left(\frac{n\mu^0}{\epsilon}\right)\right)$$

iterations to compute a strictly feasible pair $(X^*, S^*) \in \mathrm{ri}(\mathcal{P} \times \mathcal{D})$ *that satisfies*

$$\mathbf{Tr}(X^*S^*) \leq \epsilon.$$

□

7.6 PREDICTOR–CORRECTOR METHODS

Predictor–corrector methods are the most popular primal–dual methods at the moment, due to some successful implementations (see page 131). We will describe two variants, one due to Mizuno, Todd and Ye [123] for LP, and the other due to Mehrotra [121].

Mizuno–Todd–Ye predictor–corrector methods

A so-called Mizuno–Todd–Ye [123] predictor–corrector method alternates *predictor* and *corrector* steps. The predictor step is a damped step along the primal–dual affine-scaling direction (see page 100). This is followed by a corrector step defined as the full NT step with respect to $\mu = \mathbf{Tr}\,(XS)/n$, where $(X, S) \in \mathrm{ri}\,(\mathcal{P} \times \mathcal{D})$ are the current iterates. Formally, we can state the algorithm as follows.

Mizuno–Todd–Ye predictor–corrector method

Input

A centrality parameter $\tau > 0$;
A pair $\left(X^0, S^0\right) \in \mathrm{ri}\,(\mathcal{P} \times \mathcal{D})$ such that $\delta\left(X^0, S^0, \mu^0\right) \leq \tau$, where $\mu^0 := \mathbf{Tr}\,\left(X^0 S^0\right)/n$;
An accuracy parameter $\epsilon > 0$;
A parameter $0 < \theta < 1$;

begin

$X := X^0$; $S := S^0$; $\mu = \mu^0$;
while $\mathbf{Tr}\,(XS) > \epsilon$ **do**

Corrector step
Compute $\Delta X, \Delta S$ from (7.4) and (7.5) (NT direction);
$X := X + \Delta X, \quad S := S + \Delta S$ (full NT step);

Predictor step
Compute $\Delta X, \Delta S$ from (6.10) and (6.4) (primal–dual affine–scaling direction);
$X := X + \theta\Delta X, \quad S := S + \theta\Delta S$;

$\mu := (1 - \theta)\mu$;

end

end

Note that the parameter θ is used as the step length for the predictor step, as well as for the μ-update: $\mu^+ = (1 - \theta)\mu$. The following lemma gives us a dynamic way to choose θ such that we will always have $\delta(X, S, \mu) \leq \tau$. This lemma was proved for LP by Roos *et al.* [161] (Theorem II.64); their proof can be extended to SDP in a straightforward manner.

Lemma 7.6 *If $\tau = 1/3$, then the property $\delta(X, S, \mu) \leq \tau$ holds for each iterate $(X, S) \in \mathcal{P} \times \mathcal{D}$ produced by the Mizuno–Todd–Ye predictor–corrector algorithm,* *provided we use*

$$\theta = \frac{2}{1 + \sqrt{1 + 13 \left\| \frac{1}{2} \left(D_X^a D_S^a + D_X^a D_S^a \right) / \mu \right\|}}, \tag{7.9}$$

where D_X^a and D_S^a refer to the scaled primal–dual affine scaling direction ($D_X^a + D_S^a = -V$).

In particular, the lemma states implicitly that each predictor step will be feasible if the step length θ is chosen from (7.9).

The duality gap can only decrease during the predictor step — during a corrector step the duality gap stays constant, by Corollary 7.1. By Lemma 6.6 on page 110, we know that each predictor step will reduce the duality gap by a factor $(1 - \theta)$. We have to show that $\theta = \Omega\left(\frac{1}{\sqrt{n}}\right)$ in order to retain the same $O(\sqrt{n})$ worst-case iteration complexity as for the short step method. This is straightforward by recalling

$$D_X^a + D_S^a = -V, \quad \mathbf{Tr}\left(D_X^a D_S^a\right) = 0,$$

so that

$$\|D_X^a\|^2 + \|D_S^a\|^2 = \|V\|^2 = n\mu,$$

which implies

$$\|D_X^a\| \, \|D_S^a\| \leq \frac{1}{2} n\mu, \tag{7.10}$$

by using $2ab \leq a^2 + b^2$ ($a, b \in \mathbf{R}$). We now use (7.10) to give an upper bound on $\left\| \frac{1}{2} \left(D_X^a D_S^a + D_X^a D_S^a \right) / \mu \right\|$ as follows:

$$\begin{aligned} \left\| \frac{1}{2} \left(D_X^a D_S^a + D_X^a D_S^a \right) / \mu \right\| &\leq \frac{1}{2} \|D_X^a D_S^a\| / \mu + \frac{1}{2} \|D_S^a D_X^a\| / \mu \\ &\leq \|D_X^a\| \, \|D_S^a\| / \mu \leq \frac{1}{2} n, \end{aligned}$$

where the first two inequalities follow from the triangle inequality and sub-multiplicativity of the Frobenius norm respectively, and the last inequality is due to (7.10). If we look at the expression for θ in (7.9), we therefore see that

$$\theta \geq \frac{2}{1 + \sqrt{1 + \frac{13}{2} n}}.$$

This shows that the Mizuno–Todd–Ye predictor–corrector algorithm has a similar complexity as the short step method (see Theorem 7.1). Formally we have the following result.

Theorem 7.3 *If $\tau = 1/3$ and θ is given by (7.9), the Mizuno–Todd–Ye predictor–corrector algorithm stops after at most*

$$\left\lceil \left(1 + \sqrt{1 + \frac{13}{2} n} \right) \log \frac{\mathbf{Tr}\left(X^0 S^0\right)}{\epsilon} \right\rceil$$

iterations. The output is a primal-dual pair $(X,S) \in \mathrm{ri}\,(\mathcal{P}\times\mathcal{D})$ *satisfying* $\mathbf{Tr}\,(XS) \leq \epsilon$.

Superlinear convergence

Recall that a predictor step reduces the duality gap by a factor $(1-\theta)$ where θ is defined by (7.9). Also recall that the duality gap is given by $n\mu$ after the corrector step. The duality gap will therefore converge to zero at a quadratic rate if and only if $(1-\theta)\mu = O\left(\mu^2\right)$ *i.e.* $(1-\theta) = O(\mu)$.

It is easy to show that — for θ as defined in (7.9) — we have

$$1-\theta \leq \frac{13}{4}\left\|\frac{1}{2}\left(D_X^a D_S^a + D_X^a D_S^a\right)/\mu\right\|$$

(see Lemma II.65 by Roos *et al.* [161]). We will therefore have quadratic convergence asymptotically if

$$\left\|\frac{1}{2}\left(D_X^a D_S^a + D_X^a D_S^a\right)/\mu\right\| = O(\mu). \tag{7.11}$$

This bound holds in the special case of LP, and one can therefore show asymptotic quadratic convergence of the Mizuno–Todd–Ye predictor–corrector algorithm for LP (see also Roos *et al.* [161], §7.7). In the SDP case one can only prove a weaker result, under the assumption of strict complementarity.

Theorem 7.4 (Luo *et al.* [117]) *Let* $(X,S) \in \mathrm{ri}\,(\mathcal{P}\times\mathcal{D})$ *and let* D_X^a *and* D_S^a *denote the scaled primal–dual affine–scaling directions at* (X,S), i.e. $D_X^a + D_S^a = -V$. *If strict complementarity holds for* (P) *and* (D), *one has*

$$\left\|\frac{1}{2}\left(D_X^a D_S^a + D_X^a D_S^a\right)/\mu\right\| = O(\delta(X,S,\mu)+\mu). \tag{7.12}$$

The fact that one can only prove a weaker result in the SDP case is probably not due to shortcomings in the analysis, as can be seen from the following example.

Example 7.1 (Adapted from Kojima *et al.* [105]) *Consider the problems* (P) *and* (D) *in standard form with data*

$$A_1 = \begin{pmatrix} -2 & 0 \\ 0 & 0 \end{pmatrix}, \quad A_2 = \begin{pmatrix} 0 & 1 \\ 1 & -2 \end{pmatrix}, \quad C = \begin{pmatrix} 0 & 0 \\ 0 & 1 \end{pmatrix}, \quad b = \begin{pmatrix} -2 \\ 0 \end{pmatrix}.$$

The problem pair (P),(D) has a strictly complementary solution pair

$$X^* = \begin{pmatrix} 1 & 0 \\ 0 & 0 \end{pmatrix}, \quad S^* = \begin{pmatrix} 0 & 0 \\ 0 & 1 \end{pmatrix}, \quad y^* = \begin{pmatrix} 0 \\ 0 \end{pmatrix}.$$

Consider the sequence of feasible points $(X_k, S_k, y_k) \to (X^*, S^*, y^*)$ *defined by*

$$X_k := \begin{pmatrix} 1 & \epsilon_k \\ \epsilon_k & \epsilon_k \end{pmatrix}, S_k := \begin{pmatrix} (1+c)\epsilon_k & -\sqrt{c\epsilon_k} \\ -\sqrt{c\epsilon_k} & 1+2\sqrt{c\epsilon_k} \end{pmatrix}, y_k := \begin{pmatrix} (1+c)\epsilon_k/2 \\ \sqrt{c\epsilon_k} \end{pmatrix},$$

where $\epsilon_k \to 0$ *and* $c > 0$. *Here the values* $c := \frac{1}{32}$ *and* $\epsilon_k := 10^{-k/10}$ $(k = 30, \ldots, 80)$ *will be used.*

We investigate the centrality of the given sequence of feasible points. It seems clear from Figure 7.1 that $\delta(X, S, \mu) < 0.13$ *for the sequence of points. In other words, all the points lie in the region of quadratic convergence to the central path. The maximum feasible step length (denoted by* $\alpha_{\max}$*) to the boundary along the NT*

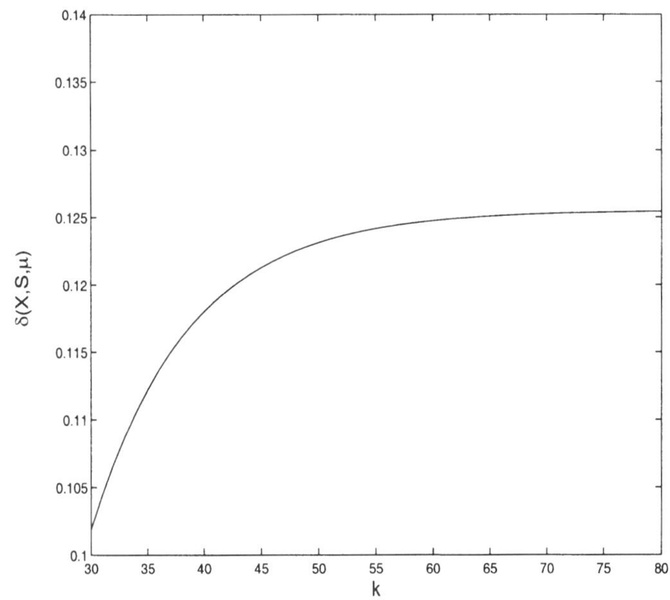

Figure 7.1. The centrality function $\delta(X_k, S_k, \mu_k)$ where $\mu_k = \frac{1}{n}\mathbf{Tr}\,(X_k S_k)$ as a function of k.

predictor direction does not approach one along the sequence of iterates, as can be seen from Figure 7.2. □

Superlinear convergence has been proved for the Mizuno-Todd-Ye predictor-corrector scheme using the NT direction under assumptions of a strictly complementary solution and increasingly centered iterates by Kojima *et al.* [105] and Luo *et al.* [117]. In particular, if we do repeated centering steps so that $\delta(X, S, \mu) = O(\mu)$ after the centering steps, then the bound (7.12) becomes the same as (7.11). The requirement for

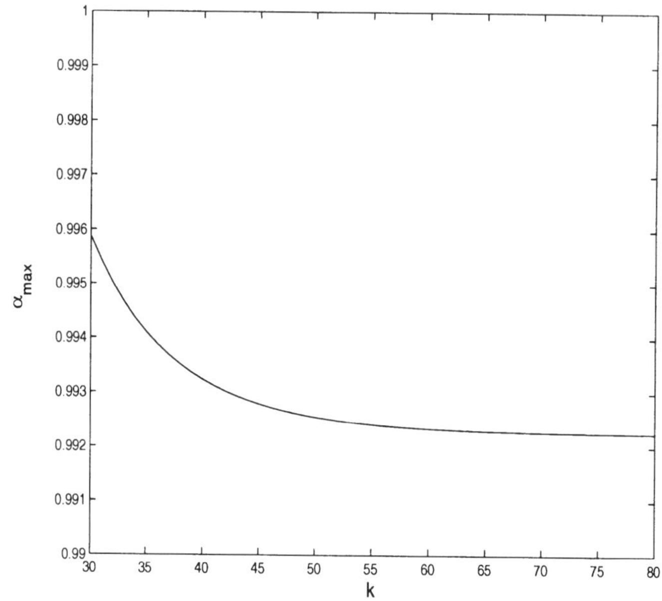

Figure 7.2. The maximal feasible step length $\alpha_{\max}$ along the NT predictor direction does not converge to one along the sequence of points (X_k, S_k).

extra centering steps is very undesirable from a practical point of view, since centering steps do not decrease the duality gap.

There is some hope that Example 7.1 does not preclude superlinear convergence of the Mizuno–Todd–Ye predictor–corrector algorithm — in the example the sequence of points is constructed in a special way, and these points do *not* correspond to actual iterates generated by the algorithm. In particular, it may be possible to prove that the centering steps become more accurate for small values of μ.

Mehrotra-type predictor–corrector methods and software

Most implementations use the so-called Mehrotra-type [121] predictor–corrector method. Here the primal–dual affine–scaling direction is also computed, but no step is made along this direction. It is used instead to find a suitable 'target value' for μ and to compute a *second order approximation* of the solution of the nonlinear system (7.2). We give a summary of the steps of the algorithm here. To this end, let $(X, S) \in \mathrm{ri}\,(\mathcal{P} \times \mathcal{D})$ denote the current iterates, as before.

- The maximal feasible step-length along the primal–dual affine–scaling direction is calculated, as well as the duality gap g^* that would result from this step.

- The value of g^* is used to choose a target value of μ via

$$\mu = \left(\frac{g^*}{\mathbf{Tr}\,(XS)} \right) \frac{g^*}{n}.$$

  Note that the value $\mu = \frac{g^*}{n}$ corresponds to the value of the duality gap that can be attained along the primal–dual affine–scaling direction. This value is now further reduced by multiplying with the factor $\frac{g^*}{\mathbf{Tr}\,(XS)}$ (the fraction by which the duality gap can be reduced).

- The search direction is now computed via a 'second order method' — one does *not* compute $\Delta X, \Delta S$ from (7.4) and (7.5), but uses the primal–dual affine–scaling directions D_X^a and D_S^a to obtain a more accurate (second order) solution of (7.2). In particular, one can solve the Lyapunov equation

$$\frac{1}{2}\left((D_X + D_S)V + V(D_X + D_S)\right) = \mu I - V^2 - \frac{1}{2}\left(D_X^a D_S^a + D_S^a D_X^a\right),$$

  instead of (7.3)[4] to obtain scaled search directions D_X and D_S.

- The maximal feasible step-lengths, say $\alpha_{\max}$ and $\beta_{\max}$ along $\Delta X = D^{\frac{1}{2}} D_X D^{\frac{1}{2}}$ and $\Delta S = D^{-\frac{1}{2}} D_S D^{-\frac{1}{2}}$ respectively, are now calculated, and the damped steps $X + (1-\epsilon)\alpha_{\max}\Delta X$ and $S + (1-\epsilon)\beta_{\max}\Delta S$ are taken for some small value $\epsilon > 0$.

See Todd *et al.* [173] for more details on the implementation.

Software Mehrotra-type predictor–corrector algorithms using the NT direction have been implemented in the software SeDuMi by Sturm [167] and SDPT3 by Toh *et al.* [175]. In the SeDuMi package only the NT direction is used, and in SDTP3 the NT direction can be specified by the user. These are two of the most successful implementations of primal–dual methods for SDP at the time of writing.

[4] Note that we are approximating the cross term $D_X D_S$ that had been neglected in (7.3), by using D_X^a and D_S^a.

8
PRIMAL–DUAL POTENTIAL REDUCTION METHODS

Preamble

The progress of a primal–dual interior point algorithm can be monitored by evaluating the so-called Tanabe–Todd–Ye potential function at each iteration. If one can guarantee that this function decreases by at least a fixed constant at each iteration, then one can immediately deduce a complexity bound for the algorithm. So-called potential reduction methods are therefore appealing from a theoretical perspective, since the analysis is conceptually simple.

Potential reduction methods were the first methods to be extended from linear programming (LP) to the more general semidefinite programming (SDP) problem; Nesterov and Nemirovski [137] and Alizadeh [3] independently analysed several potential reduction methods that have analogies in the LP literature.

A primal–dual potential reduction method suited for the structure of linear matrix inequalities arising in control theory applications was analysed by Vandenberghe and Boyd in [180]. A general potential reduction method for conic LP's involving homogeneous self–dual cones (including SDP) is presented by Nesterov and Todd [138].

Most of these methods (and some variants thereof) are described in the review paper by Vandenberghe and Boyd [181]. Two more recent surveys are by Alizadeh and S. Schmieta [7] and Tunçel [176]; these papers include potential reduction methods for more general conic LP's.

In this chapter we describe the framework of analysis for these methods, and apply the methodology to the potential reduction method of Nesterov and Todd [138].

8.1 INTRODUCTION

Primal–dual potential reduction algorithms achieve a constant decrease of the so-called Tanabe–Todd–Ye [171, 174] potential function at each iteration. This function is defined on ri $(\mathcal{P} \times \mathcal{D})$ by:

$$\Phi(X,S) := (n + \nu\sqrt{n}) \log \mathbf{Tr}\,(XS) - \log\det(XS) - n\log n,$$

where $\nu \geq 1$ is a given parameter. If one can prove that an algorithm achieves a constant reduction of Φ at each iteration, then one can immediately give a worst-case iteration complexity bound for the algorithm.

Potential reduction algorithms fit into the following framework.

Generic primal–dual potential reduction method

Input
  A strictly feasible starting pair (X^0, S^0);

Parameters
  An accuracy parameter $\epsilon > 0$;
  A parameter $\nu \geq 1$.
begin
  $X := X^0;\ S := S^0;$
  while $\mathbf{Tr}\,(XS) > \epsilon$ **do**
    Compute feasible descent directions for Φ at (X,S), say $(\Delta X, \Delta S)$;
    Find $(\alpha, \beta) =$ argmin $\Phi(X + \alpha\Delta X, S + \beta\Delta S)$ subject to $0 \leq \alpha \leq \alpha_{\max}$, $0 \leq \beta \leq \beta_{\max}$ (*plane search*), where $\alpha_{\max}$ and $\beta_{\max}$ denote the respective maximal feasible step lengths;
    $X := X + \alpha\Delta X;\ S := S + \beta\Delta S;$
  end
end

This *plane search* procedure in the algorithm is examined more closely in the next section.

8.2 DETERMINING STEP LENGTHS VIA PLANE SEARCHES OF THE POTENTIAL

Once suitable primal–dual search directions $(\Delta X, \Delta S)$ have been computed, step length parameters must be chosen to ensure feasible steps. In other words, one must find α, β such that

$$X + \alpha \Delta X \succ 0, \quad S + \beta \Delta S \succ 0.$$

This is done by performing a plane search on the potential function Φ. We briefly review this procedure here.

The intervals for feasible step lengths in both the ΔX and ΔS directions are calculated first. This is done by computing the eigenvalues of $X^{-\frac{1}{2}}\Delta X X^{-\frac{1}{2}}$ and $S^{-\frac{1}{2}}\Delta S S^{-\frac{1}{2}}$, *i.e.* the generalised eigenvalues of the pairs $(X, \Delta X)$ and $(S, \Delta S)$.

The idea is as follows. Assume that $X^{-\frac{1}{2}}\Delta X X^{-\frac{1}{2}}$ has at least one negative eigenvalue. Then

$$X + \alpha \Delta X \succeq 0 \iff I + \alpha X^{-\frac{1}{2}}\Delta X X^{-\frac{1}{2}} \succeq 0,$$

which in turn holds if and only if

$$\lambda_{\min}\left(\alpha X^{-\frac{1}{2}}\Delta X X^{-\frac{1}{2}}\right) \geq -1.$$

This is the same as

$$\alpha \leq \frac{-1}{\lambda_{\min}\left(X^{-\frac{1}{2}}\Delta X X^{-\frac{1}{2}}\right)}.$$

Thus we have

$$0 \leq \alpha \leq \alpha_{\max}, \quad 0 \leq \beta \leq \beta_{\max}, \tag{8.1}$$

where

$$\begin{aligned}
\alpha_{\max} &= \min_{i=1,\ldots,n} \left\{ \frac{-1}{\lambda_i\left(X^{-\frac{1}{2}}\Delta X X^{-\frac{1}{2}}\right)} \;\middle|\; \lambda_i\left(X^{-\frac{1}{2}}\Delta X X^{-\frac{1}{2}}\right) < 0 \right\}, \\
\beta_{\max} &= \min_{i=1,\ldots,n} \left\{ \frac{-1}{\lambda_i\left(S^{-\frac{1}{2}}\Delta S S^{-\frac{1}{2}}\right)} \;\middle|\; \lambda_i\left(S^{-\frac{1}{2}}\Delta S S^{-\frac{1}{2}}\right) < 0 \right\}.
\end{aligned}$$

Once the eigenvalues of $X^{-\frac{1}{2}}\Delta X X^{-\frac{1}{2}}$ and $S^{-\frac{1}{2}}\Delta S S^{-\frac{1}{2}}$ are known, the plane search reduces to the two dimensional minimization problem: find values (α^*, β^*) in the rectangle (8.1) that minimize

$$\begin{aligned}
f(\alpha, \beta) &:= \Phi(X + \alpha \Delta X, S + \beta \Delta S) \\
&= (n + \nu\sqrt{n}) \log(\mathbf{Tr}\,(XS) + \alpha \mathbf{Tr}\,(C\Delta X) - \beta b^T \Delta y) \\
&\quad - \sum_{i=1}^{n} \log\left[1 + \alpha \lambda_i\left(X^{-\frac{1}{2}}\Delta X X^{-\frac{1}{2}}\right)\right] \\
&\quad - \sum_{i=1}^{n} \log\left[1 + \beta \lambda_i\left(S^{-\frac{1}{2}}\Delta S S^{-\frac{1}{2}}\right)\right].
\end{aligned}$$

The function f is quasi-convex and has a unique minimizer in the interior of the rectangle of feasible step lengths defined by (8.1).[1] An important observation is that the evaluation of $f(\alpha,\beta)$, $\nabla f(\alpha,\beta)$ and $\nabla^2 f(\alpha,\beta)$ can be done in $O(n)$ operations, which means that the plane search can be done efficiently once the eigenvalues of $X^{-\frac{1}{2}}\Delta X X^{-\frac{1}{2}}$ and $S^{-\frac{1}{2}}\Delta S S^{-\frac{1}{2}}$ have been computed.

8.3 THE CENTRALITY FUNCTION Ψ

As mentioned in the introduction, the potential function Φ is composed of a 'duality gap' term and a 'centrality term'. The centrality term is given by the function:

$$\begin{aligned}\Psi(X,S) &:= n\log\frac{\frac{1}{n}\sum_{i=1}^{n}\lambda_i(XS)}{\left(\prod_{i=1}^{n}\lambda_i(XS)\right)^{1/n}}\\ &= -\log\det(XS)+n\log\mathbf{Tr}\,(XS)-n\log n.\end{aligned}$$

The function Ψ is determined by the ratio of the arithmetic and geometric means of the eigenvalues of XS. By the arithmetic–geometric mean inequality, Ψ is always non-negative and zero if and only if the pair (X,S) is centered. The following inequalities show that the centrality functions κ and Ψ are closely related:[2]

$$\log\left(\kappa(XS)\right)-2\log 2\leq\Psi(X,S)\leq(n-1)\log\left(\kappa(XS)\right). \tag{8.2}$$

As an example we show how the function Ψ may be used to explain the 'centering effect' which we observed for the primal–dual Dikin-type direction in Chapter 6 (Figure 6.1). As before we will use the NT scaling as described in Section 6.2 to obtain the scaled directions D_X and D_S, the scaled primal–dual step $D_V = D_X + D_S$, as well as V.

Example 8.1 *We show here that the primal–dual Dikin-type direction is a descent direction for Ψ at $(X,S)\in\mathrm{ri}\,(\mathcal{P}\times\mathcal{D})$, unless (X,S) are on the central path. The directional derivative of Ψ along $(\Delta X,\Delta S)$ is given by $\langle\nabla_X\Psi(X,S),\Delta X\rangle+\langle\nabla_S\Psi(X,S),\Delta S\rangle$. We will now show that the directional derivative of Ψ along the primal–dual Dikin step direction is always non-positive, and zero on the central path only. Indeed, using the expressions*[3]

$$\nabla_X\Psi(X,S)=-X^{-1}+\frac{n}{\mathbf{Tr}\,(XS)}S,\quad\nabla_S\Psi(X,S)=-S^{-1}+\frac{n}{\mathbf{Tr}\,(XS)}X,$$

---

[1] See Vandenberghe and Boyd [181], and the references therein.

[2] The inequalities in (8.2) will not be used again and only serve to show that κ is bounded in terms of Ψ. For a proof of (8.2) and an extended discussion of the function Ψ the reader is referred to Dennis and Wolkowicz [95], where other bounds are also given.

[3] The required calculus results may be found in Appendix C.

it is easy to verify that the directional derivative is given by

$$\begin{aligned} & \langle \nabla_X \Psi(X,S), \Delta X\rangle = +\langle \nabla_S \Psi(X,S), \Delta S\rangle & (8.3) \\ = & \ \mathbf{Tr}\left((D_X + D_S)\left(\frac{n}{\|V\|^2}V - V^{-1}\right)\right) \\ = & \ \mathbf{Tr}\left(D_V\left(\frac{n}{\|V\|^2}V - V^{-1}\right)\right). & (8.4) \end{aligned}$$

Substituting the primal–dual Dikin-type direction $D_V = \frac{-V^3}{\|V^2\|}$ *yields*

$$\begin{aligned} & \langle \nabla_X \Psi(X,S)\Delta X\rangle = +\langle \nabla_S \Psi(X,S)\Delta S\rangle \\ = & \ \mathbf{Tr}\left(\frac{-nV^4}{\|V\|^2\|V^2\|} + \frac{V^2}{\|V^2\|}\right) \\ = & \ \frac{\|V\|^2}{\|V^2\|}\left[1 - n\frac{\|V^2\|^2}{\|V\|^4}\right]. \end{aligned}$$

The right-hand side expression is always non-positive by the inequality

$$\|V\|^2 \le \sqrt{n}\,\|V^2\|, \tag{8.5}$$

that follows from the Cauchy–Schwartz inequality. Equality holds in (8.5) if and only if $V = \mu I$ *for some* $\mu > 0$, i.e. *if and only if* X *and* S *are on the central path.* □

8.4 COMPLEXITY ANALYSIS IN A POTENTIAL REDUCTION FRAMEWORK

The Tanabe–Todd–Ye potential function is obtained by adding an additional 'duality gap' term to the 'centrality' function Ψ as follows:

$$\begin{aligned} \Phi(X,S) &:= \nu\sqrt{n}\log \mathbf{Tr}\,(XS) + \Psi(X,S) & (8.6) \\ &= (n+\nu\sqrt{n})\log \mathbf{Tr}\,(XS) - \log\det(XS) - n\log n, \end{aligned}$$

where the (fixed) parameter $\nu \ge 1$ determines the relative 'weight' given to the duality gap term.

Using (8.6) and (8.4), it is easy to show that the directional derivative of Φ at $(X,S) \in \mathrm{ri}\ (\mathcal{P}\times\mathcal{D})$ is given by:

$$\begin{aligned} & \langle \nabla_X \Psi(X,S), \Delta X\rangle + \langle \nabla_S \Psi(X,S), \Delta S\rangle \\ = & \ \mathbf{Tr}\,(\nabla_X \Phi(X,S)\Delta X) + \mathbf{Tr}\,(\nabla_X \Phi(X,S)\Delta X) \\ = & \ (n+\nu\sqrt{n})\mathbf{Tr}\left(\frac{VD_V}{\|V\|^2}\right) + \alpha\mathbf{Tr}\left(V^{-1}D_V\right). & (8.7) \end{aligned}$$

The duality gap $\mathbf{Tr}\,(XS)$ tends to zero as Φ tends to minus infinity. In particular, we have the following bound on the duality gap $\mathbf{Tr}\,(XS)$ in terms of $\Phi(X,S)$.

Lemma 8.1 *Let $(X, S) \in \text{ri}\,(\mathcal{P} \times \mathcal{D})$. One has*

$$\mathbf{Tr}\,(XS) \leq \exp\left(\frac{\Phi(X,S)}{\nu\sqrt{n}}\right), \tag{8.8}$$

where $\Phi(X, S)$ is as defined in (8.6).

Proof:

By (8.6) we have

$$\frac{\Phi(X,s)}{\nu\sqrt{n}} = \log\left(\mathbf{Tr}\,(XS)\right) + \frac{\Psi(X,S)}{\nu\sqrt{n}} \geq \log\left(\mathbf{Tr}\,(XS)\right),$$

since $\Psi(X, S) \geq 0$. The required result follows. □

If we have an algorithm that fits in the framework on page 134, and that reduces Φ by an absolute constant at each iteration, then we can immediately give a worst-case iteration complexity bound for the algorithm. This follows from the following theorem.

Theorem 8.1 *If a given algorithm (that fits in the framework described on page 134) decreases the potential function Φ by an absolute constant c_{red} (independent of n) at each iteration, then at most*

$$\left\lceil \frac{\nu\sqrt{n}L + \Psi(X^0, S^0)}{c_{red}} \right\rceil \tag{8.9}$$

iterations are needed to satisfy the convergence condition $\mathbf{Tr}\,(XS) \leq \epsilon$, where (X^0, S^0) are the strictly feasible starting solutions and $L = \log\left(\mathbf{Tr}\,(X^0S^0)/\epsilon\right)$, as before.

Proof:

After k steps of the algorithm we will have computed a pair $(X, S) \in \text{ri}\,(\mathcal{P} \times \mathcal{D})$ such that

$$\Phi(X,S) \leq \Phi\left(X^0, S^0\right) - kc_{red}.$$

By (8.8) we therefore know that

$$\mathbf{Tr}\,(XS) \leq \exp\left(\frac{\Phi(X,S)}{\nu\sqrt{n}}\right) \leq \exp\left(\frac{\Phi(X^0,S^0) - kc_{red}}{\nu\sqrt{n}}\right).$$

We will therefore surely have $\mathbf{Tr}\,(XS) \leq \epsilon$ if

$$\exp\left(\frac{\Phi(X^0,S^0) - kc_{red}}{\nu\sqrt{n}}\right) \leq \epsilon.$$

If we rewrite this inequality we find that we will have computed a pair $(X, S) \in$ ri $(\mathcal{P} \times \mathcal{D})$ such that $\mathbf{Tr}\,(XS) \leq \epsilon$ after k steps, provided that:

$$\begin{aligned} k &\geq \frac{\nu\sqrt{n}\log\frac{1}{\epsilon} + \Phi\left(X^0, S^0\right)}{c_{red}} \\ &= \frac{\nu\sqrt{n}\log\frac{1}{\epsilon} + \nu\sqrt{n}\log\mathbf{Tr}\,\left(X^0S^0\right) + \Psi\left(X^0, S^0\right)}{c_{red}} \\ &= \frac{\nu\sqrt{n}L + \Psi\left(X^0, S^0\right)}{c_{red}}, \end{aligned}$$

where $L = \log\frac{\mathbf{Tr}\,\left(X^0S^0\right)}{\epsilon}$, and where we have used (8.6) to obtain the first equality. □

8.5 A BOUND ON THE POTENTIAL REDUCTION

Assume now that we have a given pair of feasible directions $\Delta X, \Delta S$ that are feasible descent directions for Φ at a given feasible pair $(X, S) \in$ ri $(\mathcal{P} \times \mathcal{D})$.

We first recall a useful sufficient condition for a feasible step length.

Lemma 8.2 *Assume that* $(\Delta X, \Delta S) \in \mathcal{L}^{\perp} \times \mathcal{L}$ *and* $(X, S) \in$ ri $(\mathcal{P} \times \mathcal{D})$. *We will have* $X + \alpha\Delta X \in$ ri $(\mathcal{P})$ *and* $S + \alpha\Delta S \in$ ri $(\mathcal{D})$ *if* $\alpha h < 1$, *where*

$$\begin{aligned} h^2 &:= \left\|X^{-\frac{1}{2}}\Delta X X^{-\frac{1}{2}}\right\|^2 + \left\|S^{-\frac{1}{2}}\Delta S S^{-\frac{1}{2}}\right\|^2 \\ &= \left\|V^{-\frac{1}{2}}D_X V^{-\frac{1}{2}}\right\|^2 + \left\|V^{-\frac{1}{2}}D_S V^{-\frac{1}{2}}\right\|^2. \end{aligned} \tag{8.10}$$

Proof:

We have already encountered the quantity h on page 100, where we also showed that a step length $\alpha \leq h$ is always feasible. □

The following lemma gives a bound for the change in Φ brought about by the step $(X + \alpha\Delta X, S + \alpha\Delta S)$.

Lemma 8.3 *Assume that* $(\Delta X, \Delta S) \in \mathcal{L}^{\perp} \times \mathcal{L}$ *is a strict descent direction of* Φ *at* $(X, S) \in$ ri $(\mathcal{P} \times \mathcal{D})$. *A primal–dual step* $(X + \alpha\Delta X, S + \alpha\Delta S)$ *of length* α $(\alpha h < 1)$ *is feasible and reduces the potential function* Φ *by at least*

$$\begin{aligned} \Delta\Phi &:= \Phi(X, S) - \Phi(X + \alpha\Delta X, S + \alpha\Delta S) \\ &\geq -\alpha\left(\langle\nabla_X\Psi(X, S)\Delta X\rangle + \langle\nabla_S\Psi(X, S)\Delta S\rangle\right) - \psi(-\alpha h), \end{aligned}$$

where

$$\psi(t) := t - \log(1+t), \quad t > -1 \in \mathbb{R}$$

(see Figure 3.1 on page 43).

Proof:

By the definition of the potential function one has:

$$\begin{aligned}
\Delta\Phi &\equiv \Phi(X,S) - \Phi(X+\alpha\Delta X, S+\alpha\Delta S) \\
&= (n+\nu\sqrt{n})\log\left[\frac{\mathbf{Tr}\,(XS)}{\mathbf{Tr}\,(XS)+\alpha\mathbf{Tr}\,(X\Delta S+S\Delta X)}\right] \\
&\quad + \log\left[\frac{\det(X+\alpha\Delta X)}{\det X}\right] + \log\left[\frac{\det(S+\alpha\Delta S)}{\det S}\right] \\
&= (n+\nu\sqrt{n})\log\left[\frac{\|V\|^2}{\|V\|^2+\alpha\mathbf{Tr}\,(VD_V)}\right] \\
&\quad + \log\left(\det(X^{-1})\det[X+\alpha\Delta X]\right) + \log\left(\det(S^{-1})\det[S+\alpha\Delta S]\right) \\
&= (n+\nu\sqrt{n})\log\left[\frac{1}{1+\alpha\mathbf{Tr}\,(VD_V)/\|V\|^2}\right] \\
&\quad + \log\det[I+\alpha X^{-1}\Delta X] + \log\det[I+\alpha S^{-1}\Delta S].
\end{aligned}$$

To proceed, the following inequality is needed (for a proof, see *e.g.* Roos *et al.* [161], Lemma C.1):

$$\sum_{i=1}^{k}\log(1+x_i) \geq \sum_{i=1}^{k} x_i - \psi(-\|x\|), \quad \forall x \in \mathbb{R}^k, \ \|x\| < 1, \tag{8.11}$$

where

$$\psi(t) := t - \log(1+t), \quad t > -1 \in \mathbb{R}, \tag{8.12}$$

as before (see Figure 3.1 on page 43).

We would like to apply (8.11) to the eigenvalues of $\alpha X^{-1}\Delta X$ and $\alpha S^{-1}\Delta S$. To this end, note that

$$\lambda_i\left(\alpha X^{-1}\Delta X\right) = \alpha\lambda_i\left(X^{-\frac{1}{2}}\Delta X X^{-\frac{1}{2}}\right), \ \lambda_i\left(\alpha S^{-1}\Delta S\right) = \alpha\lambda_i\left(S^{-\frac{1}{2}}\Delta S S^{-\frac{1}{2}}\right),$$

$(i = 1, \ldots, n)$, such that

$$\begin{aligned}
&\sum_{i=1}^{n}\left[\left(\lambda_i\left(\alpha X^{-1}\Delta X\right)\right)^2 + \left(\lambda_i\left(\alpha S^{-1}\Delta S\right)\right)^2\right] \\
&= \alpha^2\left\|X^{-\frac{1}{2}}\Delta X X^{-\frac{1}{2}}\right\|^2 + \alpha^2\left\|S^{-\frac{1}{2}}\Delta S S^{-\frac{1}{2}}\right\|^2 \\
&= (\alpha h)^2
\end{aligned}$$

with h as defined in (8.10). We can therefore apply (8.11) to the eigenvalues of $\alpha X^{-1}\Delta X$ and $\alpha S^{-1}\Delta S$ if $\alpha h < 1$. Thus we obtain the inequality

$$\log\det[I+\alpha X^{-1}\Delta X] + \log\det[I+\alpha S^{-1}\Delta S] \geq \alpha\mathbf{Tr}\,(X^{-1}\Delta X + S^{-1}\Delta S) - \psi(-\alpha h),$$

which holds if $\alpha h < 1$. Recall from the previous lemma that $\alpha h < 1$ is a sufficient condition for a feasible step length.

We have now obtained the relation:

$$\begin{aligned}\Delta\Phi &\geq (n+\nu\sqrt{n})\log\left[\frac{1}{1+\alpha\mathbf{Tr}\,(VD_V)/\|V\|^2}\right]+\alpha\mathbf{Tr}\,(X^{-1}\Delta X+S^{-1}\Delta S)\\ &\quad -\psi(-\alpha h)\\ &= -(n+\nu\sqrt{n})\log\left[1+\alpha\mathbf{Tr}\,(VD_V)/\|V\|^2\right]+\alpha\mathbf{Tr}\,\left(V^{-1}D_V\right)\\ &\quad -\psi(-\alpha h).\end{aligned}$$

Using the well-known inequality $-\log(1+x) \geq -x$, as well as the expression (8.7) for the directional derivative of Φ, completes the proof. □

Corollary 8.1 *Assume that D_V corresponds to a (strict) descent direction of Φ at $(X, S) \in \text{ri}\,(\mathcal{P}\times\mathcal{D})$. A primal–dual step of length α^* along this direction reduces the potential function Φ by at least*

$$\Delta\Phi \geq \psi\left(\frac{\bar{c}}{h}\right),$$

where

$$\begin{aligned}\bar{c} &:= -\left(\langle\nabla_X\Psi(X,S)\Delta X\rangle+\langle\nabla_S\Psi(X,S)\Delta S\rangle\right)\\ &= -(n+\nu\sqrt{n})\mathbf{Tr}\left(\frac{VD_V}{\|V\|^2}\right)+\mathbf{Tr}\,\left(V^{-1}D_V\right)>0,\end{aligned}$$

and

$$\alpha^* = \frac{1}{h}-\frac{1}{\bar{c}+h} < \frac{1}{h}.$$

Proof:

By Lemma 8.3 we know that

$$\Delta\Phi \geq \alpha\bar{c}-\psi(-\alpha h) = \alpha\left(\bar{c}+h\right)+\log(1-\alpha h), \tag{8.13}$$

where

$$\bar{c} := -(n+\nu\sqrt{n})\mathbf{Tr}\left(\frac{VD_V}{\|V\|^2}\right)+\mathbf{Tr}\,\left(V^{-1}D_V\right).$$

Note that $\bar{c} > 0$, because we assume that D_V corresponds to a strict descent direction of Φ at (X, S).

The function of α in (8.13) is strictly concave for $0 \leq \alpha < 1/h$, and has maximizer

$$\alpha^* = \frac{1}{h}-\frac{1}{\bar{c}+h}.$$

Thus

$$\Delta\Phi \geq \alpha^* \bar{c} - \psi(-\alpha^* h) = \psi\left(\frac{\bar{c}}{h}\right),$$

which is the required result. □

To find a lower bound on $\Delta\Phi$, we can therefore find a suitable lower bound on $\bar{c}/h$. The quantities $\bar{c}$ and h are completely determined by the current iterate V and the direction D_V that we choose. In the next section we will analyse a potential reduction method that uses the NT direction (see page 116), namely the potential reduction method of Nesterov and Todd [138].

8.6 THE POTENTIAL REDUCTION METHOD OF NESTEROV AND TODD

This method uses the NT direction in conjunction with a dynamic updating strategy for μ.

A formal statement of the algorithm is as follows.

Nesterov–Todd potential reduction method

Input
    A strictly feasible starting pair (X^0, S^0);

Parameters
    An accuracy parameter $\epsilon > 0$;
    A potential parameter $\nu \geq 1$.
begin
    $X := X^0$; $S := S^0$;
    while $\mathbf{Tr}\,(XS) > \epsilon$ **do**
        $\mu = \mathbf{Tr}\,(XS)/(n + \nu\sqrt{n})$;
        Compute ΔX, ΔS from (7.4) and (7.5);
        Find $(\alpha, \beta) = \text{argmin } \Phi(X + \alpha\Delta X, S + \beta\Delta S)$;
        $X := X + \alpha\Delta X$; $S := S + \beta\Delta S$;
    end
end

Recall that the NT direction is defined (for a given $\mu > 0$) via

$$D_V = \mu V^{-1} - V,$$

and that we have the associated centrality function

$$\delta := \delta(X,S,\mu) = \frac{1}{2}\frac{1}{\sqrt{\mu}}\|D_V\| = \frac{1}{2}\left\|\sqrt{\mu}V^{-1} - \frac{1}{\sqrt{\mu}}V\right\|. \tag{8.14}$$

We can give an upper bound for h in terms of δ by using the following lemma.

Lemma 8.4 (Jiang [96]) *Let $\mu > 0$ be given and define δ as in (8.14). One has*

$$\max\left\{\rho\left(\sqrt{\mu}V^{-1}\right), \rho\left(\frac{1}{\sqrt{\mu}}V\right)\right\} \le \delta + \sqrt{1+\delta^2},$$

where ρ denotes the spectral radius. □

Proof:

We denote the eigenvalues of $\frac{1}{\sqrt{\mu}}V$ by

$$u_i = \lambda_i\left(\frac{1}{\sqrt{\mu}}V\right) \quad (i = 1,\ldots,n).$$

The eigenvalues of $\sqrt{\mu}V^{-1}$ are therefore given by $1/u_i$ $(i = 1,\ldots,n)$. Using this notation we have

$$4\delta^2 = \sum_{i=1}^{n}\left(\frac{1}{u_i} - u_i\right)^2. \tag{8.15}$$

Now the proof proceeds exactly as the proof given in Lemma II.60 in Roos *et al.* [161] for the special case of LP. We will repeat it here to make our presentation self–contained.

Since u_i is positive for each $i = 1,\ldots,n$, we know by (8.15) that

$$-2\delta u_i \le 1 - u_i^2 \le 2\delta u_i.$$

This implies

$$u_i^2 - 2\delta u_i - 1 \le 0 \le u_i^2 + 2\delta u_i - 1.$$

Rewriting this as

$$(u_i - \delta)^2 - 1 - \delta^2 \le 0 \le (u_i + \delta)^2 - 1 - \delta^2$$

we obtain

$$(u_i - \delta)^2 \le 1 + \delta^2 \le (u_i + \delta)^2,$$

which implies

$$u_i - \delta \le |u_i - \delta| \le \sqrt{1+\delta^2} \le u_i + \delta.$$

Thus we arrive at

$$-\delta + \sqrt{1+\delta^2} \le u_i \le \delta + \sqrt{1+\delta^2}. \tag{8.16}$$

To complete the proof, we note that

$$-\delta + \sqrt{1+\delta^2} = \frac{1}{\delta + \sqrt{1+\delta^2}}$$

so that (8.16) implies $u_i \le \delta + \sqrt{1+\delta^2}$ and $1/u_i \le \delta + \sqrt{1+\delta^2}$ $(i = 1, \ldots, n)$. □

We are now ready to give an upper bound on h in terms of δ.

Lemma 8.5 *Let* $h^2 := \|V^{-\frac{1}{2}} D_X V^{-\frac{1}{2}}\|^2 + \|V^{-\frac{1}{2}} D_S V^{-\frac{1}{2}}\|^2$. *One has*

$$h \le 2\delta\left(\delta + \sqrt{1+\delta^2}\right), \tag{8.17}$$

where δ *is defined in (8.14), and* D_X, D_S *refer to the NT direction.*

Proof:

In the proof we will repeatedly use the inequality (see page 233):

$$\mathbf{Tr}\,(AB) \le \lambda_{\max}(A)\mathbf{Tr}\,(B), \quad \text{for } A, B \succeq 0.$$

By definition of h:

$$\begin{aligned}
h^2 &= \left\|V^{-\frac{1}{2}} D_X V^{-\frac{1}{2}}\right\|^2 + \left\|V^{-\frac{1}{2}} D_S V^{-\frac{1}{2}}\right\|^2 \\
&= \mathbf{Tr}\,\left(V^{-1} D_X V^{-1} D_X + V^{-1} D_S V^{-1} D_S\right) \\
&\le \lambda_{\max}(V^{-2})\mathbf{Tr}\,(D_X^2 + D_S^2) \\
&= \frac{1}{\lambda_{\min}(V^2)} \|D_V\|^2 \\
&= \rho\left(V^2\right) \|D_V\|^2 \\
&= \rho\left(\frac{1}{\mu} V^2\right) \frac{\|D_V\|^2}{\mu}.
\end{aligned}$$

Using the last lemma we now have

$$h^2 \le \left(\delta + \sqrt{1+\delta^2}\right)^2 \left(4\delta^2\right),$$

where we have also used $4\delta^2 = \frac{\|D_V\|^2}{\mu}$. □

We can now use the bound (8.17) together with Corollary 8.1 to obtain the following result.

Lemma 8.6 *Let* $(X, S) \in \mathrm{ri}\,(\mathcal{P} \times \mathcal{D})$ *be given and let* $\mu = \frac{\mathbf{Tr}\,(XS)}{\nu\sqrt{n}+n}$. *One can always find a feasible step length such that*

$$\Delta\Phi \ge \psi\left(\frac{2\delta}{\delta + \sqrt{1+\delta^2}}\right)$$

along the NT direction.

Proof:

From Corollary 8.1 we know that

$$\Delta\Phi \geq \psi\left(\frac{\bar{c}}{h}\right),$$

where

$$\bar{c} := -(n+\nu\sqrt{n})\mathbf{Tr}\left(\frac{VD_V}{\|V\|^2}\right) + \mathbf{Tr}\left(V^{-1}D_V\right).$$

Substituting $\mu = \frac{\|V\|^2}{n+\nu\sqrt{n}}$ into the expression for $\bar{c}$ we obtain

$$\begin{aligned} \bar{c} &:= -\frac{1}{\mu}\mathbf{Tr}\left(VD_V\right) + \mathbf{Tr}\left(V^{-1}D_V\right) \\ &= \frac{1}{\mu}\mathbf{Tr}\left(D_V\left(-V+\mu V^{-1}\right)\right) \\ &= \frac{1}{\mu}\|D_V\|^2 = 4\delta^2, \end{aligned}$$

where we have used $D_V = -V + \mu V^{-1}$. Thus we have

$$\frac{\bar{c}}{h} = \frac{4\delta^2}{h} \geq \frac{2\delta}{\delta + \sqrt{1+\delta^2}},$$

where the inequality follows from (8.17). The required result is now obtained by substituting the lower bound on $\bar{c}/h$ in Corollary 8.1. □

Loosely speaking, we can always reduce Φ by an absolute constant if δ is large enough. Moreover, we know from Lemma 7.5 that after an update $\mu^+ = (1-\theta)\mathbf{Tr}\,(XS)$, where $\theta \in (0,1)$, the value of $\delta\,(X,S,\mu^+)$ will be bounded from below by

$$\left(\delta(X,S,\mu^+)\right)^2 \geq \frac{n\theta^2}{4(1-\theta)}.$$

It is easy to check that the μ-updating strategy

$$\mu^+ = \frac{\mathbf{Tr}\,(XS)}{n+\nu\sqrt{n}}$$

of the Nesterov–Todd potential reduction method corresponds to

$$\theta = \frac{\nu}{\sqrt{n}+\nu},$$

and therefore by straightforward calculation one has

$$\left(\delta(X,S,\mu^+)\right)^2 \geq \frac{\nu^2 n + \nu^3\sqrt{n}}{4\left(n+\nu^2+2\nu\sqrt{n}\right)}.$$

Using $n \geq 2$ and $\nu \geq 1$ we therefore have $\delta(X, S, \mu^+) \geq 0.38$. Substituting this value into the bound from Lemma 8.6 yields $\Delta\Phi > 0.1$. This bound for $\Delta\Phi$ is rather pessimistic — it is easy to check that for $\nu \geq 10$, one has $\Delta\Phi \geq 0.27$, for example.

The Nesterov–Todd potential reduction algorithm therefore has the following worst-case complexity, by Theorem 8.1.

Theorem 8.2 *The Nesterov–Todd potential reduction algorithm requires at most*

$$\left\lceil \frac{\sqrt{n}\nu L + \Psi(X^0, S^0)}{0.1} \right\rceil$$

iterations to compute a pair $(X^*, S^*) \in \mathrm{ri}\,(\mathcal{P} \times \mathcal{D})$ *that satisfies* $\mathbf{Tr}\,(X^*S^*) \leq \epsilon$. □

Note that the method has the same $O\left(\sqrt{n}\right)$ iteration complexity as the short step method (see Theorem 7.1 on page 124), but it allows for much larger reductions of μ per iteration. For this reason potential reduction methods are more practical than short step methods. In practice, potential reduction algorithms have been replaced in implementations by the more popular predictor–corrector methods.

Software A slightly dated code by Vandenberghe and Boyd [179], called SP, uses an implementation of the Nesterov–Todd potential reduction method.

II SELECTED APPLICATIONS

9

CONVEX QUADRATIC APPROXIMATION

Preamble

Let $n \geq 1$ and let $f : \mathbb{R}^n \to \mathbb{R}$ be a convex function. Given distinct points $z_1, z_2, \cdots, z_N$ in $\mathbb{R}^n$ we consider the problem of finding a quadratic function $g : \mathbb{R}^n \to \mathbb{R}$ such that $\|[f(z_1) - g(z_1), \ldots, f(z_N) - g(z_N)]\|$ is minimal for some given norm $\| \cdot \|$. For the Euclidean norm this is the well-known *quadratic least squares* problem. (If the norm is not specified we will simply refer to g as the *quadratic approximation.*) In this chapter — that is based on Den Hertog *et al.* [51] — we show that the quadratic approximation is not necessarily convex for $n \geq 2$, even though it is convex if $n = 1$. The best *convex* quadratic approximation can be obtained in the multivariate case by using semidefinite programming.

9.1 PRELIMINARIES

Let $n \geq 1$ and let $f : \mathbb{R}^n \to \mathbb{R}$ be a convex function. Given distinct points $z_1, z_2, \cdots, z_N$ in $\mathbb{R}^n$ we consider the problem of finding a quadratic function $g : \mathbb{R}^n \to \mathbb{R}$ such that

$$f(z_i) = g(z_i) \quad i = 1, 2, \cdots, N. \tag{9.1}$$

The function g being quadratic, we can write it as

$$g(z) = z^T Q z + r^T z + \gamma \tag{9.2}$$

where $Q \in \mathcal{S}_n$, $r \in \mathbb{R}^n$ and $\gamma \in \mathbb{R}$. Hence, the problem of finding g such that (9.1) holds amounts to finding Q, r and γ such that

$$z_i^T Q z_i + r^T z_i + \gamma = f(z_i) \quad i = 1, 2, \cdots, N. \tag{9.3}$$

This is a linear system of N equations in the unknown entries of Q, r and γ. The number of unknowns in Q is equal to $\frac{1}{2}n(n+1)$, hence the total number of unknowns is given by

$$\frac{1}{2}n(n+1) + n + 1 = \frac{1}{2}(n+1)(n+2).$$

We call the points $z_1, z_2, \cdots, z_N$ *quadratically independent* if

$$z_i^T Q z_i + r^T z_i + \gamma = 0 \quad i = 1, 2, \cdots, N \quad \Rightarrow \quad Q = 0 \quad r = 0 \quad \gamma = 0. \tag{9.4}$$

Note that in this case $N \geq \frac{1}{2}(n+1)(n+2)$. Moreover, if $N = \frac{1}{2}(n+1)(n+2)$, then system (9.3) has a unique solution. We conclude that if the given points $z_1, z_2, \cdots, z_N$ are quadratically independent and $N = \frac{1}{2}(n+1)(n+2)$, then there exists a unique quadratic function g such that (9.1) holds. This is the interpolation case. When $N > \frac{1}{2}(n+1)(n+2)$, the linear system (9.3) is overdetermined and we can find a least norm solution:

$$\min_{Q,r,\gamma} \|x\|$$

where

$$x_i := z_i^T Q z_i + r^T z_i + \gamma - f(z_i), \quad i = 1, \ldots, N.$$

If the norm is the Euclidean norm, then the function g is the quadratic least squares approximation. If we do not specify the norm, we will simply refer to quadratic approximation.

9.2 QUADRATIC APPROXIMATION IN THE UNIVARIATE CASE

In this section we consider the univariate case ($n = 1$), *i.e.* f is a one-dimensional convex function. It is obvious that for any three quadratically independent points z_1, z_2, z_3 the function g will be convex. In other words, the quadratic interpolation function is convexity preserving. We proceed to show that also the quadratic approximation is convexity preserving. More precisely, we show that the quadratic approximation g of f with respect to a set of points

$$\mathcal{Z} := \{z_1, z_2, \cdots, z_N\}$$

is convex for any norm.

Theorem 9.1 *Let $z_1 < z_2 < \ldots < z_N$ and $y_i = f(z_i)$ $(i = 1, \ldots, N)$ be given, where f is a univariate convex function. The quadratic approximation g to this data set is a convex quadratic function.*

Proof:

Assume that the quadratic approximation g to the data set is strictly concave; see Figure 9.1.

Now we distinguish between two possibilities:

(i) the function g intersects f in two points;

(ii) the function g intersects f in at most one point;

Case (i) is illustrated in Figure 9.1. One can now construct the chord through the two points of intersection. This chord then defines an affine function which is clearly a better approximation to the data set at each data point in $\mathcal{Z}$.

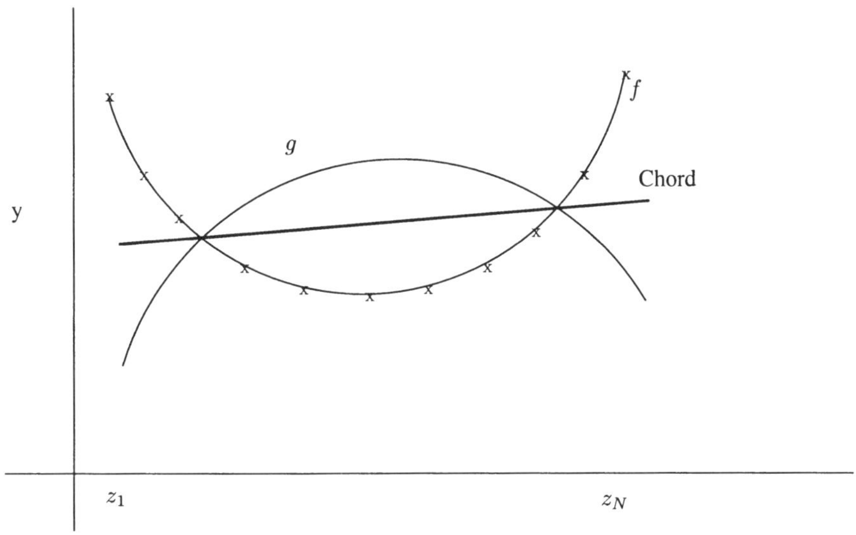

Figure 9.1. Illustration of the proof of Theorem 9.1.

In case (ii) the relative interiors of the epigraph of the function f, namely

$$\text{epi}(f) = \{(z, y) \mid y \geq f(z)\},$$

and the set

$$\{(z, y) \mid y \leq g(z)\}$$

are disjoint. These are convex sets, and therefore there exists a line separating them, by the well-known separation theorem for convex sets (see Theorem B.4 in Appendix B). This line again gives a better approximation to the data than g. □

9.3 QUADRATIC APPROXIMATION FOR THE MULTIVARIATE CASE

In this section we first show that quadratic interpolation in the multivariate case is not convexity preserving. Consequently, quadratic approximation is not convexity preserving for any norm. Subsequently we show how the best convex quadratic approximation can be obtained for the 1-norm, 2-norm, and ∞-norm by using semidefinite programming.

The following (bivariate) example shows the counterintuitive fact that quadratic interpolation is not convexity preserving in the multivariate case.

Example 9.1 *Let $f : \mathbb{R}^2 \mapsto \mathbb{R}$ be given by*

$$f(x) = -\log x_1 x_2, \qquad x_1 > 0, x_2 > 0,$$

which is clearly a convex function, $N = 6$, and the data points $z_1, \ldots, z_6 \in \mathbb{R}^2$ are the 6 columns of the matrix Z given by

$$Z = \begin{pmatrix} 1 & 2 & 3 & 2 & 4 & 6 \\ 2 & 1 & 2 & 3 & 4 & 6 \end{pmatrix}.$$

These points are quadratically independent since the coefficient matrix of the linear system (9.3) is given by

$$\begin{pmatrix} 1 & 2 & 4 & 1 & 2 & 1 \\ 4 & 2 & 1 & 2 & 1 & 1 \\ 9 & 6 & 4 & 3 & 2 & 1 \\ 4 & 6 & 9 & 2 & 3 & 1 \\ 16 & 16 & 16 & 4 & 4 & 1 \\ 36 & 36 & 36 & 6 & 6 & 1 \end{pmatrix},$$

and this matrix is nonsingular. The (unique, but rounded) solution of (9.3) is given by

$$Q = \begin{pmatrix} -0.2050 & 0.2628 \\ 0.2628 & -0.2050 \end{pmatrix}, \qquad r = \begin{pmatrix} -0.7804 \\ -0.7804 \end{pmatrix}, \qquad \gamma = 1.6219.$$

The eigenvalues of Q are approximately -0.4677 and 0.0578, showing that Q is indefinite. Hence the quadratic approximation g of f determined by the given points $z_1, z_2, \cdots, z_6$, is not convex. Figure 9.1 shows some of the level curves of f (dashed) and g (solid) as well as the points z_i $(i = 1, 2 \cdots, 6)$. The level sets of g are clearly not convex and differ substantially from the corresponding level sets of f. □

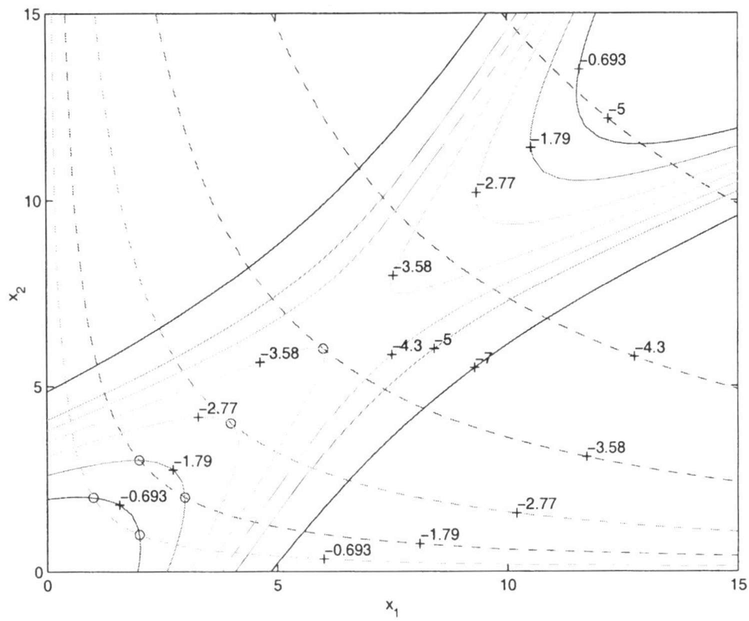

Figure 9.2. Level curves of f and g and the points where they coincide

Now we will show how to find the best *convex* quadratic approximation using SDP. Our aim is to obtain a good convex quadratic approximation g of f on the points in the finite set

$$\mathcal{Z} := \{z_1, z_2, \cdots, z_N\},$$

with respect to different norms. Convexity of g is equivalent to the matrix Q in (9.2) being positive semidefinite.

The ∞-norm

To this end, one can minimize the infinity norm of $f - g$ at $\mathcal{Z}$, yielding the objective

$$\min \max_{z \in \mathcal{Z}} |f(z) - g(z)|. \tag{9.5}$$

It will be convenient to use the notation

$$s(z) := f(z) - g(z) = f(z) - z^T Q z - r^T z - \gamma \qquad z \in \mathcal{Z}.$$

Thus we can rewrite problem (9.5) as

$$\min \{t \mid -t \leq s(z) \leq t \ (\forall z \in \mathcal{Z}), \quad Q \succeq 0\}, \tag{9.6}$$

where the variables in the optimization problem are

$$t \in \mathbf{R},\ Q \in \mathcal{S}_n^+,\ r \in \mathbf{R}^n,\ \gamma \in \mathbf{R}. \tag{9.7}$$

The **1**-*norm*

Minimization of the 1-norm of $f - g$ at $\mathcal{Z}$, yields the problem

$$\min \sum_{z \in \mathcal{Z}} |f(z) - g(z)| . \tag{9.8}$$

This is equivalent to solving

$$\min \left\{ \sum_{z \in \mathcal{Z}} t_z \;\middle|\; -t_z \leq s(z) \leq t_z \; (\forall z \in \mathcal{Z}), \quad Q \succeq 0 \right\}, \tag{9.9}$$

where the variables in the optimization problem are

$$t_{z_1} \in \mathbb{R}, \ldots, t_{z_N} \in \mathbb{R}, \; Q \in \mathcal{S}_n^+, \; r \in \mathbb{R}^n, \; \gamma \in \mathbb{R}.$$

The **2**-*norm*

Finally, we can minimize the 2-norm of $f - g$ at $\mathcal{Z}$ (least squares), as follows.

$$\min \sum_{z \in \mathcal{Z}} (f(z) - g(z))^2 , \tag{9.10}$$

and this can be rewritten as

$$\min \left\{ t \;\middle|\; \sqrt{\sum_{z \in \mathcal{Z}} s(z)^2} \leq t, \quad Q \succeq 0 \right\}, \tag{9.11}$$

where the variables in the optimization problem are the same as in (9.7).

For the ∞ and 1-norms, the resulting problems (9.6) and (9.9) have linear constraints and the semidefinite constraint $Q \succeq 0$. Problem (9.11) (for the 2-norm) has an additional second order cone (Lorentz cone) constraint, and is therefore also an SDP problem (see Section 1.3 on page 3).

Example 9.2 *For the bivariate example given above we calculated the least squares (2-norm) quadratic approximation while preserving convexity. We solved problem (9.11) using the SDP solver SeDuMi [167], to obtain the following (rounded) solution:*

$$Q = 0.02750 \begin{pmatrix} 1 & 1 \\ 1 & 1 \end{pmatrix}, \quad r = -0.7287 \begin{pmatrix} 1 \\ 1 \end{pmatrix}, \quad \gamma = 1.2196.$$

The eigenvalues of Q are approximately 0.55 *and* 0, *showing that Q is positive semidefinite. Hence the best convex quadratic approximation g of f determined by the given* *points $z_1, z_2, \cdots, z_6$, is convex, but degenerate. Note that Q is not positive definite because the constraint $Q \succeq 0$ is binding at the optimal solution of problem (9.11). (If we remove the constraint $Q \succeq 0$, then we get the non-convex interpolation function of the previous example.)*

Figure 9.3 shows some of the level curves of f (dashed) and g (solid) as well as the points z_i ($i = 1, 2 \cdots, 6$). Comparing with Figure 9.2 we see that the convex

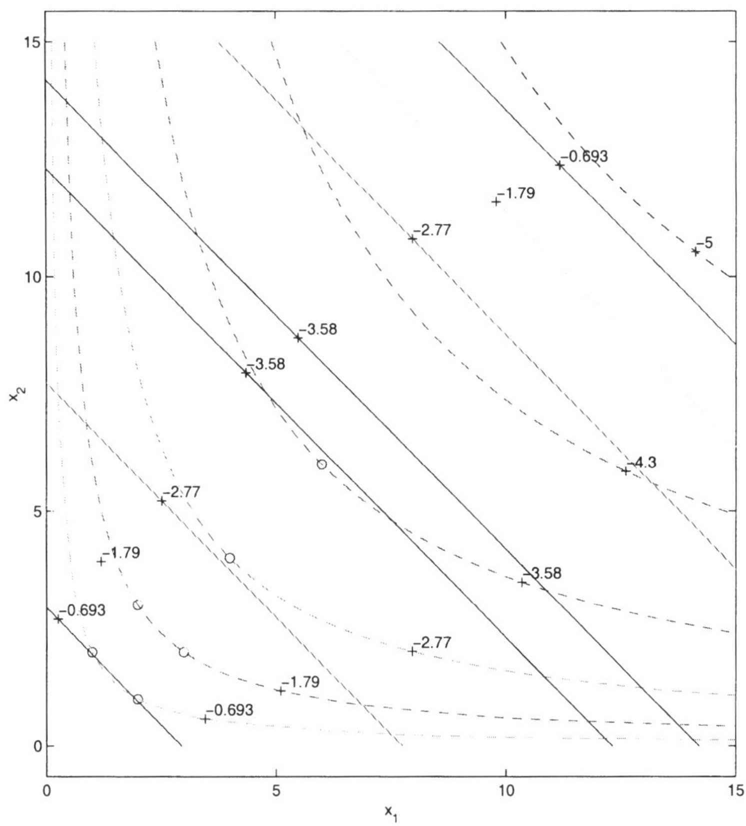

Figure 9.3. Level curves of f and g

approximation approximates f much better within the convex hull of the six specified points, if the measure of quality is the maximum error or integral of the error function

$$err(z) = |f(z) - g(z)|$$

over the convex hull. (The convex hull defines a natural 'trust region' for the approximation.) □

10

THE LOVÁSZ ϑ-FUNCTION

Preamble

The Lovász ϑ-function maps a graph $G = (V, E)$ to $\mathbb{R}_+$, and $\vartheta(G)$ is given as the optimal value of a certain SDP problem. In particular, $\vartheta(\bar{G})$ gives an upper bound on the clique number of G, where $\bar{G}$ denotes the complement of G. The upper bound is no worse than the chromatic number of G. This result is known as the 'sandwich theorem' (a name coined by Knuth [103]); it gives a polynomial-time approximation to both the clique and the chromatic numbers (these numbers cannot be computed in polynomial time, unless $P = NP$). In this chapter we will give a proof of the sandwich theorem, and derive some alternative formulations for the ϑ-function as SDP's.

10.1 INTRODUCTION

The ϑ-function was introduced by Lovász [115] to give a polynomial time lower bound on the so-called *Shannon capacity* of a graph (see Section 10.4). It is interesting to note that the definition of $\vartheta(\bar{G})$ as an SDP problem dates back to 1979. As such it is a nice example of a relatively old problem that has benefited from the emergence of efficient solution algorithms in the 1990's. (In 1979 it was known that SDP problems could be solved in polynomial time via the ellipsoid method, but no practical algorithms were available.) The ϑ-function has proved to be of great importance in combinatorial optimization. The sandwich theorem states that

$$\omega(G) \le \vartheta(\bar{G}) \le \chi(G),$$

which means that $\vartheta(\bar{G})$ can be seen as a polynomial time approximation to both $\omega(G)$ and $\chi(G)$. In a sense one cannot find better approximations to the clique or chromatic number polynomial time: neither $\omega(G)$ nor $\chi(G)$ can be approximated within a factor $|V|^{1-\epsilon}$ for any $\epsilon > 0$ in polynomial time, unless $NP = ZPP$[1] (see Håstad [81] and Feige and Kilian [57]).

The ϑ-function also forms the basis for a number of approximation algorithms, that will be described in the next chapter. A detailed exposition of the properties of the ϑ-function is given by Knuth [103].

10.2 THE SANDWICH THEOREM

The sandwich theorem relates three properties of a graph $G(V, E)$: the chromatic number $\chi(G)$, the clique number $\omega(G)$, and the Lovász number $\vartheta\left(\bar{G}\right)$, of the complementary graph $\bar{G}$ which can be defined as the optimal value of the following SDP problem:

$$\vartheta\left(\bar{G}\right) := \max_X \mathbf{Tr}\left(ee^T X\right) = e^T X e \tag{10.1}$$

subject to

$$\left.\begin{array}{rcl} x_{ij} & = & 0,\ (i,j) \notin E\ (i \neq j) \\ \mathbf{Tr}\,(X) & = & 1 \\ X & \succeq & 0. \end{array}\right\} \tag{10.2}$$

The 'sandwich theorem' states the following.[2]

Theorem 10.1 (Lovász's Sandwich Theorem [115]) *For any graph $G = (V, E)$ one has*

$$\omega(G) \le \vartheta\left(\bar{G}\right) \le \chi(G).$$

Proof:

In order to prove the first inequality of the theorem, we show that problem (10.1)–(10.2) is a relaxation of the maximum clique problem.

Let x_C denote the incidence vector of a clique C of size k in G, i.e:

$$(x_C)_i = \begin{cases} 1 & \text{if } i \in C \\ 0 & \text{otherwise.} \end{cases}$$

---

[1]The complexity class $ZPP \subset NP$ is a generalization of the complexity class P, and is defined as the class of languages that have Las Vegas algorithms (randomized algorithms with zero sided error) running in expected polynomial time; for more information, see Motwani and Raghavan [128], page 22.

[2]The proof given here is due to De Klerk *et al.* [45]; another proof is given by Knuth [103].

It is easy to check that the rank 1 matrix

$$X := \frac{1}{k} x_C x_C^T$$

is feasible in (10.2) with objective value

$$e^T X e = \frac{1}{k} \left(e^T x_C\right)^2 = \frac{k^2}{k} = k.$$

We therefore have $\omega(G) \leq \vartheta\left(\bar{G}\right)$, which is the first part of the sandwich theorem.

The second part is to prove $\vartheta\left(\bar{G}\right) \leq \chi(G)$. To this end, we write down the Lagrangian dual of the SDP relaxation (10.2) to obtain

$$\vartheta\left(\bar{G}\right) = \min_{\lambda, Y} \lambda \tag{10.3}$$

subject to

$$\left.\begin{array}{rcl} Y + ee^T & \preceq & \lambda I \\ y_{ij} & = & 0,\ (i,j) \in E\ (i \neq j) \\ y_{ii} & = & 0,\ i \in V. \end{array}\right\} \tag{10.4}$$

Note that both the primal problem (10.1) and dual problem (10.3) satisfy the Slater constraint qualification (Assumption 2.2 on page 23), and therefore have the same optimal value, namely $\vartheta\left(\bar{G}\right)$.

Given a colouring of G with k colours, we must construct a feasible solution for (10.4) with $\lambda \leq k$. Such a colouring defines a partition $V = \cup_{i=1}^k C_i$ where the C_i's are subsets of nodes sharing the same colour. In other words, the C_i's must be disjoint stable sets (co-cliques). Now let $\gamma_i = |C_i|$ and define

$$M_i := k\left(I_{\gamma_i} - E_{\gamma_i}\right), \quad i = 1, \ldots, k,$$

where I_{γ_i} is the $(\gamma_i \times \gamma_i)$ identity matrix, and E_{γ_i} the all-1 matrix of the same size.

We will show that the block diagonal matrix

$$Y = \begin{pmatrix} M_1 & 0 & \ldots & 0 \\ 0 & M_2 & \ldots & 0 \\ \vdots & \vdots & \ddots & 0 \\ 0 & 0 & \ldots & M_k \end{pmatrix} \tag{10.5}$$

is feasible in (10.4) if $\lambda = k$. By construction, Y satisfies the last two constraints in (10.4). We must still show that $Y + ee^T \preceq kI$, *i.e.* the largest eigenvalue of $Y + ee^T$ must be at most k.

The Raleigh–Ritz theorem (see Appendix A) states that for any symmetric matrix A, one has:

$$\lambda_{\max}(A) = \max\left\{x^T Ax \mid \|x\| = 1\right\}. \tag{10.6}$$

It follows that the maximal eigenvalue of Y is given by

$$\lambda_{\max}(Y) = \max_{\alpha}\left\{\sum_{i=1}^{k}\alpha_i\lambda_{\max}(M_i),\ \sum_{i=1}^{k}\alpha_i = 1,\ \alpha_i \geq 0\quad \forall i\right\}. \tag{10.7}$$

Moreover one has $\lambda_{\max}(M_i) = k$, so that (10.7) yields $\lambda_{\max}(Y) = k$. The eigenvector corresponding to k is orthogonal to the all-1 vector e. To see this, note that $Yx = \lambda x$ implies

$$-k(\gamma_i - 1)\sum_{j\in C_i} x_j = \lambda \sum_{j\in C_i} x_j, \quad i = 1,\ldots,k,$$

so that $\sum_{j\subset C_i} x_j = 0$ $(i = 1,\ldots,k)$ if $\lambda > 0$. In particular, $e^T x = 0$ from which it follows that k is also an eigenvalue of $Y + ee^T$. Assuming that k is not the largest eigenvalue of $Y + ee^T$, then the largest eigenvalue must have an eigenspace orthogonal to the eigenspace of k. The orthogonal complement of the eigenspace of k is spanned by the vectors

$$(x_{C_i})_j := \begin{cases} 1 & \text{if } j \in C_i \\ 0 & \text{otherwise,} \end{cases}$$

where $i = 1,\ldots,k$. The maximal eigenvalue of $Y + ee^T$ can therefore be computed from (10.6):

$$\begin{aligned}\lambda_{\max}\left(Y + ee^T\right) &= \max_{\|x\|=1}\left\{x^T\left(Y + ee^T\right)x\ :\ x \in \text{span}\{x_{C_1},\ldots,x_{C_k}\}\right\}\\ &= \max_{\alpha}\left\{x^T Yx + \left(e^T x\right)^2\ :\ x = \sum_{i=1}^{k}\alpha_i x_{C_i},\ \sum_{i=1}^{k}\gamma_i\alpha_i^2 = 1\right\}.\end{aligned}$$

Substituting the expression for x, and using the construction of Y simplifies this to

$$\begin{aligned}&\lambda_{\max}\left(Y + ee^T\right)\\ &= \max_{\alpha}\left\{-k\sum_{i=1}^{k}\alpha_i^2\left(\gamma_i^2 - \gamma_i\right) + \left(\sum_{i=1}^{k}\gamma_i\alpha_i\right)^2 \,\middle|\, \sum_{i=1}^{k}\gamma_i\alpha_i^2 = 1\right\}\\ &= k + \max_{\alpha}\left\{-k\sum_{i=1}^{k}(\alpha_i\gamma_i)^2 + \left(\sum_{i=1}^{k}\gamma_i\alpha_i\right)^2 \,\middle|\, \sum_{i=1}^{k}\gamma_i\alpha_i^2 = 1\right\}.\end{aligned}$$

The function to be maximized is always non-positive, since it is of the form

$$-kz^T z + (e^T z)^2 \leq -kz^T z + (\|e\|\|z\|)^2 = -kz^T z + k\|z\|^2 = 0,$$

where $z_i = \alpha_i\gamma_i$, $(i = 1,\ldots,k)$. This leads to the contradiction $\lambda_{\max}\left(Y + ee^T\right) \leq k$.

We conclude that $\lambda_{\max}\left(Y + ee^T\right) = k$. □

Example 10.1 ($\vartheta(\bar{G})$ of the pentagon) *Let $G = (V, E)$ be the graph of a pentagon. The adjacency matrix*[3] *of G is given by*

$$A = \begin{bmatrix} 0 & 1 & 0 & 0 & 1 \\ 1 & 0 & 1 & 0 & 0 \\ 0 & 1 & 0 & 1 & 0 \\ 0 & 0 & 1 & 0 & 1 \\ 1 & 0 & 0 & 1 & 0 \end{bmatrix}. \tag{10.8}$$

We have $\omega(G) = 2$, $\chi(G) = 3$. We will show that $\vartheta(\bar{G}) = \sqrt{5}$ in this case. To this end, note that the matrix

$$\begin{aligned} X &= \frac{1}{10}\left(2I + (\sqrt{5}-1)A\right) \\ &= \frac{1}{10}\begin{bmatrix} 2 & \sqrt{5}-1 & 0 & 0 & \sqrt{5}-1 \\ \sqrt{5}-1 & 2 & \sqrt{5}-1 & 0 & 0 \\ 0 & \sqrt{5}-1 & 2 & \sqrt{5}-1 & 0 \\ 0 & 0 & \sqrt{5}-1 & 2 & \sqrt{5}-1 \\ \sqrt{5}-1 & 0 & 0 & \sqrt{5}-1 & 2 \end{bmatrix}, \end{aligned}$$

is positive semidefinite and feasible in (10.2), with corresponding objective value $\mathbf{Tr}\left(ee^TX\right) = \sqrt{5}$. *We want to prove that X is in fact an optimal solution. This can be done by finding a dual feasible solution* — i.e. *a solution satisfying (10.4)* — *with the same objective value. Such a solution is given by*

$$\lambda = \sqrt{5} \text{ and } Y = \frac{\sqrt{5}-5}{2}\left(ee^T - A - I\right).$$

Note that Y is feasible in (10.4) by construction. It is also easy to verify that $Y + ee^T \preceq \lambda I$, as required. □

[3]The adjacency matrix $A = A(G)$ of a graph $G = (V, E)$ has entries

$$a_{ij} = \begin{cases} 1 & \text{if } (i,j) \in E \\ 0 & \text{otherwise.} \end{cases}$$

The sandwich theorem can equivalently be stated as

$$\alpha(G) \le \vartheta(G) \le \chi(\bar{G}),$$

where $\alpha(G)$ is the stability number of G.

An immediate consequence is that $\vartheta(G) = \alpha(G)$ if G is a so-called *perfect graph.*

Definition 10.1 (Perfect graph) *A graph $G = (V, E)$ is called perfect if $\alpha(G') = \omega(G')$ for every induced subgraph G' of G.*

10.3 OTHER FORMULATIONS OF THE ϑ-FUNCTION

We have given a proof of the equivalence of two different definitions of $\vartheta\left(\bar{G}\right)$ via (10.1) and (10.3). These and other equivalent definitions of $\vartheta\left(\bar{G}\right)$ are discussed in Grötschel *et al.* [72].

We will derive one more formulation of the ϑ-function in this section. This formulation will be used extensively in the next chapter.

Theorem 10.2 (Karger *et al.* [98]) *The ϑ-function of the complement of a graph $G = (V, E)$ is given by*

$$\vartheta(\bar{G}) = \min_{U,t} t$$

subject to

$$\begin{aligned} u_{ij} &= \frac{-1}{t-1}, \quad (i,j) \in E \\ u_{ii} &= 1, \quad i = 1, \ldots, n \\ U &\succeq 0. \end{aligned}$$

Proof:

From (10.3) and (10.4) we can deduce that

$$\vartheta\left(\bar{G}\right) - 1 = \min t$$

subject to

$$\begin{aligned} Y + ee^T &\preceq tI + I \\ y_{ij} &= 0, \ (i,j) \in E \quad (i \neq j) \\ y_{ii} &= 0, \ i \in V. \end{aligned}$$

This implies that

$$\frac{-1}{\vartheta(\bar{G}) - 1} = -\max \frac{1}{t}$$

subject to

$$\begin{aligned} \tfrac{1}{t}\left(Y + ee^T - I\right) &\preceq I \\ y_{ij} &= 0,\ (i,j) \in E \quad (i \neq j) \\ y_{ii} &= 0,\ i \in V. \end{aligned}$$

By introducing the new variables $p = 1/t$ and $\bar{Y} = \frac{1}{t}Y$ we obtain

$$\frac{-1}{\vartheta(\bar{G}) - 1} = -\max p$$

subject to

$$\begin{aligned} \bar{Y} + p\left(ee^T - I\right) &\preceq I \\ \bar{y}_{ij} &= 0,\ (i,j) \in E \quad (i \neq j) \\ \bar{y}_{ii} &= 0,\ i \in V. \end{aligned}$$

The dual problem of this last formulation takes the form

$$\frac{-1}{\vartheta(\bar{G}) - 1} = -\min \mathbf{Tr}\, Z$$

subject to

$$\begin{aligned} \mathbf{Tr}\left(\left(ee^T - I\right) Z\right) &= 1 \\ z_{ij} &= 0,\ (i,j) \notin E \quad (i \neq j) \\ Z &\succeq 0. \end{aligned}$$

Note that the optimal value of the last formulation is also $-1/(\vartheta(\bar{G}) - 1)$, by the strong duality theorem (Theorem 2.2 on page 26), since the Slater regularity condition (Assumption 2.2 on page 23) is satisfied. We can rewrite the last formulation as

$$\frac{-1}{\vartheta(\bar{G}) - 1} = \max(-\mathbf{Tr}\, Z)$$

subject to

$$\begin{aligned} \textstyle\sum_{(i,j) \in E} z_{ij} &= 1 \\ z_{ij} &= 0,\ (i,j) \notin E \quad (i \neq j) \\ Z &\succeq 0. \end{aligned}$$

Taking the dual of this formulation yields the SDP problem:

$$\frac{-1}{\vartheta(\bar{G}) - 1} = \min p$$

subject to

$$\begin{aligned} u_{ii} &= 1, \quad i \in V \\ u_{ij} &= p,\ (i,j) \in E \quad (i \neq j) \\ U &\succeq 0, \end{aligned}$$

where we have once again used the strong duality theorem. It is easy to see that this last problem is equivalent to the one in the statement of the theorem. □

10.4 THE SHANNON CAPACITY OF A GRAPH

The ϑ-function was actually introduced by Lovász [115] in order to study the so-called Shannon capacity of a graph — a quantity that arises naturally in applications in coding theory. To define the Shannon capacity, we need to introduce the *strong product* of graphs.

Definition 10.2 (Strong product of graphs) *The strong product $G_1 * G_2$ of two graphs $G_1 = (V_1, E_1)$ and $G = (V_2, E_2)$ is defined as the graph with vertex set $V := V_1 \times V_2$ and edge set:*

$$E := \{((\bar{v}_i, v_j), (\bar{v}_k, v_l)) \mid [(\bar{v}_i, \bar{v}_k) \in E_1 \text{ or } i = k] \text{ and } [(v_j, v_l) \in E_2 \text{ or } j = l]\},$$

where at least one of the two conditions $i \neq k$ or $j \neq l$ holds.

It is easy to see that $\alpha(G_1 * G_2) \geq \alpha(G_1)\alpha(G_2)$. Indeed, if $S_1 \subset V_1$ and $S_2 \subset V_2$ are stable sets of G_1 and G_2 respectively, then $S_1 \times S_2$ is a stable set of $G_1 \times G_2$. Thus

$$\alpha(G)^r \leq \alpha\left(\underbrace{G * \ldots * G}_{r \text{ times}}\right) := \alpha\left(G^r\right).$$

Example 10.2 *We consider the problem of transmitting data via a communication channel. The data is coded as words consisting of the letters of an alphabet. During transmission, it may happen that a letter is changed to an 'adjacent' letter. We associate a set of vertices V with the letters of the alphabet, and join two vertices by an edge if the two corresponding letters are adjacent. Let us call the resulting graph $G = (V, E)$. For a given integer $r > 0$, we would now like to construct the largest possible dictionary of r-letter words with the property that one word in the dictionary cannot be changed to another word in the same dictionary during transmission. Two r-letter words*

$$(l_1, \ldots, l_r) \text{ and } (\hat{l}_1, \ldots, \hat{l}_r)$$

correspond to the endpoints of an edge in G^r if and only if for each $i = 1, \ldots, r$, either $l_i = \hat{l}_i$, or the letters l_i and $\hat{l}_i$ are adjacent. It is therefore easy to see that the maximal number of words in this dictionary is $\alpha\left(G^r\right)$. □

The Shannon capacity can now be defined using the strong product.

Definition 10.3 (Shannon capacity of a graph) *The Shannon capacity of a graph $G = (V, E)$ is defined as*

$$\Theta(G) := \lim_{r\to\infty} \alpha\left(G^r\right)^{\frac{1}{r}}.$$

Note that this limit is finite, since $\alpha\left(G^r\right) \leq |V|^r$. Also note that $\Theta(G) \geq \alpha(G)$ and

$$\Theta(G) = \lim_{r\to\infty} \alpha\left(G^{2r}\right)^{\frac{1}{2r}} = \lim_{r\to\infty} \left(\alpha\left(\left(G^2\right)^r\right)^{\frac{1}{r}}\right)^{\frac{1}{2}} = \sqrt{\Theta\left(G^2\right)}. \qquad (10.9)$$

The ϑ-function is multiplicative with respect to the strong graph product.

Theorem 10.3 (Lovász [115]) *Let two graphs $G_1 = (V_1, E_1)$ and $G_2 = (V_2, E_2)$ be given. Then*

$$\vartheta\left(G_1 * G_2\right) = \vartheta(G_1)\vartheta(G_2).$$

Proof:

The idea of the proof is simple — we will use optimal solutions of the optimization problems that yield $\vartheta(G_1)$ and $\vartheta(G_2)$ to construct an optimal solution of the problem that yields $\vartheta(G_1 * G_2)$. This construction is done by taking the Kronecker product of optimal solutions.[4]

To start the proof, let two graphs $G_1 = (V_1, E_1)$ and $G = (V_2, E_2)$ be given, and let X_1 and X_2 be optimal solutions of the problem

$$\vartheta\left(G\right) := \max_X \mathbf{Tr}\left(ee^TX\right) = e^TXe \qquad (10.10)$$

subject to

$$\left.\begin{array}{rcl} x_{ij} & = & 0, \quad (i,j) \in E \ (i \neq j) \\ \mathbf{Tr}\left(X\right) & = & 1 \\ X & \succeq & 0, \end{array}\right\} \qquad (10.11)$$

for $G = G_1$ and $G = G_2$ respectively. Let $X^* := X_1 \otimes X_2$. By the properties of the Kronecker product (see Appendix E), we have $X^* \succeq 0$ and $\mathbf{Tr}\, X^* = 1$. Moreover, we can label the vertices of $G_1 * G_2 := (V', E')$ in such a way that $\left(X^*\right)_{ij} = 0$ if $(i,j) \in E'$. Indeed, if we denote

$$V_1 = \left\{v_1, \ldots, v_{|V_1|}\right\}, \quad V_2 = \left\{\bar{v}_1, \ldots, \bar{v}_{|V_2|}\right\},$$

then a suitable labeling of $V' \equiv V_1 \times V_2$ is

$$V' = \left\{\left(v_1, \bar{v}_1\right), \ldots, \left(v_1, \bar{v}_{|V_2|}\right), \left(v_2, \bar{v}_1\right), \ldots, \left(v_{|V_1|}, \bar{v}_{|V_2|}\right)\right\}.$$

---

[4]The approach used here is based on a proof by Pasechnik [142].

In other words, X^* is a feasible solution of problem (10.10)–(10.11) for $G = G_1 * G_2$.

Finally the objective value of problem (10.10)–(10.11) for $G = G_1 * G_2$ at X^* is

$$\begin{aligned}
e_{|V'|}^T X^* e_{|V'|} &= e_{|V'|}^T (X_1 \otimes X_2) e_{|V'|} \\
&= \left(e_{|V_1|}^T \otimes e_{|V_2|}^T\right) (X_1 \otimes X_2) \left(e_{|V_1|} \otimes e_{|V_2|}\right) \\
&= \left(e_{|V_1|}^T \otimes e_{|V_2|}^T\right) \left((X_1 e_{|V_1|}) \otimes (X_2 e_{|V_2|})\right) \\
&= \left(e_{|V_1|}^T X_1 e_{|V_1|}\right) \otimes \left(e_{|V_2|}^T X_2 e_{|V_2|}\right) \\
&= \left(e_{|V_1|}^T X_1 e_{|V_1|}\right) \left(e_{|V_2|}^T X_2 e_{|V_2|}\right) \\
&= \vartheta(G_1)\vartheta(G_2),
\end{aligned}$$

where we have used the property $(AB) \otimes (CD) = (A \otimes C)(B \otimes D)$ of the Kronecker product repeatedly (see page 250).

We conclude that $\vartheta(G_1 * G_2) \geq \vartheta(G_1)\vartheta(G_2)$. To prove that equality holds, we must consider the dual formulation of the ϑ-function for a graph $G = (V, E)$. Note that the dual formulation, namely (10.3)–(10.4), can be rewritten as:

$$\vartheta(G) = \min_{\lambda, S} \lambda$$

subject to

$$\begin{aligned}
S &\succeq ee^T \\
s_{ij} &= 0, \quad (i,j) \notin E \ (i \neq j) \\
s_{ii} &= \lambda, \quad i \in V.
\end{aligned}$$

By the Schur complement theorem (Theorem A.9 on page 235), $S \succeq ee^T$ is equivalent to

$$Z := \begin{bmatrix} S & e \\ e^T & 1 \end{bmatrix} \succeq 0.$$

We can therefore give yet another formulation for $\vartheta(G)$, namely:

$$\vartheta(G) = \min_{\lambda, Z} \lambda \tag{10.12}$$

subject to

$$\left.\begin{aligned}
Z &\in \mathcal{S}_{|V|+1}^+ \\
z_{ij} &= 0, \quad (i,j) \notin E \ (i \neq j) \\
z_{ii} &= \lambda, \quad i \in V. \\
z_{|V|+1,i} &= 1, \quad i = 1, \ldots, |V| + 1.
\end{aligned}\right\} \tag{10.13}$$

Now let $(\vartheta(G_1), Z_1)$ and $(\vartheta(G_2), Z_2)$ denote the optimal solutions of (10.12)–(10.13) for G_1 and G_2 respectively. Similarly to what we did for the primal problem, we will construct a feasible solution for the problem (10.12)–(10.13) associated with $G_1 * G_2$ by using $(\vartheta(G_1), Z_1)$ and $(\vartheta(G_2), Z_2)$.

To this end, let $Z^* = Z_1 \otimes Z_2$. In the same way as before, it is straightforward to show that Z^* is feasible in (10.13) (for $G = G_1 * G_2$) and the associated objective value at Z^* is $\lambda = \vartheta(G_1)\vartheta(G_2)$. This completes the proof. □

By using the sandwich theorem (Theorem 10.1) and Theorem 10.3 successively, we have

$$\alpha\left(G^r\right)^{\frac{1}{r}} \leq \left(\vartheta\left(G^r\right)\right)^{\frac{1}{r}} = \left(\left[\vartheta\left(G\right)\right]^r\right)^{\frac{1}{r}} = \vartheta(G),$$

so that $\Theta(G) \leq \vartheta(G)$. This is an important result, since it is not known whether there exists an algorithm (polynomial or not) to compute $\Theta(G)$. To summarize, we have the following theorem.

Theorem 10.4 (Lovász [115]) *Let $G = (V, E)$ be given. Then*

$$\alpha(G) \leq \Theta(G) \leq \vartheta(G) \leq \chi(\bar{G}).$$

Example 10.3 (Shannon capacity of the pentagon) *Let $G = (V, E)$ be the graph of a pentagon. The complement G is isomorphic to G in this case, and therefore has the same ϑ-number. Its Shannon capacity $\Theta(G)$ is upper bounded by $\vartheta(G) = \vartheta(\bar{G}) = \sqrt{5}$ (see Example 10.1 on page 161).*

Now us denote the vertices of G by $V = \{1, 2, 3, 4, 5\}$, and connect the vertices such that the adjacency matrix of G is given by (10.8) on page 161. Now $G * G$ has an independent set of size 5, given by

$$\{(1,2),\ (2,4),\ (3,1),\ (4,3),\ (5,5)\}.$$

Thus $\Theta(G)$ is lower bounded by $\sqrt{5}$ because by (10.9) we have

$$\Theta(G) = \sqrt{\Theta\left(G^2\right)} \geq \sqrt{\alpha\left(G^2\right)} \geq \sqrt{5}.$$

It follows that $\Theta(G) = \sqrt{5}$. □

The last example may seem simple enough at first glance, but the Shannon capacity of the pentagon was unknown until Lovász introduced the ϑ-function. In fact, in a recent talk entitled *The combinatorial optimization top 10 list*, W.R. Pulleyblank [151] reserved a place for this result as one of the ten most important results in combinatorial optimization.

It is worth mentioning that the Shannon capacity of the graph of a heptagon (7-cycle) is still unknown.

11

GRAPH COLOURING AND THE MAX-*K*-CUT PROBLEM

Preamble

Given a simple[1] graph $G = (V, E)$, the (unweighted)[2] MAX-k-CUT problem is to assign k colours to the vertices V in such a way that the number of non-defect edges (with endpoints of different colours) is maximal. Any assignment of k colours to all the vertices V is called a k-cut, and the number of non-defect edges is called the weight of the k-cut.

Example 11.1 *Let us consider the Petersen graph which has* $|V| = 10$, $|E| = 15$, *and chromatic number* $\chi(G) = 3$. *In Figure 11.1 we show a* 2*-cut for this graph. (The vertices are partitioned into two sets indicated by triangular and square markers.) Note that there are three defect edges for this* 2*-cut (these edges are denoted by dashed lines). The* 12 *remaining edges are non-defect and we conclude that the optimal value of the MAX-2-CUT (MAX-CUT) problem is at least* 12 *for this graph. In Figure 11.2 we show a maximum* 3*-cut for the same graph. Note that there are no defect edges*

[1] A graph is called simple if there is at most one edge between a given pair of vertices and if there is no loop (an edge which connects a vertex to itself).

[2] One can generalize the problem by adding nonnegative weights on the edges. The (weighted) MAX-k-problem is then to assign k colours to the vertices in such a way that the total weight of the non-defect edges is as large as possible. The results in this chapter can easily be extended to include nonnegative edge weights.

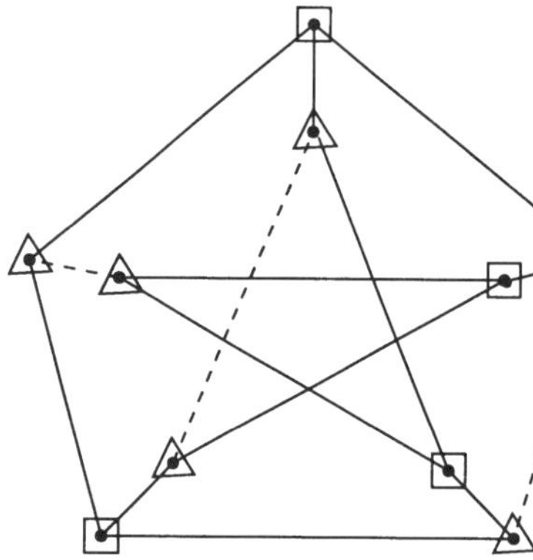

Figure 11.1. A maximum 2-cut of the Petersen graph

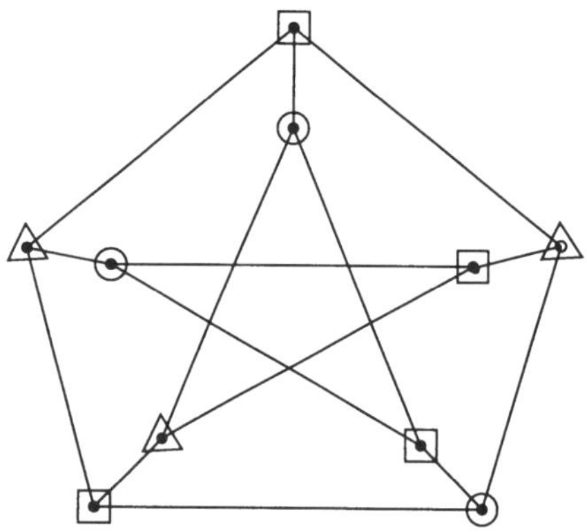

Figure 11.2. A maximum 3-cut of the Petersen graph.

now — the 3-cut gives a legal colouring of the Petersen graph using three colours (indicated by circular, square, and triangular markers respectively). □

MAX-k-CUT is an NP-complete problem for $k \geq 2$. In the case $k = 2$ the problem is usually referred to as MAX-CUT. In a seminal work, Goemans and Williamson [66] first used SDP to formulate an approximation algorithm for MAX-CUT which produces a 2-cut with weight within a factor 0.878 of the optimal weight. This work was extended for $k \geq 3$ by Frieze and Jerrum [59] and their results were further refined by De Klerk *et al.* [42]. In this chapter we will review all these results as well as related results by Karger *et al.* [98] for the approximate colouring problem. ('How many colours do you need to colour a κ-colourable graph correctly in polynomial time?') We will show that the Lovaśz ϑ-function forms the basis for all these approaches.

The algorithms in this chapter are so-called *randomized algorithms*. For an excellent introductory text on this topic the reader is referred to Motwani and Raghavan [128].

11.1 INTRODUCTION

The Lovász ϑ-function [115] of a graph $G = (V, E)$ forms the base for many semidefinite programming (SDP) relaxations of combinatorial optimization problems. Karger *et al.* [98] devised approximation algorithms for colouring κ-colourable graphs, *i.e.* graphs G with chromatic number $\chi(G) = \kappa$. Their so-called vector chromatic number is closely related to — and bounded from above by — the ϑ-function. The authors of [98] proved that k-colourable graphs can be coloured in polynomial time by using at most[3] $\min\left\{\tilde{O}\left(n^{1-3/(\kappa+1)}\right), \tilde{O}\left(\Delta^{1-2/\kappa}\right)\right\}$ colours, where $n = |V|$ and Δ is the valency of the graph; their paper also contains a review of earlier results. The work by Karger *et al.* employs the ideas of semidefinite approximation algorithms and associated randomized rounding, as introduced in the seminal work of Goemans and Williamson [66] on the MAX-CUT and other problems. The results by Karger *et al.* [98] cannot be improved much for general κ, since approximating the chromatic number of a general graph to within a factor $n^{1-\epsilon}$ for some $\epsilon > 0$ would imply $ZPP = NP$ (see Feige and Kilian [57]).[4] There is still some room for improvement of the results of Karger *et al.* for fixed values of κ — the best related hardness result states that all 3-colourable graphs cannot be coloured in polynomial time using 4 colours, unless P = NP (see Khanna *et al.* [102]). The best known (exponential) algorithm for exact 3-colouring runs in $O(1.3446^n)$ time (see Beigel and Eppstein [17]).

The graph colouring problem for k-colourable graphs can be seen as a special case of the (unweighted) MAX-k-CUT problem. Approximation algorithms for the MAX-k-CUT problem assign a colour from a set of k colours to each vertex in polynomial time so as to maximize the number of non-defect edges. For a survey of heuristics for MAX-k-CUT and applications in VLSI design see Cho and Sarrafzadeh [33]. The approximation guarantee of a MAX-k-CUT approximation algorithm is the ratio of the number of non-defect edges in the approximate solution to the maximum number of non-defect edges. The fraction $1 - 1/k$ is achieved by a random assignment of colours to the vertices (see Example 11.2). This can be slightly improved to $\left(1 - \frac{1}{k} + 2\log(k)/k^2\right)$ for sufficiently large values of k, as shown by Frieze and Jerrum [59]; there is very little room for further improvement of this result, since the attainable approximation guarantee is upper bounded by $1 - 1/(34k)$, unless P=NP (see Kann *et al.* [97]). For fixed values of k the approximation guarantee can be improved. For example, a guarantee of 0.836 is attainable for MAX-3-CUT, as was independently shown by De Klerk and Pasechnik [42] and Goemans and Williamson [67]. The approach By Frieze and Jerrum [59] is closely linked to that by Karger *et al.*

---

[3]The $\tilde{O}$ notation means that powers of logarithms may be suppressed.

[4]For a review of the complexity classes P, NP, ZPP, *etc.* the reader is referred to Motwani and Raghavan [128].

in the sense that both underlying semidefinite programming relaxations are related to the same formulation of the Lovász ϑ-function.

11.2 THE ϑ-APPROXIMATION OF THE CHROMATIC NUMBER

The Lovász ϑ–function has many representations (see Chapter 10); the representation we will use in this chapter is by Karger *et al.* [98]:

$$(*) \quad \vartheta(\bar{G}) = \min_{U,t} t$$

subject to

$$\begin{aligned} u_{ij} &= \frac{-1}{t-1}, \quad \text{if } (i,j) \in E \\ u_{ii} &= 1, \quad i = 1, \ldots, n \\ U &\succeq 0. \end{aligned}$$

Recall that the 'sandwich' theorem of Lovász ensures that $\omega(G) \leq \vartheta(\bar{G}) \leq \chi(G)$, where $\omega(G)$ and $\chi(G)$ denote the clique and chromatic numbers of $G = (V, E)$ respectively, and $\bar{G}$ is the complementary graph of G.

11.3 AN UPPER BOUND FOR THE OPTIMAL VALUE OF MAX-K-CUT

Our goal is to assign k colours to the $|V| = n$ vertices of a given graph $G = (V, E)$ such that the number of non-defect edges is as large as possible. We will give a mathematical formulation for this problem where we will use the following result from linear algebra.

Lemma 11.1 *If $n \geq k$, one can find k vectors $r_1, \ldots, r_k$ in $\mathbb{R}^n$ such that*

$$r_i^T r_j = \begin{cases} -\frac{1}{k-1} & \text{if } i \neq j \\ 1 & \text{if } i = j. \end{cases} \tag{11.1}$$

Proof:

Let $U \in \mathcal{S}_k$ be defined by

$$u_{ij} = \begin{cases} -\frac{1}{k-1} & \text{if } i \neq j \\ 1 & \text{if } i = j, \end{cases}$$

where $i, j = 1, \ldots, k$. The matrix U is now positive semidefinite, since it is diagonally dominant (see Theorem A.6 in Appendix A). We can therefore obtain a factorization $U = R^T R$ where the columns of R are given by $[r_1, \ldots, r_k]$ say. The vectors

$r_1, \ldots, r_k$ satisfy condition (11.1) by construction. □

We are now ready to give a mathematical formulation of the MAX-k-CUT problem. To this end, let $r_1, \ldots, r_k$ denote a set of vectors in $\mathbb{R}^n$ $(n > k)$ which satisfies condition (11.1). We will associate these vectors with k different colours. Similarly we will associate n unit vectors y_i $(i = 1, \ldots, n)$ with the set of vertices V. Thus we assign r_j to y_i if we wish to assign colour j to vertex i.

$$\text{(MAX-}k\text{-CUT)} \quad OPT := \max_{y_1, \ldots, y_n} \frac{k-1}{k} \sum_{(i,j) \in E} (1 - y_i^T y_j)$$

subject to

$$y_j \in \{r_1, \ldots, r_k\} \quad (j = 1, \ldots, n). \tag{11.2}$$

After an assignment of colours to the endpoints of the edge $(i, j) \in E$ we have

$$\frac{k-1}{k}(1 - y_i^T y_j) = \begin{cases} 1 & \text{if } (i,j) \text{ is non-defect} \\ 0 & \text{if } (i,j) \text{ is defect.} \end{cases}$$

Thus it is easy to see that we have given a valid mathematical formulation of MAX-k-CUT.

We obtain an SDP relaxation of this problem if we weaken the requirement (11.2) to

$$\|y_j\| = 1 \ (i = 1, \ldots, n), \quad 0 \geq y_i^T y_j \geq \frac{-1}{k-1} \quad (i,j) \in E, \tag{11.3}$$

to obtain the problem

$$\max_{y_1, \ldots, y_n} \frac{k-1}{k} \sum_{(i,j) \in E} (1 - y_i^T y_j) \tag{11.4}$$

subject to (11.3).

If we define the matrices $Y = [y_1, \ldots, y_n]$ and $U = Y^T Y$, we can rewrite the relaxation as

$$\max_{U} \frac{k-1}{k} \sum_{(i,j) \in E} (1 - u_{ij}) \tag{11.5}$$

subject to

$$\begin{aligned} 0 \geq u_{ij} &\geq \frac{-1}{k-1}, \quad \text{if } (i,j) \in E \\ u_{ii} &= 1, \quad i = 1, \ldots, n \\ U &\succeq 0. \end{aligned}$$

Note the very close similarity between this problem and the $(*)$ formulation of $\vartheta(\bar{G})$.

Note that the optimal value in (11.5) is an upper bound on the optimal value of MAX-k-CUT, because (11.5) is a relaxation of MAX-k-CUT.

11.4 APPROXIMATION GUARANTEES

We are going to formulate a randomized approximation algorithm for the MAX-k-CUT problem presently. First we will review the concepts of approximation algorithms and approximation guarantees.

Definition 11.1 *We call a algorithm which runs in (randomized) polynomial time an α_k-approximation algorithm for MAX-k-CUT if — for any graph $G = (V, E)$ — the algorithm produces a k-cut with weight at least $\alpha_k * OPT$, where OPT denotes the optimal value of the MAX-k-CUT problem, as before. We refer to the value α_k as the approximation guarantee.*

Before proceeding we give a simple example of a $\frac{k-1}{k}$-approximation algorithm for MAX-k-CUT.

Example 11.2 *Let us consider perhaps the simplest possible randomized algorithm for MAX-k-CUT, where we assign a colour (from our set of k colours) randomly to each vertex in turn. Consider an edge (i, j). The probability that both endpoints are assigned the same colour is $1/k$. The expected number of non-defect edges is therefore $\frac{k-1}{k}|E|$. In other words, our algorithm yields a k-cut containing at least $\frac{k-1}{k}|E|$ edges, in expectation. Since the optimal value of MAX-k-CUT cannot exceed $|E|$, its approximation guarantee is $\alpha_k \geq \frac{k-1}{k}$.* □

Let $y_1, \ldots, y_n$ denote an optimal solution of the SDP problem (11.4). Recall that the optimal value of problem (11.4) is an upper bound on OPT.

Let us now consider a family of approximation algorithms where we assign k colours to the vertices in such a way that the probability that edge (i, j) is defect after the assignment is a known function of $y_i^T y_j$, say $p(y_i^T y_j)$.

The expected number of edges in a k-cut generated by such an algorithm is simply $\sum_{(i,j)\in E}\left(1 - p\left(y_i^T y_j\right)\right)$. Thus the approximation guarantee is given by

$$\begin{aligned}
\alpha_k &= \frac{\sum_{(i,j)\in E}\left(1 - p\left(y_i^T y_j\right)\right)}{OPT} \\
&\geq \frac{\sum_{(i,j)\in E}\left(1 - p\left(y_i^T y_j\right)\right)}{\frac{k-1}{k}\sum_{(i,j)\in E}\left(1 - y_i^T y_j\right)} \\
&\geq \min_{(i,j)\in E} \frac{1 - p\left(y_i^T y_j\right)}{\frac{k-1}{k}\left(1 - y_i^T y_j\right)} \\
&\geq \min_{0\geq\rho\geq -1/(k-1)} \frac{1 - p(\rho)}{\frac{k-1}{k}(1 - \rho)}.
\end{aligned}$$

To obtain the second inequality we have used the following.

$$\frac{\sum_{i=1}^{n} x_i}{\sum_{i=1}^{n} y_i} \geq \min_i \frac{x_i}{y_i} \quad \text{for } x_i > 0,\ y_i > 0\ (i = 1, \ldots, n).$$

Thus we have obtained the bound

$$\alpha_k \geq \min_{0 \geq \rho \geq -1/(k-1)} \frac{k(1 - p(\rho))}{(k-1)(1-\rho)} \tag{11.6}$$

on the value of the approximation guarantee. We will use it to derive upper bounds on the approximation guarantees of the algorithms in the next section.

Note that the expression (11.6) no longer depends on the optimal solution of the SDP problem (11.4). The bound is therefore valid if we use any set of unit vectors $y_1, \ldots, y_n$ in $\mathbb{R}^n$ in the formulation of an algorithm instead of the optimal set for problem (11.4). The only requirement is that $0 \geq y_i^T y_j \geq -1/(k-1)$ if $(i, j) \in E$. We will therefore not use problem (11.4) again — we only needed it to derive the upper bound on α_k. Instead we will use a set of vectors that corresponds to the $(*)$ formulation of $\vartheta(\bar{G})$.

11.5 A RANDOMIZED MAX-K-CUT ALGORITHM

We now propose a randomized MAX-k-CUT algorithm based on the Lovász ϑ-function. The algorithm assigns k colours to the vertices of the graph in such a way that the expected fraction of defect edges is provably small. Let

$$t = \max\left\{k, \lceil \vartheta(\bar{G}) \rceil\right\}$$

and let (U, t) be a feasible solution of problem $(*)$. The basic idea of the algorithm is now as follows: define vectors $v_1, \ldots, v_n$ as the columns of a Choleski factorization of U. These vectors are associated with the vertices of G, as before. We now generate k vectors in $\mathbb{R}^n$ randomly and assign colour j to vertex i if the jth random vector is 'closest' to v_i (in a sense to be made precise).

Formally, we can state the algorithm as follows.

Randomized MAX-k-CUT Algorithm

1. Let $t = \max\left\{k, \lceil \vartheta(\bar{G}) \rceil\right\}$, and let (U, t) be a feasible solution of problem $(*)$;
2. Take the Choleski factorization $U = \bar{V}^T \bar{V}$, and denote $\bar{V} = [v_1, \ldots, v_n]$;
3. Generate k different random vectors[5] $r^{(1)}, \ldots, r^{(k)} \in \mathbb{R}^n$; Each vector $r^{(i)}$ is associated with a colour i;

[5]Random vectors here means that each component of each vector is drawn independently from the standard normal distribution with mean zero and variance one.

4. Assign colour i to vertex j if $i = \arg\max_t v_j^T r^{(t)}$, *i.e.* assign the colour corresponding to the 'closest' random vector.

We first show that this algorithm can be restated in an equivalent way, where we work in a higher dimension, but only need one random $r \in \mathbb{R}^{kn}$ instead of k different random vectors in $\mathbb{R}^n$. This alternative formulation is easier to analyze for small, fixed values of k. The first formulation of the algorithm is essentially due to Frieze and Jerrum [59] and the alternative formulation is due to De Klerk *et al.* [42].

Randomized MAX-k-CUT Algorithm (Alternative formulation)

1. Let $t = \max\left\{k, \lceil \vartheta(\bar{G}) \rceil\right\}$, and let (U, t) be a feasible solution of problem $(*)$. Define
$$Y = U \otimes \frac{k}{k-1}\left(I_k - \frac{1}{k}c_k c_k^T\right). \tag{11.7}$$

2. *Rounding scheme:*

   Perform a factorization $Y = V^T V$ and denote
$$V = [v_1^1, v_1^2 \ldots, v_1^k, \ldots, v_n^k];$$

   Choose a random unit vector $r \in \mathbf{R}^{kn}$ from the uniform distribution on the unit ball in $\mathbb{R}^{kn}$.

   Assign colour i to vertex j if $i = \arg\max_t {v_j^t}^T r$.

The vector v_j^t corresponds to the situation where colour t is assigned to vertex j. In particular, vertex j is assigned colour t by the algorithm if v_j^t is the 'closest' to the random vector r of all the vectors $v_j^1, \ldots, v_j^k$.

Theorem 11.1 *The two statements of the randomized MAX-k-CUT algorithm are equivalent.*

Proof:

Let U, t again denote a solution of problem $(*)$. The first step in the alternative formulation is taking the Kronecker product
$$Y = U \otimes c\left(I_k - \frac{1}{k}e_k e_k^T\right) = \bar{V}^T \bar{V} \otimes c\left(I_k - \frac{1}{k}e_k e_k^T\right)$$
where $c = \frac{k}{k-1}$, and subsequently obtaining the Choleski factor $\left[v_1^1, v_1^2, \ldots, v_n^k\right]$ of Y. Vertex i is then assigned colour p if and only if
$$p = \arg\max_t {v_i^t}^T r \tag{11.8}$$

for a random $r \in \mathbb{R}^{kn}$. The identity (see Appendix E):

$$\left(B^TB\right) \otimes \left(G^TG\right) = \left(B^T \otimes G^T\right)(B \otimes G) = (B \otimes G)^T (B \otimes G),$$

implies

$$\bar{V}^T\bar{V} \otimes c\left(I_k - \frac{1}{k}e_ke_k^T\right) = \left(\bar{V} \otimes \sqrt{c}\left(I_k - \frac{1}{k}e_ke_k^T\right)\right)^T \bar{V} \otimes \sqrt{c}\left(I_k - \frac{1}{k}e_ke_k^T\right), \tag{11.9}$$

where we have used the fact that the matrix $\left(I_k - \frac{1}{k}e_ke_k^T\right)$ is idempotent[6]. We now define the matrices:

$$X_i = \left[v_i^1,\, v_i^2, \ldots,\, v_i^k\right]^T, \quad i = 1, \ldots, n. \tag{11.10}$$

Note that by (11.8), vertex i is assigned colour p if and only if

$$p = \arg\max_t (X_i r)_t.$$

By (11.9) and (11.10) we have

$$X_i = v_i^T \otimes \sqrt{c}I_k - v_i^T \otimes \frac{\sqrt{c}}{k}e_ke_k^T,$$

so that

$$X_i r = \left(v_i^T \otimes \sqrt{c}I_k\right) r - \left(v_i^T \otimes \frac{\sqrt{c}}{k}e_ke_k^T\right) r = \left(v_i^T \otimes \sqrt{c}I_k\right) r - c_i e_k$$

where c_i is a scalar depending on r and v_i. Note that we can ignore the last term when finding the largest component of $X_i r$. In other words,

$$\arg\max_p (X_i r)_p = \arg\max_p \left(v_i^T \otimes \sqrt{c}I_k r\right)_p. \tag{11.11}$$

Finally, we construct a set of k random vectors in $\mathbb{R}^n$ from r as follows:

$$r^{(i)} := \left[r_i, r_{i+k}, r_{i+2k}, \ldots, r_{i+(n-1)k}\right]^T, \quad i = 1, \ldots, k.$$

Note that, by construction,

$$\left(v_i^T \otimes \sqrt{c}I_k\right) r = \sqrt{c}\left[v_i^T r^{(1)}, \ldots, v_i^T r^{(k)}\right]^T, \quad i = 1, \ldots, n,$$

so that

$$\arg\max_p (X_i r)_p = \arg\max_p v_i^T r^{(p)}, \quad i = 1, \ldots, n,$$

by (11.11). This completes the proof. □

[6]Recall that a matrix A is called idempotent if $A^2 = A$.

11.6 ANALYSIS OF THE ALGORITHM

We proceed to give a simple proof — using geometric arguments only — which establishes the probability that a given edge is defect after running the randomized MAX-k-cut algorithm (alternative formulation). In particular, we wish to know what the probability (say p_1) is that both endpoints of a given edge are assigned colour 1. The probability that the edge is defect is then simply kp_1 since the number of colours used equals k.

Note that both endpoints of an edge (i, j) have been assigned colour 1 if and only if:

$$r^T v_i^1 \geq r^T v_i^q, \quad q = 2, \ldots, k,$$

and

$$r^T v_j^1 \geq r^T v_j^q, \quad q = 2, \ldots, k.$$

In other words, r must lie in the dual cone of the cone generated by the vectors

$$(v_i^1 - v_i^2), \ldots, (v_i^1 - v_i^k), (v_j^1 - v_j^2), \ldots, (v_j^1 - v_j^k). \tag{11.12}$$

An alternative geometrical interpretation is that the half space with outward pointing normal vector $-r$ must contain the vectors in (11.12), *i.e.* the vectors (11.12) must lie on a specific side of a random hyperplane with normal vector r (the same side as r).[7]

For convenience of notation we define the unit vectors

$$w_i^q = \frac{v_i^1 - v_i^q}{\|v_i^1 - v_i^q\|}, \quad q = 2, \ldots, k, \quad i = 1, \ldots, n.$$

The 'w-vectors' can be viewed as a set of $(2k-2)$ points on the $(2k-3)$-dimensional unit hypersphere

$$\mathcal{S}^{(2k-3)} := \left\{ x \in \mathbb{R}^{2k-2} \mid \|x\| = 1 \right\},$$

and thus define a so-called *spherical simplex*[8] (say $\mathbf{S}$) in the space $\mathcal{S}^{(2k-3)}$.

The *Gram matrix* of the w vectors (which has the inner products of the w-vectors – *i.e.* the cosines of the edge lengths of $\mathbf{S}$ – as entries) is known explicitly, since the corresponding entries in the matrix Y in (11.7) are known. In particular, it is easy to show that the Gram matrix is given by:

$$\text{Gram}(\mathbf{S}) := \frac{1}{2} \begin{bmatrix} 1 & \rho \\ \rho & 1 \end{bmatrix} \otimes \left(I_{k-1} + e_{k-1} e_{k-1}^T \right), \tag{11.13}$$

[7] The probability that a given set of vectors lie on the same side of a random hyperplane has been investigated recently by Karloff and Zwick [99] (at most 4 vectors) and Zwick [194] (general case) in the context of MAX-SAT approximation algorithms; see Chapter 13 of this monograph. In what follows, we employ the same approach as these authors.

[8] A one dimensional spherical simplex is simply a segment of the unit circle, a two dimensional spherical simplex is a triangle on a unit sphere, *etc.*

where $\rho = \frac{-1}{t-1}$, and $t = \max\left\{\lceil\vartheta(\bar{G})\rceil, k\right\}$, as before.

From a geometrical viewpoint, we are interested in the volume of the spherical simplex (say $\mathbf{S}^*$) which is dual to the spherical simplex $\mathbf{S}$, as a fraction of the total volume of the unit hypersphere $\mathcal{S}^{(2k-3)}$. This dual spherical simplex is given by:

$$\mathbf{S}^* = \left\{x \in \mathcal{S}^{(2k-3)} \;\middle|\; x^T z \geq 0 \quad \forall z \in \mathbf{S}\right\}.$$

The Gram matrix associated with $\mathbf{S}^*$ is given by taking the inverse of Gram($\mathbf{S}$) in (11.13) and subsequently normalizing its diagonal. Straightforward calculations show that this matrix takes the form

$$\mathrm{Gram}(\mathbf{S}^*) := \frac{k}{k-1}\begin{bmatrix} 1 & -\rho \\ -\rho & 1 \end{bmatrix} \otimes \left(I_{k-1} - \frac{1}{k}e_{k-1}e_{k-1}^T\right). \tag{11.14}$$

The volume of a spherical simplex is completely determined by the off-diagonal entries of its Gram matrix. Unfortunately, there is no closed form expression available for the volume function in general, and it must be evaluated by numerical integration. The integral which yields p_1 is given by (see Alekseevskij *et al.* [2]):

$$\begin{aligned} p_1 &= \frac{\mathrm{vol}(\mathbf{S}^*)}{\mathrm{vol}\left(\mathcal{S}^{(2k-3)}\right)} \qquad\qquad (11.15) \\ &= \frac{1}{\sqrt{\det(\mathrm{Gram}(\mathbf{S}))\pi^{2k-2}}}\int_0^\infty \cdots \int_0^\infty e^{-y^T \mathrm{Gram}(\mathbf{S})^{-1} y} dy_1 \ldots dy_{2k-2}. \end{aligned}$$

We will write $p_1(\rho)$ from now on to emphasize that p_1 is a function of ρ only (for a given k).

11.7 APPROXIMATION RESULTS FOR MAX-K-CUT

In the next two subsections we will calculate $p_1(\rho)$ for $k = 2$ (MAX-CUT) and $k = 3$ (MAX-3-CUT) after which we will consider general $k > 3$. The cases $k = 2$ and $k = 3$ are special because the integral in (11.15) can then be solved analytically.

RESULTS FOR MAX-CUT

In the MAX-CUT case ($k = 2$), the integral in (11.15) can be solved analytically. In this case the spherical simplex $\mathbf{S}^*$ is merely a circular line segment of length $\arccos(-\rho)$ on the unit circle. This follows from the fact that the Gram matrix of $\mathbf{S}^*$ is given by

$$\mathrm{Gram}(\mathbf{S}^*) = \begin{bmatrix} 1 & -\rho \\ -\rho & 1 \end{bmatrix}$$

in this case, by (11.14). Thus we have

$$p_1(\rho) = \frac{\mathrm{vol}(\mathbf{S}^*)}{\mathrm{vol}\left(\mathcal{S}^{(1)}\right)} = \frac{\arccos(-\rho)}{2\pi},$$

so that the probability of an edge becoming defect is simply

$$p(\rho) = 2p_1(\rho) = \arccos(-\rho)/\pi.$$

Thus the upper bound (11.6) on the performance guarantee α_2 becomes

$$\begin{aligned} \alpha_2 &\geq \min_{0\geq\rho\geq-1} \frac{2(1-p(\rho))}{(1-\rho)} \\ &= \min_{0\geq\rho\geq-1} \frac{1-\frac{1}{\pi}\arccos(-\rho)}{\frac{1}{2}(1-\rho)} \\ &\geq 0.878. \end{aligned}$$

The last inequality is illustrated in Figure (11.3).

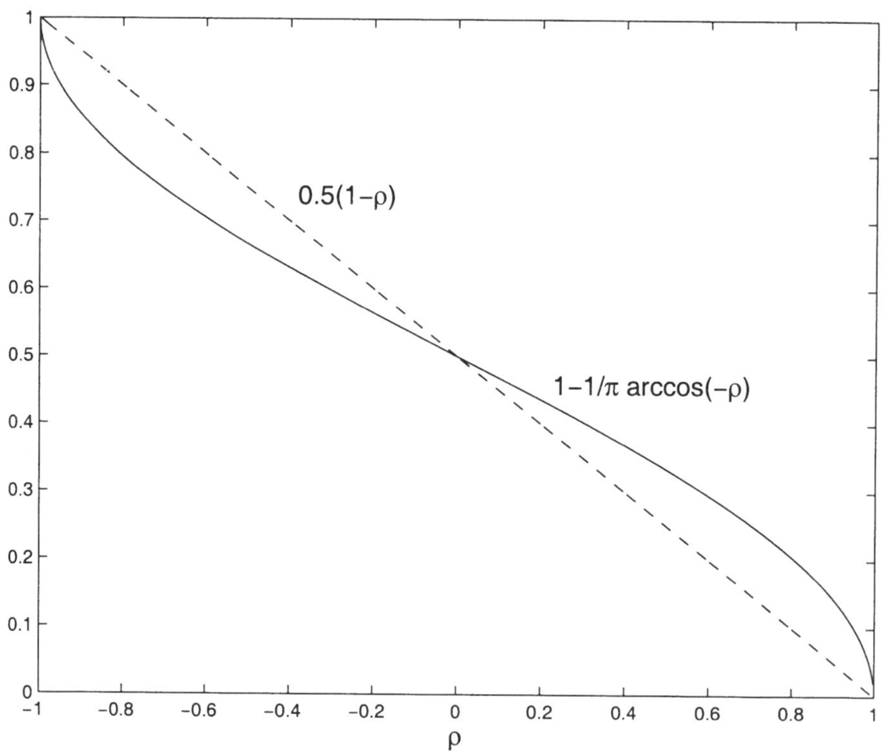

Figure 11.3. The approximation of the MAX-CUT algorithm is given by the worst case ratio of the values $1 - \frac{1}{\pi}\arccos(-\rho)$ (solid line) and $\frac{1}{2}(1-\rho)$ (dashed line). The worst case ratio is roughly 0.878.

The $\alpha_2 \geq 0.878$ bound is the celebrated result by Goemans and Williamson [66]; in a recent talk by W.R. Pulleyblank [151] it was given the status of one of the ten most important results in combinatorial optimization.

RESULTS FOR MAX-3-CUT

We will now calculate the MAX-3-CUT guarantee of our algorithm.

In this case, the integral (11.15) can in fact be solved analytically to obtain:

$$p_1(\rho) = \frac{1}{9} + \frac{\arccos(-\rho) - \arccos^2(\rho/2)}{4\pi^2}.$$

This was first shown by Goemans and Williamson [67] and subsequently by De Klerk and Pasechnik [40] in a different way.

The resulting approximation guarantee from (11.6) is shown in Figure 11.4 as a function of $\vartheta(\bar{G})$. Note that the worst-case approximation guarantee occurs where $\vartheta(\bar{G}) = 3$, and is approximately given by $\alpha_3 = 0.836008$.

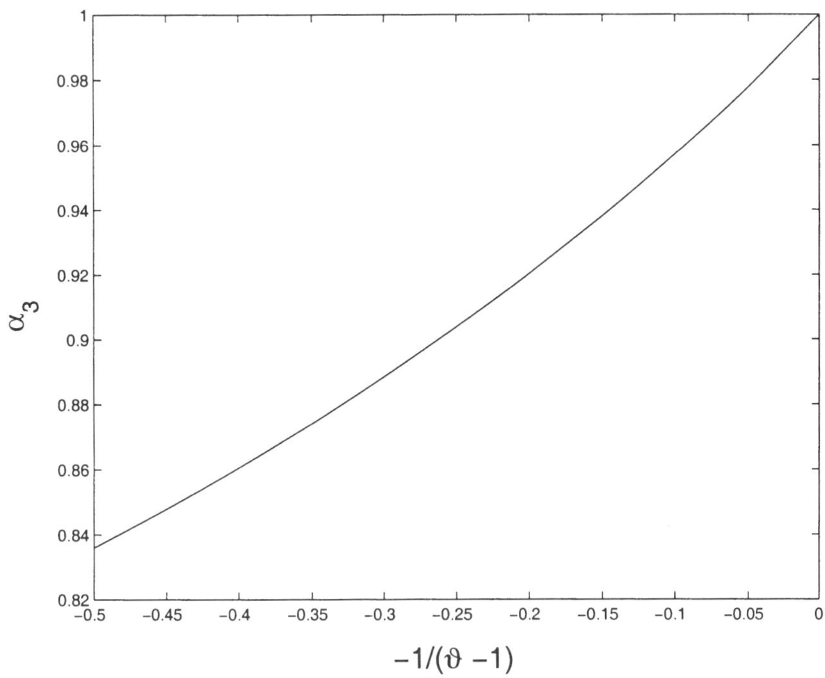

Figure 11.4. The MAX-3-CUT guarantee as a function of $\vartheta(\bar{G})$.

RESULTS FOR GENERAL MAX-K-CUT

Recall from (11.6) that the approximation guarantee of our randomized MAX-k-CUT algorithm is given by

$$\alpha_k := \min_{-1/(k-1) \leq \rho \leq 0} \frac{k(1 - kp_1(\rho))}{(k-1)(1-\rho)}. \tag{11.16}$$

Also recall that the worst-case for the approximation guarantee for MAX-3-CUT occurs for graphs G where $\vartheta(\bar{G}) \leq 3$. We cannot prove the analogous result for all $k > 3$, but will show that it is true asymptotically (for $k \gg 0$). For small fixed values of k we can give similar numerical proofs as for $k = 3$ (*cf.* Figure 11.4).

We have done this for $k = 4, \ldots, 15$ using the software MVNDST (for calculating multivariate normal probabilities) by Genz [62], and the approximation guarantees are plotted in Figure 11.5 using '+' markers. The plotted values lie above the curve $1-1/k$

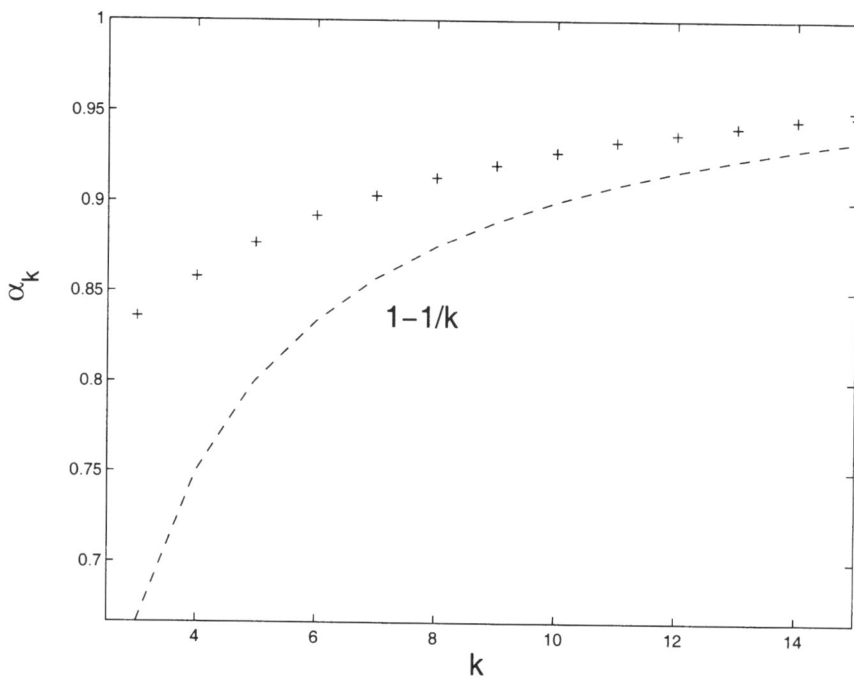

Figure 11.5. The MAX-k-CUT approximation guarantees of the algorithm are indicated by '+' markers.

(which in turn corresponds to the random assignments of k colours to the vertices, *i.e.* the algorithm in Example 11.2). We will see that this is true in general (at least for $k \gg 0$).

The approximate numerical values of α_k ($k = 1, \ldots, 10$) are shown in Table 11.1. These are the best known approximation results for MAX-k-CUT ($k \leq 10$) at the time of writing.

k:	2	3	4	5	6	7	8	9	10
α_k:	0.878	0.836	0.858	0.877	0.892	0.903	0.913	0.920	0.927

Table 11.1. The MAX-k-CUT approximation guarantees for $2 \leq k \leq 10$

ASYMPTOTIC ANALYSIS

Now we investigate the asymptotic behaviour of $p_1(\rho)$ in (11.15) for fixed $\rho < 0$ as $k \to \infty$. One can show the following.

Theorem 11.2 (De Klerk *et al.* [42])

$$p_1(\rho) \sim \frac{\Gamma(\frac{1}{1+\rho})^2}{\sqrt{1-\rho^2}} \frac{(4\pi \log(k-1))^{\frac{-\rho}{1+\rho}}}{(k-1)^{\frac{2}{1+\rho}}} \tag{11.17}$$

as $k \to \infty$, where Γ denotes the gamma function.

If we substitute (11.17) into (11.16) and differentiate with respect to ρ, we find that the minimum in (11.16) is attained when $\rho = -1/(k-1)$, *i.e.* for graphs G with $\vartheta(\bar{G}) \leq k$.

Now it is easy to show that the performance guarantee becomes

$$\alpha_k \sim 1 - k^{\frac{-k}{k-2}} \sim 1 - \frac{1}{k} + \frac{2\log k}{k^2},$$

as $k \to \infty$. The $\alpha_k \sim 1 - \frac{1}{k} + \frac{2\log k}{k^2}$ result was first shown by Frieze and Jerrum [59], but the proof which was sketched here is due to De Klerk *et al.* [42].

11.8 APPROXIMATE COLOURING OF κ-COLOURABLE GRAPHS

We now turn our attention to a different, but related problem, namely *approximate graph colouring*. Given is a κ-colourable graph $G = (V, E)$, *i.e.* $\chi(G) \leq \kappa$ for some fixed $\kappa \geq 3$. We call an assignment of colours to *all* the vertices of G a *legal colouring* if there are no defect edges, and ask how many colours we need to do a legal colouring of G in polynomial time.

In this section we will sketch a proof of the following result.

Theorem 11.3 (Karger *et al.* [98]) *A κ-colourable graph with maximum degree Δ can be legally coloured in polynomial time using $\tilde{O}\left(\Delta^{1-2/\kappa}\right)$ colours.*

We can use our approximation algorithm for MAX-k-CUT to assign k colours to the vertices of G. If k is 'large enough', then fewer than $\frac{1}{2}n$ edges will be defect. Such an assignment of colours to the vertices corresponds to a so-called *semi–colouring* of G.

Definition 11.2 (Semi–colouring) *An assignment of k colours to at least half the vertices of G such that there are no defect edges is called a semi–colouring.*

An assignment of k colours to all the vertices of G such that at most $\frac{1}{2}n$ edges are defect yields a semi-colouring, by simply removing the colour from one endpoint of each defect edge.

We can obtain a legal colouring via successive semi-colourings, as follows.

(1) Choose a 'sufficiently large' value of k and use the randomized MAX-k-CUT approximation algorithm to obtain a semi-colouring of G using k colours;

(2) Remove the vertices that have been coloured in step (1);

(3) Replace G by the induced subgraph on the remaining vertices;

(4) Repeat steps (1) to (3) by introducing a new set of k colours each time step (1) is performed, until no vertices are left.

Assume for the moment that we know how to choose k such that we can always obtain a semi-colouring. The following lemma gives a bound on the number of colours required by the above scheme to give a legal colouring of G.

Lemma 11.2 (Karger *et al.* [98]) *If one can obtain a semi-colouring (using k colours) of any graph with at most n vertices in polynomial time, then any graph on n vertices can be legally coloured in polynomial time using $O(k \log n)$ colours.*

Proof:

Let $G = (V, E)$ be given and perform a semi-colouring using k colours. Remove the legally coloured vertices from the graph and find a new semi-colouring using k new colours on the subgraph induced on the remaining vertices. After repeating this process $\hat{t}$ times we have an induced subgraph on at most $\left(\frac{1}{2}\right)^{\hat{t}} n$ vertices. If $\hat{t} = \Omega(\log(n))$, then this subgraph has only $O(1)$ vertices which we can colour using $O(1)$ different colours. □

We return to the problem of finding a suitable value for k, and prove that a semi-colouring will be obtained with high probability if we can find a k-cut such that the (expected) number of defect edges is at most $\frac{1}{4}n$.

Lemma 11.3 *Let $G = (V, E)$ be given. Assume that we have a randomized procedure — say PROC — that assigns k-colours to V in polynomial time such that the expected* *number of defect edges does not exceed* $\frac{1}{4}n$. *Then a semi-colouring of* G *can be obtained in polynomial time with probability* $1 - \epsilon$ *for any given* $\epsilon > 0$.

Proof:

By Markov's inequality (see *e.g.* Theorem 3.2 in Motwani and Raghavan [128]), the probability that the number of defect edges is more than twice the expected value is at most $\frac{1}{2}$. Repeating the procedure PROC $\hat{t}$ times, the probability that we have not found a k-cut with at most $\frac{1}{2}n$ defect edges is upper bounded by $\left(\frac{1}{2}\right)^{\hat{t}}$. Once we have obtained a k-cut with at most $\frac{1}{2}n$ defect edges, we simply remove a colour from one endpoint of each defect edge to obtain a semi-colouring. □

The total number of edges in a graph is upper bounded by $|E| \leq \frac{1}{2}n\Delta$, where Δ is the maximum degree of any vertex.

By Theorem 11.2, the expected number of defect edges is therefore bounded via:

$$kp_1(\rho)|E| \leq \frac{1}{2}kp_1(\rho)n\Delta \sim \frac{1}{2}k^{\frac{\rho-1}{\rho+1}}n\Delta. \tag{11.18}$$

The expected number of defect edges will be at most $\frac{1}{4}n$ if

$$k \geq (2\Delta)^{\frac{1+\rho}{1-\rho}}.$$

This last inequality gives us a suitable way to choose k.

For κ-colourable graphs we have $\rho = -1/(\kappa - 1)$, and consequently we obtain a semi-colouring using $O\left(\Delta^{1-2/\kappa}\right)$ colours, and subsequently a complete colouring using $\tilde{O}\left(\Delta^{1-2/\kappa}\right)$ colours. Thus we have sketched a proof of Theorem 11.3.

12

THE STABILITY NUMBER OF A GRAPH AND STANDARD QUADRATIC OPTIMIZATION

Preamble

In this chapter we consider the class of conic optimization problems obtained when we replace the positive semidefinite cone with the cone of copositive matrices.[1] The resulting optimization problems are called copositive programs. An important application of copositive programming is that the stability number of a graph is given as the optimal value of a suitable copositive program. The only problem is that — unlike for the positive semidefinite cone — one cannot optimize over the copositive cone in polynomial time, unless $P = NP$. However, one can approximate the copositive cone arbitrarily well by using linear matrix inequalities. The maximum stable set problem can in turn be seen as a special case of the standard quadratic optimization problem (where one optimizes a quadratic function over the standard simplex). One can apply the same ideas as for the stable set problem to the more general case. We will review all these results in this chapter. The presentation is based on the papers De Klerk and Pasechnik [41] and Bomze and De Klerk [27].

[1] Recall that a matrix $A \in \mathcal{S}_n$ is called copositive if $x^T A x \geq 0$ for all $x \in \mathbb{R}^n_+$.

12.1 PRELIMINARIES

The maximum stable set problem Recall that for a graph $G = (V, E)$, a subset $V' \subset V$ is called a stable set of G if the induced subgraph on V' contains no edges. The maximum stable set problem is to find the stable set of maximal cardinality.

Example 12.1 *Let us consider the Petersen graph again. In Figure 12.1 we show a maximum stable set for this graph. (The vertices in the stable set are marked with square markers.)* □

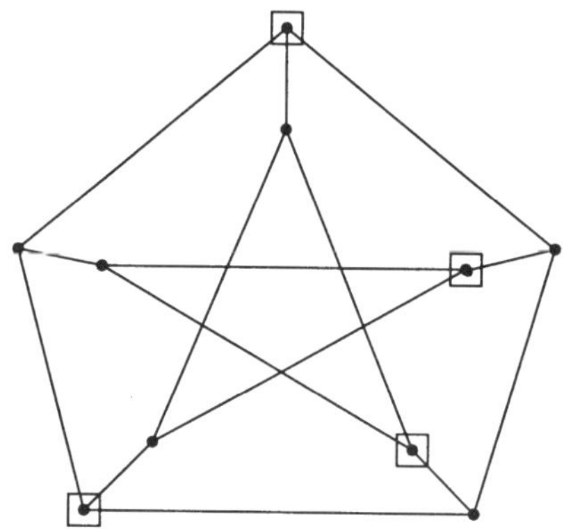

Figure 12.1. A maximum stable set of the Petersen graph

The maximum stable set problem is equivalent to finding the largest clique in the complementary graph, and cannot be approximated within a factor $|V|^{\frac{1}{2}-\epsilon}$ for any $\epsilon > 0$ unless $P = NP$, or within a factor $|V|^{1-\epsilon}$ for any $\epsilon > 0$ unless $NP = ZPP$ (Håstad [81]). The best known approximation guarantee for this problem is $O\left(\frac{|V|}{(\log|V|)^2}\right)$ (Boppana and Halldórsson [29]). For a survey of the maximum clique problem, see Bomze *et al.* [26].

Copositive programming As on page 28, we again consider a generic primal and dual pair of conic linear programs:

$$(P_{\mathcal{K}}) \quad p^* := \inf_X \left\{ \mathbf{Tr}\,(CX) \mid \mathbf{Tr}\,(A_i X) = b_i \ (i = 1, \ldots, m), \ X \in \mathcal{K} \right\}$$

$$(D_{\mathcal{K}}) \quad d^* := \sup_{y \in \mathbb{R}^m} \left\{ b^T y \;\middle|\; \sum_{i=1}^{m} y_i A_i + S = C, \ S \in \mathcal{K}^* \right\},$$

where $\mathcal{K}$ is a closed convex cone. We define the following convex cones of matrices:

- the copositive cone:

$$\mathcal{C}_n := \left\{A \in \mathcal{S}_n \mid x^T A x \geq 0 \quad \text{for all } x \in \mathbb{R}^n_+\right\};$$

- the cone of completely positive matrices:

$$\mathcal{C}_n^* = \left\{A \in \mathcal{S}_n \;\middle|\; A = \sum_{i=1}^{k} x_i x_i^T,\ x_i \in \mathbb{R}^n_+ \ (i = 1, \ldots, k) \ \text{ for any } k\right\};$$

- the cone of nonnegative matrices:

$$\mathcal{N}_n = \{A \in \mathcal{S}_n \mid a_{ij} \geq 0 \ (i,j = 1, \ldots, n)\}.$$

The completely positive cone is the dual of the copositive cone for the usual inner product $\langle X, Y\rangle := \mathbf{Tr}\,(XY)$.

Optimization over the cones $\mathcal{S}_n^+$ and $\mathcal{N}_n$ correspond to LP and SDP respectively, and can therefore be done in polynomial time (to compute an ϵ-optimal solution) using interior point methods. On the other hand, copositive programming (where $\mathcal{K} = \mathcal{C}_n$) is reducible to some NP-hard problems, like the maximum stable set problem, and can therefore not be solved in polynomial time, unless $P = NP$.

12.2 THE STABILITY NUMBER VIA COPOSITIVE PROGRAMMING

From the sandwich theorem (Theorem 10.1) we know that $\alpha(G) \leq \vartheta(G)$. We now show that we can actually obtain the stability number $\alpha(G)$ by replacing the semidefinite cone by the completely positive cone in the definition of the Lovász ϑ-function (see (10.1) on page 158).

Theorem 12.1 *Let $G = (V, E)$ be given with $|E| = n$. The stability number of G is given by:*

$$\alpha(G) = \max \mathbf{Tr}\left(ee^T X\right) \tag{12.1}$$

subject to

$$\left.\begin{aligned} X_{ij} &= 0,\ \{i,j\} \in E\ (i \neq j) \\ \mathbf{Tr}\,(X) &= 1 \\ X &\in \mathcal{C}_n^*. \end{aligned}\right\} \tag{12.2}$$

Proof:

Consider the convex cone:

$$\mathcal{C}_G := \{X \in \mathcal{C}_n^* \ : \ X_{ij} = 0,\ \{i,j\} \in E\}.$$

The extreme rays of this cone are of the form xx^T where $x \in \mathbb{R}^n$ is nonnegative and its support corresponds to a stable set of G. This follows from the fact that all extreme rays of $\mathcal{C}_n^*$ are of the form xx^T for nonnegative $x \in \mathbb{R}^n$ (see page 239). Therefore, the extreme points of the set defined by (12.2) are given by the intersection of the extreme rays with the hyperplane defined by $\mathbf{Tr}\,(X) = 1$.

Since the optimal value of problem (12.1) is attained at an extreme point, there is an optimal solution of the form:

$$X^* = x^* x^{*T}, \; x^* \in \mathbb{R}^n, \; x^* \geq 0, \; \|x^*\| = 1,$$

and where the support of x^* corresponds to a stable set, say S^*. Denoting the optimal value of problem (12.1) by λ, we therefore have

$$\lambda = \max_{\|x\|=1} (e^T x)^2, \; x \geq 0, \; \text{support}(x) = \text{support}(x^*).$$

The optimality conditions of this problem imply

$$x^* = \frac{1}{\sqrt{|S^*|}} x_{S^*},$$

and therefore

$$\lambda = (e^T x^*)^2 = \frac{|S^*|^2}{|S^*|} = |S^*|.$$

This shows that S^* must be the maximum stable set, and consequently $\lambda = \alpha(G)$. □

Note that – since $X \in \mathcal{C}_n^*$ is always nonnegative – we can simplify (12.1) and (12.2) to

$$\alpha(G) = \max\left\{\mathbf{Tr}\,\left(ee^T X\right) \; : \; \mathbf{Tr}\,(AX) = 0, \; \mathbf{Tr}\,(X) = 1, \; X \in \mathcal{C}_n^*\right\}, \qquad (12.3)$$

where A is the adjacency matrix of G. The dual problem of (12.3) is given by

$$\inf_{\lambda, y \in \mathbb{R}} \left\{\lambda \; : \; Q := \lambda I + yA - ee^T \in \mathcal{C}_n\right\}. \qquad (12.4)$$

The primal problem (12.3) is not strictly feasible (some entries of X must be zero), even though the dual problem (12.4) is strictly feasible (set $Q = (n+1)I - ee^T$)). By the conic duality theorem, we can therefore only conclude that the primal optimal set is nonempty and *not* that the dual optimal set is nonempty. We will prove now, however, that $Q = \alpha(I + A) - ee^T$ is always a dual optimal solution. This result follows from the next lemma.

Lemma 12.1 *For a given graph $G = (V, E)$ with adjacency matrix A and stability number $\alpha(G)$, and a given parameter $\epsilon \geq 0$, the matrix*

$$Q_\epsilon^* = (1 + \epsilon)\alpha(I + A) - ee^T$$

is copositive.

Proof:

Let $\epsilon \geq 0$ be given. We will show that Q^*_ϵ is copositive.

To this end, denote the standard simplex by

$$\Delta := \left\{ x \in \mathbf{R}^n \ : \ \sum_{i=1}^n x_i = 1, \ x \geq 0 \right\},$$

and note that:

$$\begin{aligned} \min_{x \in \Delta} x^T Q^*_\epsilon x &= \min_{x \in \Delta} (1+\epsilon)\alpha \left(x^T x + x^T A x \right) - x^T e e^T x \\ &= (1+\epsilon)\alpha \min_{x \in \Delta} \left(x^T x + x^T A x \right) - 1. \end{aligned}$$

We now show that the minimum is attained at $x^* = \frac{1}{|S^*|} x_{S^*}$ where S^* denotes the maximum stable set, as before. In other words, we will show that

$$\min_{x \in \Delta} x^T Q^*_\epsilon x = \epsilon. \tag{12.5}$$

Let $x^* \in \Delta$ be a minimizer of $x^T Q^*_\epsilon x$ over Δ.

If the support of x^* corresponds to a stable set, then the proof is an easy consequence of the inequality:

$$\text{argmax} \left\{ \|x\| \ : x \in \Delta, \ (e^T x)^2, \ x \geq 0, \ \text{support}(x) = S \right\} = \frac{1}{|S|} x_S \quad \forall S \subset V,$$

which can readily be verified via the optimality conditions.

Assume therefore that the support of x^* does not correspond to a stable set, *i.e.* $x^*_i > 0$ and $x^*_j > 0$ where $\{i, j\} \in E$.

Now we fix all the components of x to the corresponding values of x^* except components i and j. Note that, defining $c_0 := \sum_{k \neq i,j} x^*_k$, one can find constants c_1, c_2 and c_3 such that

$$\begin{aligned} {x^*}^T Q^*_\epsilon x^* &= \min_{x_i + x_j = 1 - c_0, x_i \geq 0, x_j \geq 0} (1+\epsilon)\alpha \left(x_i^2 + 2x_i x_j + x_j^2 \right) + \\ &\quad + x_i c_1 + x_j c_2 + c_3 \\ &= \min_{x_i + x_j = 1 - c_0, x_i \geq 0, x_j \geq 0} (1+\epsilon)\alpha \left(x_i + x_j \right)^2 + x_i c_1 + x_j c_2 + c_3 \\ &= \min_{x_i + x_j = 1 - c_0, x_i \geq 0, x_j \geq 0} (1+\epsilon)\alpha \left(1 - c_0 \right)^2 + x_i c_1 + x_j c_2 + c_3. \end{aligned}$$

The final optimization problem is simply an LP in the two variables x_i and x_j and attains its minimal value in an extremal point where $x_i = 0$ or $x_j = 0$. We can therefore replace x^* with a vector $\bar{x}$ such that ${x^*}^T Q x^* = \bar{x}^T Q \bar{x}$ and $\bar{x}_i \bar{x}_j = 0$.

By repeating this process we obtain a minimizer of $x^T Q_\epsilon^* x$ over Δ with support corresponding to a stable set. □

The lemma shows that Q_ϵ^* is copositive and therefore ϵ-optimal in (12.4). For $\epsilon = 0$ we have the following result.

Corollary 12.1 *For any graph $G = (V, E)$ with adjacency matrix A one has*

$$\alpha(G) = \min_{\lambda} \left\{ \lambda \, : \, \lambda(I + A) - ee^T \in \mathcal{C}_n \right\}.$$

Remark 12.1 *The result of Corolary 12.1 is also a consequence of a result by Motzkin and Straus [129], who proved that*

$$\frac{1}{\alpha(G)} = \min_{x \in \Delta} x^T (A + I) x,$$

where A is the adjacency matrix of G. To see the relationship between the two results we also need the known result (see e.g. *Bomze* et al. *[28]) that minimization of a quadratic function over the simplex is equivalent to a copositive programming problem:*

$$\begin{aligned} \min_{x \in \Delta} x^T Q x &= \min_{X \in (\mathcal{C}_n)^*} \left\{ Tr(QX) \, : \, Tr\left(ee^T X\right) = 1 \right\} \\ &= \max_{\lambda \in \mathbb{R}} \left\{ \lambda \, : \, Q - \lambda ee^T \in \mathcal{C}_n \right\} \end{aligned}$$

for any $Q \in \mathcal{S}_n$, where the second inequality follows from the strong duality theorem.

Corollary 12.1 implies that we can simplify our conic programs even further to obtain:

$$\alpha(G) = \max \left\{ \mathbf{Tr}\left(ee^T X\right) \, : \, \mathbf{Tr}\left((A + I)X\right) = 1, \; X \in \mathcal{C}_n^* \right\}, \tag{12.6}$$

with associated dual problem:

$$\alpha(G) = \min_{\lambda \in \mathbb{R}} \left\{ \lambda \, : \, Q := \lambda(I + A) - ee^T \in \mathcal{C}_n \right\}. \tag{12.7}$$

Note that both these problems are strictly feasible and the conic duality theorem now guarantees complementary primal-dual optimal solutions.

12.3 APPROXIMATIONS OF THE COPOSITIVE CONE

The reformulation of the stable set problem as a conic copositive program makes it clear that copositive programming is not tractable (see also Quist *et al.* [152]). In fact, even the problem of determining whether a matrix is not copositive is NP-complete (see Murty and Kabadi [132]).

Although we have obtained a nice convex reformulation of the stable set problem, there is no obvious way of solving this reformulation.[2] A solution to this problem was recently proposed by Parillo [141], who showed that one can approximate the copositive cone to any given accuracy by a sufficiently large set of linear matrix inequalities. In other words, each copositive programming problem can be approximated to any given accuracy by a sufficiently large SDP. Of course, the size of the SDP can be exponential in the size of the copositive program. In the next subsection we will review the approach of Parillo, and subsequently work out the implications for the copositive formulation of the maximum stable set problem. We will also look at a weaker, LP-based approximation scheme.

REPRESENTATIONS AS SUM-OF-SQUARES AND POLYNOMIALS WITH NONNEGATIVE COEFFICIENTS

We can represent the copositivity requirement for an $(n \times n)$ symmetric matrix M as

$$P_M(x) := (x \circ x)^T M (x \circ x) = \sum_{i,j=1}^{n} M_{ij} x_i^2 x_j^2 \geq 0, \ \forall x \in \mathbb{R}^n, \tag{12.8}$$

where '$\circ$' indicates the componentwise (Hadamard) product. We therefore wish to know whether the polynomial $P_M(x)$ is nonnegative for all $x \in \mathbb{R}^n$. Although one cannot answer this question in polynomial time in general, one can decide in polynomial time whether $P_M(x)$ can be written as a sum of squares. Before we give a formal exposition of the methodology, we give an example which illustrates the basic idea.

Example 12.2 (Parillo [141]) *We show how to obtain a sum of squares decomposition for the polynomial* $h(x) := 2x_1^4 + 2x_1^3x_2 - x_1^2x_2^2 + 5x_2^4$*:*

$$\begin{aligned} h(x) &= 2x_1^4 + 2x_1^3x_2 - x_1^2x_2^2 + 5x_2^4 \\ &= \begin{bmatrix} x_1^2 \\ x_2^2 \\ x_1x_2 \end{bmatrix}^T \begin{bmatrix} 2 & 0 & 1 \\ 0 & 5 & 0 \\ 1 & 0 & -1 \end{bmatrix} \begin{bmatrix} x_1^2 \\ x_2^2 \\ x_1x_2 \end{bmatrix} \end{aligned}$$

[2] In Bomze *et al.* [28], some ideas from interior point methods for semidefinite programming are adapted for the copositive case, but convergence cannot be proved. The absence of a computable self-concordant barrier for this cone basically precludes the application of interior point methods to copositive programming.

$$= \begin{bmatrix} x_1^2 \\ x_2^2 \\ x_1x_2 \end{bmatrix}^T \begin{bmatrix} 2 & -\lambda & 1 \\ -\lambda & 5 & 0 \\ 1 & 0 & -1+2\lambda \end{bmatrix} \begin{bmatrix} x_1^2 \\ x_2^2 \\ x_1x_2 \end{bmatrix} \quad \forall \lambda \in \mathbf{R}.$$

For $\lambda = 3$ the coefficient matrix is positive semidefinite. Denoting the coefficient matrix by $\tilde{M}$, we have

$$\tilde{M} = L^T L, \quad L = \frac{1}{\sqrt{2}} \begin{bmatrix} 2 & -3 & 1 \\ 0 & 1 & 3 \end{bmatrix},$$

and consequently

$$2h(x) = \left(2x_1^2 - 3x_2^2 + x_1x_2\right)^2 + \left(x_2^2 + 3x_1x_2\right)^2.$$

Thus we have obtained a sum of squares decomposition for $h(x)$. □

Following the idea in the example, we represent $P_M(x)$ via

$$P_M(x) = \tilde{x}^T \tilde{M} \tilde{x} \tag{12.9}$$

where $\tilde{x} = [x_1^2, \ldots, x_n^2, x_1x_2, x_1x_3, \ldots, x_{n-1}x_n]^T$, and $\tilde{M}$ is a symmetric matrix of order $n + \frac{1}{2}n(n-1)$.

Note that — as in the example — $\tilde{M}$ is not uniquely determined. The non-uniqueness follows from the identities:

$$\begin{aligned} (x_ix_j)^2 &= (x_i^2)(x_j)^2 \\ (x_ix_j)(x_ix_k) &= (x_i^2)(x_jx_k) \\ (x_ix_j)(x_kx_l) &= (x_ix_k)(x_jx_l) = (x_ix_l)(x_jx_k). \end{aligned}$$

It is easy to see that the possible choices for $\tilde{M}$ define an affine space.

Condition (12.8) will certainly hold if at least one of the following two conditions holds:

1. A representation of $P_M(x) = \tilde{x}^T \tilde{M} \tilde{x}$ exists with $\tilde{M}$ symmetric positive semidefinite. In this case we obtain the sum of squares decomposition $P_M(x) = \|L\tilde{x}\|^2$ where $L^T L = \tilde{M}$ denotes the Choleski factorization of $\tilde{M}$;

2. All the coefficients of $P_M(x)$ are nonnegative.

Note that the second condition implies the first.

Parillo showed that $P_M(x)$ in (12.8) allows a sum of squares decomposition if and only if $M \in \mathcal{S}_n^+ + \mathcal{N}_n$, which is a well-known sufficient condition for copositivity. Let us define the cone $\mathcal{K}_n^0 := \mathcal{S}_n^+ + \mathcal{N}_n$. Similarly, $P_M(x)$ has only nonnegative coefficients if and only if $M \in \mathcal{N}_n$ which is a weaker sufficient condition for copositivity, and we define $\mathcal{C}_n^0 = \mathcal{N}_n$.

Higher order sufficient conditions can be derived by considering the polynomial

$$P_M^{(r)}(x) = P_M(x)\left(\sum_{i=1}^{n} x_i^2\right)^r = \left(\sum_{i,j=1}^{n} M_{ij}x_i^2x_j^2\right)\left(\sum_{i=1}^{n} x_i^2\right)^r, \tag{12.10}$$

and asking whether $P_M^{(r)}(x)$ — which is a homogeneous polynomial of degree $2(r+2)$ — has a sum of squares decomposition, or whether it only has nonnegative coefficients.

For $r = 1$, Parillo showed that a sum of squares decomposition exists if and only if[3] the following system of linear matrix inequalities has a solution:

$$\begin{aligned} M - M^{(i)} &\in \mathcal{S}_n^+, \quad i = 1, \ldots, n & (12.11)\\ M_{ii}^{(i)} &= 0, \quad i = 1, \ldots, n & (12.12)\\ M_{jj}^{(i)} + 2M_{ij}^{(j)} &= 0, \quad i \neq j & (12.13)\\ M_{jk}^{(i)} + M_{ik}^{(j)} + M_{ij}^{(k)} &\geq 0, \quad i < j < k, & (12.14) \end{aligned}$$

where $M^{(i)}$ $(i = 1, \ldots, n)$ are symmetric matrices. Bomze and De Klerk [27] have shown that this system of LMI's is the same as

$$\begin{aligned} M - M^{(i)} &\in \mathcal{S}_n^+ + \mathcal{N}_n, \quad i = 1, \ldots, n\\ M_{ii}^{(i)} &= 0, \quad i = 1, \ldots, n\\ M_{jj}^{(i)} + 2M_{ij}^{(j)} &= 0, \quad i \neq j\\ M_{jk}^{(i)} + M_{ik}^{(j)} + M_{ij}^{(k)} &\geq 0, \quad i < j < k. \end{aligned}$$

Similarly, $P_M^{(1)}(x)$ has only nonnegative coefficients if M satisfies the above system, but with $\mathcal{S}_n^+ + \mathcal{N}_n$ replaced by $\mathcal{N}_n$. (This shows that $\mathcal{K}_n^0 \subset \mathcal{C}_n^0$, as it should be.)

Note that the sets of matrices which satisfy these respective sufficient conditions for copositivity define two respective convex cones. In fact, this is generally true for all r.

Definition 12.1 ($\mathcal{C}_n^r$ and $\mathcal{K}_n^r$) *Let any integer $r \geq 0$ be given. The convex cone $\mathcal{K}_n^r$ consists of the matrices $M \in \mathcal{S}_n$ for which $P_M^{(r)}(x)$ in (12.10) has a sum of squares*

[3] In fact, Parillo [141] only proved the 'if'-part; the proof of the converse is given in Bomze and De Klerk [27].

decomposition; similarly we define the cone $\mathcal{C}_n^r$ *as the cone of matrices* $M \in \mathcal{S}_n$ *for which* $P_M^{(r)}(x)$ *in (12.10) has only nonnegative coefficients.*

Note that $\mathcal{C}_n^r \subset \mathcal{K}_n^r$ for all $r = 0, 1, \ldots$. (If $P_M(x)$ has only nonnegative coefficients, then it obviously has a sum of squares decomposition. The converse is not true in general.)

The systems of linear (in)equalities (resp. LMI's) which define $\mathcal{C}_n^r$ (resp. $\mathcal{K}_n^r$) are given in Bomze and De Klerk [27] for $r > 1$.

UPPER BOUNDS ON THE ORDER OF APPROXIMATION

Every strictly copositive M lies in some cone $\mathcal{C}_n^r$ for r sufficiently large; this follows from the celebrated theorem of Pólya.

Theorem 12.2 (Pólya [146]) *Let f be a homogeneous polynomial that is positive on the simplex*

$$\Delta = \left\{ z \in \mathbf{R}^n \;\middle|\; \sum_{i=1}^{n} z_i = 1,\ z \geq 0 \right\}.$$

For sufficiently large N all the coefficients of the polynomial

$$\left(\sum_{i=1}^{n} z_i \right)^N f(z)$$

are positive.

One can apply this theorem to the copositivity test (12.8) by letting $f(z) = z^T M z$ and associating $x \circ x$ with z.

To summarize, we have the following theorem.

Theorem 12.3 *Let M be strictly copositive. One has*

$$\mathcal{N}_n = \mathcal{C}_n^0 \subset \mathcal{C}_n^1 \subset \ldots \subset \mathcal{C}_n^N \ni M$$

and consequently

$$\mathcal{S}_n^+ + \mathcal{N}_n = \mathcal{K}_n^0 \subset \mathcal{K}_n^1 \subset \ldots \subset \mathcal{K}_n^N \ni M$$

for some sufficiently large N.

We can derive an upper bound on N by calculating the coefficients $P_M^{(r)}(x)$ explicitly. This is done in the next theorem.

Theorem 12.4 (Bomze and De Klerk [27]) *Let $M \in \mathcal{S}_n$ be given. One has $M \in \mathcal{C}_n^r$ if*

$$m^T M m - m^T \operatorname{diag} M \geq 0 \quad \forall m \in I_n(r+2),$$

where

$$I_n(r) := \left\{ m \in \mathbb{Z}_+^n \;\middle|\; \sum_{i=1}^n m_i = r \right\}. \tag{12.15}$$

Proof:

We only sketch the proof here; the interested reader is referred to Bomze and De Klerk [27] for more details.

By definition, we have $M \in \mathcal{C}_n^r$ if and only if the coefficients of

$$P_M^{(r)}(x) = \left(M_{ij} x_i^2 x_j^2\right) \left(\sum_{i=1}^n x_i^2\right)^r$$

are nonnegative. Using the multinomial theorem,[4] we can rewrite $P_M^{(r)}(x)$ as

$$P_M^{(r)}(x) = \sum_{m \in I_n(r+2)} a_m x_1^{2m_1} \cdots x_n^{2m_n},$$

where the coefficient a_m of the monomial $x_1^{2m_1} \cdots x_n^{2m_n}$ is given by

$$a_m = \frac{c(m)}{(r+2)(r+1)} \left[m^T M m - m^T \operatorname{diag} M\right], \tag{12.16}$$

where $c(m)$ is the multinomial coefficient of m, namely

$$c(m) := \frac{(\sum_i m_i)!}{m_1! \cdots m_n!}. \tag{12.17}$$

Since, $c(m) > 0$, the required result follows from (12.16). □

Corollary 12.2 *If $M \in \mathcal{S}_n$ is strictly copositive, then $M \in \mathcal{C}_n^r \subset \mathcal{K}_n^r$ if $r \geq L/\kappa - 2$, where*

$$L = \max_i (M)_{ii}, \tag{12.18}$$

---

[4]The multinomial theorem states that

$$(z_1 + \ldots + z_n)^r = \sum_{m \in I_n(r)} c(m) z_1^{m_1} \cdots z_n^{m_n},$$

where $I_n(r)$ and $c(m)$ are defined in (12.15) and (12.17) respectively.

and

$$\kappa = \min_{z \in \Delta} z^T M z. \tag{12.19}$$

Proof:

As before, we will have $M \in \mathcal{C}_n^r$ if and only if all the coefficients (again denoted by a_m) of the polynomial

$$P_M^r(x) = \left(\sum_{i,j=1}^{n} M_{ij} x_i^2 x_j^2 \right) \left(\sum_{i=1}^{n} x_i^2 \right)^r$$

are nonnegative. By Theorem 12.4 we have

$$\begin{aligned}
\min_{m \in I_n(r+2)} a_m &= \min_{m \in I_n(r+2)} \left(m^T M m - m^T \operatorname{diag} M \right) \\
&\geq \min_{m \in I_n(r+2)} m^T M m - \max_{m \in I_n(r+2)} m^T \operatorname{diag} M \\
&= \min_{m \in I_n(r+2)} m^T M m - (r+2)L \\
&\geq (r+2)^2 \min_{z \in \Delta} z^T M z - (r+2)L \\
&= (r+2)^2 \left(\kappa - \frac{L}{r+2} \right).
\end{aligned}$$

In other words, all coefficients of $P_M^r(x)$ will be nonnegative if

$$\kappa - \frac{L}{r+2} \geq 0.$$

The required result follows. □

Note that κ can be arbitrarily small, and cannot be computed in polynomial time in general; indeed, it follows from Remark 12.1 that one cannot minimize a quadratic function over the simplex in polynomial time (see also Section 12.7).

Remark 12.2 *A tight upper bound on the size of N in Theorem 12.2 has recently been given by Powers and Reznik [150] (for general homogeneous polynomials positive on the simplex). Their general bound can be used to derive a slightly weaker result than stated in Corollary 12.2.*

12.4 APPLICATION TO THE MAXIMUM STABLE SET PROBLEM

We can now define successive approximations to the stability number. In particular, we define successive LP-based approximations via

$$\zeta^{(r)}(G) = \min_{\lambda} \left\{ \lambda \;:\; Q = \lambda(I + A) - ee^T \in \mathcal{C}_n^r \right\}, \tag{12.20}$$

for $r = 0, 1, 2, \ldots$, where we use the convention that $\zeta^{(r)}(G) = \infty$ if the problem is infeasible.

Similarly, we define successive SDP-based approximations via:

$$\vartheta^{(r)}(G) = \min_{\lambda} \left\{ \lambda \; : \; Q = \lambda(I + A) - ee^T \in \mathcal{K}_n^r \right\}, \tag{12.21}$$

for $r = 0, 1, 2, \ldots$ Note that we have merely replaced the copositive cone $\mathcal{C}_n$ in (12.7) by its respective approximations $\mathcal{C}_n^r$ and $\mathcal{K}_n^r$. We will refer to r as the *order of approximation.*

The minimum in (12.21) is always attained. The proof follows directly from the conic duality theorem, if we note that $\lambda = n + 1$ always defines a matrix Q in the interior of $\mathcal{K}_n^0$ (and therefore in the interior of $\mathcal{K}_n^r \supset \mathcal{K}_n^0$ for all $r = 1, 2, \ldots$) via (12.21), and that

$$X^0 := \frac{1}{n^2 + n + |E|}(nI + ee^T)$$

is always strictly feasible in the associated primal problem:

$$\vartheta^{(r)}(G) = \max \left\{ \mathbf{Tr}\left(ee^T X\right) \; : \; \mathbf{Tr}\left((A + I)X\right) = 1, \; X \in (\mathcal{K}_n^r)^* \right\}.$$

The strict feasibility of X^0 follows from the fact that it is in the interior of $\mathcal{C}_n^*$: for any copositive matrix $Y \in \mathcal{C}_n$ we have

$$\mathbf{Tr}\left(X^0 Y\right) = \frac{1}{n^2 + n + |E|} \left(n\mathbf{Tr}\left(Y\right) + e^T Y e\right).$$

This expression can only be zero if Y is the zero matrix. In other words, $\mathbf{Tr}\left(X^0 Y\right) > 0$ for all nonzero $Y \in \mathcal{C}_n$, which means that X^0 is in the interior of $\mathcal{C}_n^*$. Consequently X^0 is also in the interior of $(\mathcal{K}_n^r)^*$ for all r, since $\mathcal{C}_n^* \subset (\mathcal{K}_n^r)^*$ $(r = 0, 1, \ldots)$.

Note that

$$\alpha(G) \leq \vartheta^{(r)}(G) \leq \zeta^{(r)}(G), \quad r = 0, 1, \ldots$$

since $\mathcal{C}_n^r \subset \mathcal{K}_n^r \subset \mathcal{C}_n$.

AN UPPER BOUND ON THE ORDER OF APPROXIMATION

We can now prove the following.

Theorem 12.5 (De Klerk and Pasechnik [41]) *Let a graph $G = (V, E)$ be given with stability number $\alpha(G)$, and let $\zeta^{(i)}$ $(i = 0, 1, 2, \ldots)$ be defined as in (12.20).*

One has:

$$\zeta^{(0)} \geq \zeta^{(1)} \geq \ldots \geq \lfloor \zeta^{(r)} \rfloor = \alpha(G),$$

for $r \geq \alpha(G)^2$. Consequently, also $\lfloor \vartheta^{(r)} \rfloor = \alpha(G)$ for $r \geq \alpha(G)^2$.

Proof:

Denote, as in the proof of Lemma 12.1,

$$Q_\epsilon^* = (1+\epsilon)\alpha(G)(I+A) - ee^T,$$

for a given $\epsilon \geq 0$.

We will now prove that $Q_\epsilon^* \in \mathcal{C}_n^r$ for $r \geq \alpha(G)^2 - \alpha(G) - 2$ if

$$\epsilon := \frac{1}{\alpha(G) + 1/[\alpha(G)-1]}. \tag{12.22}$$

Note that if we choose ϵ in this way, then Q_ϵ^* corresponds to a feasible solution of (12.20) where $\lambda = (1+\epsilon)\alpha(G) < 1 + \alpha(G)$, and we can therefore round this value of λ down to obtain $\alpha(G)$.

We proceed to bound the parameters κ and L in Corollary 12.2 for the matrix Q_ϵ^*.

- The value L is given by $L = (1+\epsilon)\alpha(G) - 1$.
- The condition number κ is given by $\kappa = \epsilon$, by (12.5).

Now we have

$$L/\kappa = \frac{(1+\epsilon)\alpha(G) - 1}{\epsilon} = \alpha(G)^2 + 1. \tag{12.23}$$

From Corollary 12.2 now follows that $Q_\epsilon^* \in \mathcal{C}_n^r$ for $r \geq \alpha(G)^2$. □

Example 12.3 *Consider the case where $G = (V, E)$ is the graph of the pentagon. It is well known that $\alpha(G) = 2$ and $\vartheta(G) = \vartheta'(G) = \sqrt{5}$ in this case (see Example 10.1).*

We will show that $\vartheta^{(1)}(G) = 2$; to this end, note that the matrix

$$Q = \begin{pmatrix} 1 & -1 & 1 & 1 & -1 \\ -1 & 1 & -1 & 1 & 1 \\ 1 & -1 & 1 & -1 & 1 \\ 1 & 1 & -1 & 1 & -1 \\ -1 & 1 & 1 & -1 & 1 \end{pmatrix} \tag{12.24}$$

corresponds to a feasible solution of (12.21) for $r = 1$, with $\lambda = 2$. The feasibility follows from the known fact that Q in (12.24) is in $\mathcal{K}_n^1$ (but not in $\mathcal{K}_n^0 = \mathcal{S}_n^+ + \mathcal{N}_n$). To verify this, we have to show that

$$(x \circ x)^T Q (x \circ x) \left(x_1^2 + \ldots + x_4^2\right)$$

has a sum of squares decomposition. Denoting $z = x \circ x$, one can show

$$\begin{aligned}
z^T Q z \left(\sum_{i=1}^{4} z_i \right) &= z_1(z_1 - z_2 + z_3 + z_4 - z_5)^2 \\
&+ z_2(z_2 - z_3 + z_4 + z_5 - z_1)^2 \\
&+ z_3(z_3 - z_4 + z_5 + z_1 - z_2)^2 \\
&+ z_4(z_4 - z_5 + z_1 + z_2 - z_3)^2 \\
&+ z_5(z_5 - z_1 + z_2 + z_3 - z_4)^2 \\
&+ 4(z_1 z_2 z_4 + z_2 z_3 z_5 + z_3 z_4 z_1 + z_4 z_5 z_2 + z_5 z_1 z_3).
\end{aligned}$$

By using $z_i = x_i^2$ $(i = 1, \ldots, 4)$ we obtain the required sum of squares decomposition. □

Example 12.4 *Let $G = (V, E)$ be the complement of the graph of an icosahedron (see Figure 12.2). In this case $n = 12$ and $\alpha(G) = 3$. One can solve the relevant SDP*

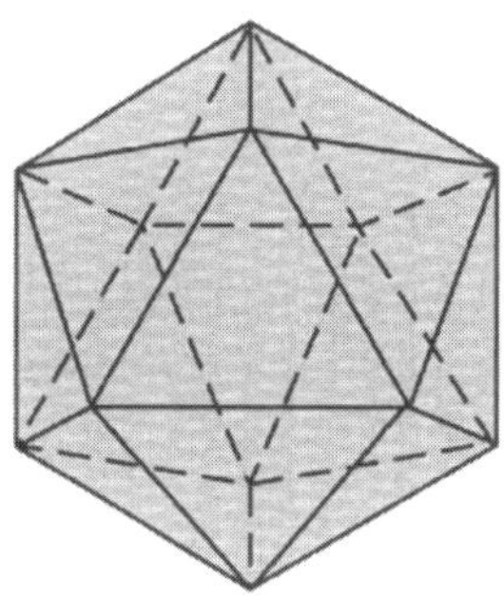

Figure 12.2. An icosahedron is a regular polyhedron with 12 vertices and 20 faces each of which is an equilateral triangle.

problem to obtain: $\vartheta^{(1)}(G) = 1 + \sqrt{5} \approx 3.236068$.

Although $\lfloor \vartheta^{(1)}(G) \rfloor = \alpha(G)$, one has $Q := \alpha(G)(A + I) - ee^T \notin \mathcal{K}_n^1$. Thus Q gives an example of a 12×12 copositive matrix which is not in $\mathcal{K}_n^1$. □

12.5 THE STRENGTH OF LOW-ORDER APPROXIMATIONS

In this section we investigate the strength of the approximations $\vartheta^{(r)}$ to $\alpha(G)$ for $r = 0$ and $r = 1$.

First we show that $\vartheta^{(0)}$ coincides with the ϑ'-function of Schrijver [163], which in turn can be seen as a strengthening of Lovász ϑ-approximation to $\alpha(G)$. To be precise,

$\vartheta'(\bar{G})$ is obtained by adding a nonnegativity restriction to X in the formulation of $\vartheta(\bar{G})$ (see (10.2) on page 158).

Lemma 12.2 *Let a graph $G = (V, E)$ be given with adjacency matrix A and let ϑ' denote the Schrijver ϑ'-function [163]:*

$$\vartheta'(G) = \max\left\{\mathbf{Tr}\left(ee^TX\right) \;:\; \mathbf{Tr}\,(AX) = 0,\; \mathbf{Tr}\,(X) = 1,\; X \in (\mathcal{K}_n^0)^*\right\}.$$

Then

$$\vartheta'(G) = \vartheta^{(0)}(G).$$

Proof:

Recall that

$$\vartheta^{(0)}(G) = \min_{\lambda}\left\{\lambda \;:\; \lambda(I + A) - ee^T \in \mathcal{K}_n^0\right\}, \tag{12.25}$$

whereas the dual formulation for $\vartheta'(G)$ is:

$$\vartheta'(G) = \min_{\lambda, y}\left\{\lambda \;:\; \lambda I + yA - ee^T \in \mathcal{K}_n^0\right\}. \tag{12.26}$$

Further recall that $\mathcal{K}_n^0 = \mathcal{S}_n^+ + \mathcal{N}_n$ and let

$$\lambda I + yA - ee^T = S + N, \text{ where } S \in \mathcal{S}_n^+ \text{ and } N \in \mathcal{N}_n. \tag{12.27}$$

Without loss of generality we assume $N_{ii} = 0$ for all $i \in \{1, \ldots, n\}$, as the sum of two positive semidefinite matrices is positive semidefinite and thus the diagonal part of N can be added to S and subtracted from N.

Assume $A_{ij} \neq 0$. Note that our choice of S and N is such that $S_{ii} = \lambda - 1$. Thus, as $S_{ij} + N_{ij} = y - 1$ and $S_{ii} \geq S_{ij}$,[5] one obtains $\lambda - 1 + N_{ij} \geq y - 1$, so $N_{ij} \geq y - \lambda$. Hence $N + (\lambda - y)A \in \mathcal{N}_n$. Therefore $\lambda(I + A) - ee^T \in \mathcal{K}_n^0$ as long as (12.27) holds. Hence we can always assume that $y = \lambda$. □

Let us restate the definition of $\vartheta^{(1)}(G)$ by using (12.11)–(12.14) as follows.

$$\begin{aligned}
\vartheta^{(1)}(G) &:= \min \beta & &\text{subject to} & (12.28)\\
\beta(I + A) - ee^T - M^{(i)} &\in \mathcal{S}_n^+, & &i = 1, \ldots, n & (12.29)\\
M_{ii}^{(i)} &= 0, & &i = 1, \ldots, n & (12.30)\\
M_{jj}^{(i)} + 2M_{ij}^{(j)} &= 0, & &i \neq j & (12.31)\\
M_{jk}^{(i)} + M_{ik}^{(j)} + M_{ij}^{(k)} &\geq 0, & &i < j < k, & (12.32)
\end{aligned}$$

where $M^{(i)}$ $(i = 1, \ldots, n)$ are symmetric matrices.

[5] Here we use the fact that $S \in \mathcal{S}_n^+$ and has a constant diagonal.

For $v \in V$, denote by $v^{\perp}$ the union of the neighbourhood[6] of v with v itself, and for $D \subseteq V$ denote by $G(D)$ the subgraph of G induced on D (that is $G(D) = (D, \{(x,y) \in E \mid x, y \in D\})$.

Using the above notation, it is possible to show the following.

Theorem 12.6 (De Klerk and Pasechnik [41]) *The system of LMi's* (12.28)–(12.32) *has a feasible solution with* $\beta = 1 + \max_{k \in V}(\vartheta'(G(V - k^{\perp})))$ *and* $M_{ij}^{(i)} = 0$ *for all* i, j. *Thus*

$$\vartheta^{(1)}(G) \leq 1 + \max_{k \in V}(\vartheta'(G(V - k^{\perp}))).$$

In particular, if $G(V - k^{\perp})$ *is perfect for all* $k \in V$ *where* $k^{\perp} \neq V$, *then* $\vartheta^{(1)}(G) = \beta = \alpha(G)$.

Thus, for instance, the pentagon example of the previous section can be generalized to all cycles.

Corollary 12.3 *Let* $G = (V, E)$ *be a cycle of length* n. *One has* $\vartheta^{(1)}(G) = \alpha(G)$. *Similarly* $\alpha(G) = \vartheta^{(1)}(G)$ *if G is a wheel.*

Proof:

Let $G = (V, E)$ be a cycle of length n. The required result now immediately follows from Theorem 12.6 by observing that $G(V - v^{\perp})$ is an (n-3)-path for all $v \in V$. The proof for wheels is similar. □

Also, complements of triangle-free graphs are recognized.

Corollary 12.4 *If* $G = (V, E)$ *has stability number* $\alpha(G) = 2$, *then* $\vartheta^{(1)}(G) = 2$.

Proof:

Immediately follows from Theorem 12.6 by observing that $G(V - v^{\perp})$ is a clique (or the empty graph) for all $v \in V$. □

As a consequence the complements of cycles or wheels are also recognized. The proof proceeds in the same way as before and is therefore omitted.

Corollary 12.5 *Let* $G = (V, E)$ *be the complement of a cycle or of a wheel. In both cases one has* $\vartheta^{(1)}(G) = \alpha(G)$.

[6]By neighbourhood of v we mean the set of vertices adjacent to v in G.

We conjecture that the result of Corollary 12.4 can be extended to include all values of α.

Conjecture 12.1 (De Klerk and Pasechnik [41]) *If $G = (V, E)$ has stability number $\alpha(G)$, then $\vartheta^{(\alpha(G)-1)}(G) = \alpha(G)$.*

12.6 RELATED LITERATURE

There have been several – seemingly different – so-called lift-and-project strategies for approximating combinatorial optimization problems. The term 'lifting' refers to the fact that one works in a higher dimensional space, and 'project' refers to a projection back to the original space.

Sherali and Adams [164] first showed how to describe the stable set polytope as the projection of a polytope in a higher dimension. Lovász and Schrijver [116] also showed how to obtain an exact description of the stable set polytope via LP or SDP. They provided upper bounds on the number of liftings (the order of approximation). As such, the results that have been presented in this chapter are similar to their results, but are derived from a different perspective.

Anjos and Wolkowicz [11] have introduced a technique of successive Lagrangian relaxations for the MAX-CUT problem, which also leads to SDP relaxations of size $(n^r \times n^r)$ after $r - 1$ steps. Recently, Laserre [111, 110] has introduced yet another lift-and-project approach, based on the properties of moment matrices. Most recently, Laurent [113, 112] has investigated the relationships between these approaches.

12.7 EXTENSIONS: STANDARD QUADRATIC OPTIMIZATION

We have already mentioned in Remark 12.1 that the maximum stable set problem is a special case of a so-called *standard quadratic optimization* problem.[7] The standard quadratic optimization problem (standard QP) is to find the global minimizers of a quadratic form over the standard simplex, *i.e.* we consider the global optimization problem

$$\underline{p} := \min_{x \in \Delta} x^T Q x \tag{12.33}$$

where Q is an arbitrary symmetric $n \times n$ matrix, and Δ is the standard simplex in $\mathbb{R}^n$,

$$\Delta = \left\{ x \in \mathbb{R}^n_+ : e^T x = 1 \right\},$$

as before. We assume that the objective is not constant over Δ, which means that Q and ee^T are linearly independent.

Note that the minimizers of (12.33) remain the same if Q is replaced with $Q + \gamma ee^T$ where γ is an arbitrary constant. So without loss of generality assume henceforth that

[7] For a review on standard quadratic optimization and its applications, see Bomze [25].

all entries of Q are non-negative. Furthermore, we can minimize the general quadratic function $x^TAx + 2c^Tx$ over Δ by setting $Q = A + ec^T + ce^T$ in (12.33).

Recall from Remark 12.1 that problem (12.33) can be rewritten as the copositive programming problem:

$$\underline{p} = \min\left\{\mathbf{Tr}\, QX \;:\; \mathbf{Tr}\, ee^TX = 1,\; X \in \mathcal{C}_n^*\right\}$$

with associated dual problem

$$\underline{p} = \max\left\{\lambda \;:\; Q - \lambda ee^T \in \mathcal{C}_n,\; \lambda \in \mathbf{R}\right\}. \tag{12.34}$$

LP-BASED APPROXIMATIONS

Let us define:

$$p_{\mathcal{C}}^{(r)} = \min\left\{\mathbf{Tr}\, QX \;:\; \mathbf{Tr}\, ee^TX = 1,\; X \in (\mathcal{C}_n^r)^*\right\}, \tag{12.35}$$

for $r = 0, 1, \ldots$ which has dual formulation

$$p_{\mathcal{C}}^{(r)} = \max\left\{\lambda \;:\; Q - \lambda ee^T \in \mathcal{C}_n^r,\; \lambda \in \mathbf{R}\right\}. \tag{12.36}$$

Note that problem (12.36) is a relaxation of problem (12.34) where the copositive cone is approximated by $\mathcal{C}_n^r$. It therefore follows that $p_{\mathcal{C}}^{(r)} \le \underline{p}$ for all r. We now provide an alternative representation of $p_{\mathcal{C}}^{(r)}$. This representation uses the following rational grid which approximates the standard simplex:

$$\Delta(r) := \{y \in \Delta : (r+2)y_i \in \{0, 1, \ldots, n\} \quad (i = 1, \ldots, n)\}\,. \tag{12.37}$$

A naive approximation of problem (12.33) would be

$$p_{\Delta(r)} := \min\{y^TQy : y \in \Delta(r)\} \ge \underline{p}. \tag{12.38}$$

The next theorem shows that there is a close link between $p_{\mathcal{C}}^{(r)}$ and the naive approximation $p_{\Delta(r)}$. In particular, one can obtain $p_{\mathcal{C}}^{(r)}$ in a similar way as the naive approximation $p_{\Delta(r)}$ is obtained, *i.e.* by only doing function evaluations at points on the grid $\Delta(r)$.

Theorem 12.7 (Bomze and De Klerk [27]) *For any given integer $r > 0$, one has*

$$p_{\mathcal{C}}^{(r)} = \tfrac{r+2}{r+1} \min_{y \in \Delta(r)} \left(y^TQy - \tfrac{1}{r+2}\mathrm{diag}\, Q^Ty\right). \tag{12.39}$$

Proof:

The proof is an easy consequence of Theorem 12.4. Substituting $M = Q - \lambda ee^T$ in (12.16), it follows from (12.36) that $p_{\mathcal{C}}^{(r)}$ is the largest value of λ such that

$$\frac{c(m)}{(r+2)(r+1)}\left[m^T Q m - m^T \operatorname{diag} Q - \lambda\left((r+2)^2 - (r+2)\right)\right] \geq 0 \qquad (12.40)$$

for all $m \in I_n(r+2)$, where

$$c(m) := \frac{(\sum_i m_i)!}{m_1! \cdots m_n!},$$

and

$$I_n(r) := \left\{ m \in \mathbb{Z}_+^n \;\middle|\; \sum_{i=1}^n m_i = r \right\},$$

as before. In other words,

$$(r+1)(r+2)p_{\mathcal{C}}^{(r)} = \min_{m \in I_n(r+2)m} m^T Q m - m^T \operatorname{diag} Q. \qquad (12.41)$$

The required result now follows by setting $y = \frac{1}{r+2}m$ in (12.41). □

Given the result in Theorem (12.7), it is straightforward to derive the following approximation guarantee.

Theorem 12.8 (Bomze and De Klerk [27]) *Let* $\overline{p} := \max_{x \in \Delta} x^T Q x$. *One has*

$$\underline{p} - p_{\mathcal{C}}^{(r)} \leq \frac{1}{r+1}\left(\overline{p} - \underline{p}\right)$$

as well as

$$p_{\Delta(r)} - \underline{p} \leq \frac{1}{r+2}\left(\overline{p} - \underline{p}\right).$$

Proof:

By Theorem 12.7 we have

$$\begin{aligned}
p_{\mathcal{C}}^{(r)} &= \tfrac{r+2}{r+1} \min\left\{ y^T Q y - \tfrac{1}{r+2} \operatorname{diag} Q^T y : y \in \Delta(r) \right\} \\
&\geq \tfrac{r+2}{r+1}\left(\underline{p} - \max_{y \in \Delta(r)} \tfrac{1}{r+2}\left(\operatorname{diag} Q\right)^T y\right) \\
&= \tfrac{r+2}{r+1}\left(\underline{p} - \tfrac{1}{r+2} \max_i Q_{ii}\right) \\
&\geq \tfrac{r+2}{r+1}\left(\underline{p} - \tfrac{1}{r+2}\overline{p}\right) \\
&= \underline{p} + \tfrac{1}{r+1}\left(\underline{p} - \overline{p}\right).
\end{aligned}$$

The first result follows. The second relation is derived in the same way: by Theorem 12.7 we have

$$\begin{aligned}\min_{y\in\Delta(r)} y^TQy &\le \tfrac{r+1}{r+2}p_{\mathcal{C}}^{(r)} + \tfrac{1}{r+2}\max_i Q_{ii}\\ &\le \tfrac{r+1}{r+2}\underline{p} + \tfrac{1}{r+2}\overline{p},\end{aligned}$$

which implies the second statement. □

SDP-BASED APPROXIMATIONS

Similarly as in the definition of $p_{\mathcal{C}}^{(r)}$, we can define SDP-based approximations to $\underline{p}$ using the cones $\mathcal{K}_n^r$ instead of $\mathcal{C}_n^r$, namely:

$$p_{\mathcal{K}}^{(r)} = \min\left\{\mathbf{Tr}\,QX \ :\ \mathbf{Tr}\,ee^TX = 1,\ X \in (\mathcal{K}_n^r)^*\right\}, \tag{12.42}$$

for $r = 0, 1, \ldots$ which has dual formulation

$$p_{\mathcal{K}}^{(r)} = \max\left\{\lambda \ :\ Q - \lambda ee^T \in \mathcal{K}_n^r,\ \lambda \in \mathbf{R}\right\}. \tag{12.43}$$

It may not immediately be clear that the optimal values in (12.42) and (12.43) are attained and equal. However, this follows from Theorem 2.3 by noting that $X = \frac{1}{n^2+n}(nI_n + ee^T)$ is feasible in problem (12.42) as well as strictly completely positive and therefore in the interior of $(\mathcal{K}_n^r)^*$ for all $r = 0, 1, \ldots$; similarly, $\lambda = -1$ defines a feasible solution of (12.43) in the interior of $\mathcal{K}_n^0$, and consequently in the interior of $\mathcal{K}_n^r$ for all $r = 0, 1, \ldots$.

Note that $p_{\mathcal{K}}^{(r)} \ge p_{\mathcal{C}}^{(r)}$ for all $r = 0, 1, \ldots$, since $\mathcal{C}_n^r \subset \mathcal{K}_n^r$.

It follows that Theorem 12.8 also holds if $p_{\mathcal{C}}^{(r)}$ is replaced by $p_{\mathcal{K}}^{(r)}$.

COMPARISON WITH KNOWN APPROXIMATION RESULTS

Consider the generic optimization problem:

$$\phi^* := \max\left\{f(x) \ :\ x \in S\right\},$$

for some nonempty, bounded convex set S, and let

$$\phi_* := \min\left\{f(x) \ :\ x \in S\right\}.$$

Definition 12.2 (Nesterov *et al.* [139]) *A value ψ approximates ϕ_* with relative accuracy $\mu \in [0,1]$ if*

$$|\psi - \phi_*| \le \mu(\phi^* - \phi_*).$$

We say ψ approximates ϕ_* with relative accuracy $\mu \in [0,1]$ in the weak sense *if*

$$|\psi - \phi_*| \le \mu.$$

The approximation is called implementable if $\psi \ge \phi_*$.

Note that if ψ is an *implementable* approximation, then $\psi = f(x)$ for some $x \in S$.

It is known from Bellare and Rogaway [18] that there is no polynomial-time μ-approximation of the optimal value of the problem

$$\min\left\{x^TQx \;:\; Bx = b,\; 0 \le x \le e\right\}$$

for some $\mu \in \left(0, \frac{1}{3}\right)$ in the weak sense, unless $P = NP$.

Using semidefinite programming techniques, the problem

$$\min\left\{x^TQx \;:\; Bx = \tfrac{1}{2}\,e,\; 0 \le x \le e\right\}$$

can be approximated in polynomial time with the relative accuracy $(1-O(1/\log(n)))$, see Nesterov *et al.* [139] and the references therein.

Note the the standard quadratic optimization problem (12.33) is a special case of this problem where $B = \frac{1}{2}e^T$. Nesterov [134] first showed that problem (12.33) allows a 2/3 polynomial-time, implementable approximation. The result in Theorem 12.8 improves on this result, since it can be restated as follows.

Theorem 12.9 (Bomze and De Klerk [27]) *The value $p_{\Delta(r)}$ as defined in (12.38) is a polynomial-time, implementable, $\frac{1}{r+2}$-approximation of $\underline{p}$.*

In other words, for any given $\epsilon > 0$ we obtain an implementable ϵ-approximation for problem (12.33).

Remark 12.3 *At first glance the complexity result for standard quadratic optimization seems to be in conflict with the in-approximability results for the maximum stable set problem (see page 188). Recall from Remark 12.1 that the stability number can be obtained by solving the following standard quadratic optimization problem:*

$$\frac{1}{\alpha(G)} = \min_{x \in \Delta} x^T(A+I)x$$

where A is the adjacency matrix of $G = (V, E)$.

The complexity result in Theorem 12.8 therefore only guarantees that we can approximate $\frac{1}{\alpha(G)}$ arbitrarily well in polynomial time, and this is not the same as approximating $\alpha(G)$. In particular, we may have to use $r = \alpha(G)^2$ in order to determine $\alpha(G)$. To see this, denote $Q = A + I$ and define $\alpha^{(r)} = 1/p_{\mathcal{C}}^{(r)}$ where $p_{\mathcal{C}}^{(r)}$ is defined in (12.35). Now we may assume without loss of generality $\alpha^{(r)} \ge 2$ (assume G is not a complete graph). This means $p_{\mathcal{C}}^{(r)} \le \frac{1}{2}$.

Now we apply Theorem 12.8, to arrive at

$$\frac{1}{\alpha(G)} - \frac{1}{\alpha^{(r)}} \leq \frac{1 - \frac{1}{\alpha(G)}}{r+1} \leq \frac{1 - \frac{1}{\alpha^{(r)}}}{r+1}. \tag{12.44}$$

If we set $r = \alpha(G)^2$ *in (12.44), then we can rewrite (12.44) as the equivalent inequality*

$$\alpha^{(r)} - \alpha(G) \leq \frac{\alpha(G)^2 - \alpha(G)}{\alpha(G)^2 - \alpha(G) + 1}.$$

Hence we get $\alpha^{(r)} - \alpha(G) < 1$ *which implies* $\lfloor \alpha^{(r)} \rfloor = \alpha(G)$ *if* $r \geq \alpha(G)^2$. *In other words, Theorem 12.8 is a generalization of Theorem 12.5, as it should be.*

13

THE SATISFIABILITY PROBLEM

Preamble

In this chapter we investigate relaxations of the satisfiability problem (SAT) via semi-definite programming. Loosely speaking, the SAT problem is to determine whether one can assign values to a set of logical variables in such a way that a given set of logical expressions (clauses) in these variables are satisfied.

Example 13.1 *Let p_1, p_2, p_3 denote logical variables that can take the values TRUE or FALSE, say. Further let $\neg p_i$ denote the negation of p_i, and let '$\vee$' denote the logical 'OR' operator. The set of clauses*

$$\begin{array}{ccccc} p_1 & \vee & p_2 & \vee & \neg p_3 \\ & & \neg p_2 & \vee & \neg p_3 \\ \neg p_1 & & & \vee & p_3 \end{array}$$

has a truth assignment $p_1 = TRUE$, $p_2 = FALSE$, $p_3 = TRUE$, i.e. *each of the three clauses are satisfied by this assignment.* □

A satisfiability problem can be relaxed to a set of LMI's in different ways. If the resulting SDP problem is infeasible, then a certificate of unsatisfiability of the SAT instance is obtained. We will look at two types of relaxations in this chapter, one due to Karloff and Zwick [99] and the other due to De Klerk *et al.* [50].[1]

13.1 INTRODUCTION

The satisfiability problem (SAT) is the original NP-complete problem (see Cook [35]). A formal problem statement requires the notion of a propositional CNF formula.

Definition 13.1 (Propositional CNF formula) *A propositional formula* $\mathbf{F}$ *in* conjunctive normal form *(CNF) is defined as a conjunction of clauses, where each clause* $\mathbf{C}_i$ *is a disjunction of* literals. *Each literal is a* logical variable *or its negation* ($\neg$)*. Let* m *be the number of clauses and* n *the number of logical variables of* $\mathbf{F}$*. A clausal propositional formula is denoted as*

$$\mathbf{F} = \mathbf{C}_1 \wedge \mathbf{C}_2 \wedge \ldots \wedge \mathbf{C}_m,$$

where '$\wedge$' denotes the logical 'AND' operator, each clause $\mathbf{C}_i$ *is of the form*

$$\mathbf{C}_i = \bigvee_{j \in I_i} p_j \vee \bigvee_{j \in J_i} \neg p_j,$$

with $I_i,\ J_i \subseteq \{1, \ldots, n\}$ *disjoint.*

Definition 13.2 (SAT) *The satisfiability (SAT) problem of propositional logic is to determine whether or not a truth assignment to the logical variables exists such that each clause evaluates to true* (i.e. *at least one of its literals is true) and thus the formula is true (satisfied).*

For a lengthy survey of algorithms for SAT, see Gu *et al.* [74].

The more general MAX-k-SAT problem is to find the maximum number of clauses that can be simultaneously satisfied, where the clause length is at most k; the MAX-$\{k\}$-SAT problem involves clauses of length *exactly* k (similar definitions hold for k-SAT and $\{k\}$-SAT). A clause of length k is called a k-clause. We will not consider clauses of length one (unit clauses) for the satisfiability problem, since these can be removed in an obvious way (unit resolution).

Reduction to the maximum stable set problem

SAT can be polynomially reduced to the maximum stable set problem in the following way: given a CNF formula $\mathbf{F}$ of the above form, we construct the following graph, say $G(\mathbf{F})$. For each literal in each clause we define a vertex of $G(\mathbf{F})$, and we connect all

[1] The presentation in this chapter is based on De Klerk *et al.* and De Klerk and Van Maaren [49].

the vertices corresponding to any given clause. In other words, each k-clause defines a clique of size k in $G(\mathbf{F})$. We further connect any pair of vertices that correspond to the literals p_i and $\neg p_i$ for all i. It is easy to see that a stable set of size m in $G(\mathbf{F})$ corresponds to a truth assignment of **F**, by setting the literals corresponding to the vertices in the stable set to TRUE. Conversely, any truth assignment of **F** corresponds to a stable set of size at least m in $G(\mathbf{F})$. We can therefore apply the methodology in Chapter 12 to decide if a given formula is satisfiable, by answering the question: 'is $\alpha(G(\mathbf{F})) \geq m$?', where $\alpha(G(\mathbf{F}))$ denotes the stability number of $G(\mathbf{F})$, as before.

Approaches using SDP

A more direct approach was explored by De Klerk *et al.* [50], where a simple SDP relaxation (called the *gap* relaxation) of propositional formulae was proposed. If the gap relaxation of a given formula is infeasible, then a certificate of unsatisfiability is obtained. The authors showed that unsatisfiability can be detected in this way for several polynomially solvable subclasses of SAT problems. The gap relaxation in De Klerk *et al.* [50] is closely related to the types of MAX-k-SAT relaxations studied earlier by Karloff and Zwick [99], Zwick [194], and Halperin and Zwick [79].

The work of these authors in turn employs the ideas of semidefinite approximation algorithms and associated randomized rounding, as introduced in the seminal work of Goemans and Williamson [66] on approximation algorithms for the MAX-CUT and other problems (see page 179).

If the SDP relaxation is feasible, then one can still 'round' a solution of the relaxation in an attempt to generate a truth assignment for the SAT-formula in question (rounding schemes). In this chapter we will also explore the theoretical properties of rounding schemes. A $7/8$ guarantee is obtained for satisfiable instances of MAX-3-SAT by rounding a solution of the relaxation due to Karloff and Zwick [99] (see Section 13.2). This result is tight (unless $P = NP$) in view of a recent in-approximability result; Håstad [80] has shown that — even for satisfiable $\{3\}$-SAT formulae — one cannot find a polynomial time algorithm that satisfies a fraction $7/8 + \epsilon$ of the clauses for any $\epsilon > 0$, unless $P = NP$.[2] The negative result therefore also holds for satisfiable instances of MAX-3-SAT. As such, the results by Karloff and Zwick [99] and Håstad [80] give a beautiful example of the recent progress made in approximation algorithms on the one side, and in-approximability results on the other. One can also not help but feel that these results support the $P \neq NP$ conjecture!

[2]Note that the trivial randomized $\{3\}$-SAT algorithm that sets p_i to TRUE with probability $\frac{1}{2}$ and to FALSE with probability $\frac{1}{2}$ (for all i independently), will satisfy $7/8$ of the clauses in a $\{3\}$-SAT formula (in expectation).

13.2 BOOLEAN QUADRATIC REPRESENTATIONS OF CLAUSES

Associating a $\{-1,1\}$–variable x_j with each logical variable p_j, a clause $\mathbf{C}_i$ can be written as a linear inequality in the following way.

$$C_i(x) = \sum_{j\in I_i} x_j - \sum_{j\in J_i} x_j \geq 2 - \ell(\mathbf{C}_i), \tag{13.1}$$

where $\ell(\mathbf{C}_i)$ denotes the length of clause i, i.e. $\ell(\mathbf{C}_i) = |I_i \cup J_i|$. Using matrix notation, the integer linear programming formulation of the satisfiability problem can be stated as

$$\text{find } x \in \{-1,1\}^n \text{ such that } Ax \geq r,$$

where the vector r has components $r_k = 2 - \ell(\mathbf{C}_k)$, and the matrix $A \in \mathbf{R}^{m\times n}$ is the so-called *clause–variable* matrix: $a_{ki} = 1$ if $i \in I_k$, $a_{ki} = -1$ if $i \in J_k$, while $a_{ki} = 0$ for any $i \notin I_k \cup J_k$.

To derive an SDP relaxation for a CNF formula, one can represent each clause $\mathbf{C}_i$ as one or more Boolean quadratic (in)equalities, that are satisfied if x corresponds to a truth assignment for this clause.

Consider a k-clause $\mathbf{C}_i$, and let x refer to the k $\{-1,1\}$-variables that appear in $\mathbf{C}_i$. There are 2^k possible choices for the values of the variables that appear in x, and all but one correspond to truth assignments of $\mathbf{C}_i$. Assume that we wish to find a quadratic inequality

$$x^T A_i x + 2b_i^T x + c_i \leq 0, \quad x \in \{-1,1\}^k, \tag{13.2}$$

that is a valid Boolean quadratic representation of $\mathbf{C}_i$. In other words, we want to find values for A_i, b_i and c_i such that (13.2) holds if x corresponds to any one of the $2^k - 1$ truth assignments for $\mathbf{C}_i$. We can view the requirement

$$x^T A_i x + 2b_i^T x + c_i \leq 0 \quad \text{for all truth assignments } x \text{ of } \mathbf{C}_i,$$

as $2^k - 1$ linear inequality constraints in the variables $A_i \in \mathcal{S}_k$, $b_i \in \mathbf{R}^k$ and $c_i \in \mathbf{R}$; *i.e.* each truth assignment defines one inequality. These inequalities define a polyhedral cone in $\mathcal{S}_k \times \mathbf{R}^k \times \mathbf{R}$. Using techniques from LP, we can find the extreme rays of this cone. Each extreme ray corresponds to a valid quadratic inequality. Since any point in a convex cone is a convex combination of points on extreme rays of the cone, any valid quadratic inequality for $\mathbf{C}_i$ is a nonnegative aggregation of the quadratic inequalities corresponding to the extreme rays.

In this way, Karloff and Zwick [99] derived a set of seven valid quadratic functions that can be used to derive all possible valid quadratic inequalities for 3-clauses; see also De Klerk *et al.* [50].

Example 13.2 *Consider the clause* $p_1 \vee p_2 \vee p_3$. *All valid Boolean quadratic inequalities for this clause are nonnegative aggregations of the following seven quadratic* *inequalities:*

$$\left.\begin{array}{rcl} x_1x_2 + x_1x_3 - x_2 - x_3 & \leq & 0 \\ x_1x_2 + x_2x_3 - x_1 - x_3 & \leq & 0 \\ x_1x_3 + x_2x_3 - x_1 - x_2 & \leq & 0 \\ -x_1x_2 - x_1x_3 - x_2x_3 - 1 & \leq & 0 \\ -x_1x_2 + x_1 + x_2 - 1 & \leq & 0 \\ -x_1x_3 + x_1 + x_3 - 1 & \leq & 0 \\ -x_2x_3 + x_2 + x_3 - 1 & \leq & 0 \\ x_1, x_2, x_3 & \in & \{-1, 1\}. \end{array}\right\} \quad (13.3)$$

Note that the first three inequalities in (13.3) only hold if $x \in \{-1,1\}^3$ corresponds to a truth assignment of $p_1 \vee p_2 \vee p_3$. The remaining four inequalities are valid for all $x \in \{-1,1\}^3$. These four inequalities are called triangle inequalities.

The analogous inequalities for all other possible 3-clauses are obtained by replacing x_i by $-x_i$ in (13.3) if p_i appears negated in the clause. □

More recently, Halperin and Zwick [79] derived a similar set of generic quadratic inequalities for 4-clauses. One can do this for clauses of any length; the only problem is that the number of quadratic inequalities increases exponentially. For 4-clauses we already have 23 quadratic inequalities, and for 5-clauses there are 694. For this reason, Van Maaren [178] and De Klerk *et al.* [50] considered so-called *elliptic representations of clauses*, where one uses only one quadratic inequality per clause. For the k-clause $p_1 \vee \ldots \vee p_k$ the elliptic representation takes the form:

$$(x_1 + \ldots + x_k - 1)^2 \leq (k-1)^2, \quad x_1, \ldots, x_n \in \{-1, 1\}. \qquad (13.4)$$

For $k = 3$ we can use $x_i^2 = 1$ to derive

$$\sum_{i \neq j} x_i x_j - \sum_{i=1}^{3} x_i \leq 0. \qquad (13.5)$$

This quadratic inequality is simply the sum of the first three of the seven generic inequalities in (13.3).

13.3 FROM BOOLEAN QUADRATIC INEQUALITY TO SDP

There is a standard way to relax Boolean quadratic inequalities to LMI's, that is originally due to Shor [166].

The Boolean quadratic inequality (13.2) can be rewritten as

$$\mathbf{Tr}\begin{bmatrix} A_i & b_i \\ b_i^T & c_i \end{bmatrix}\begin{bmatrix} xx^T & x \\ x^T & 1 \end{bmatrix} \leq 0, \quad x \in \{-1,1\}^n, \tag{13.6}$$

by using the properties of the trace operator (see Appendix A). The matrix

$$\begin{bmatrix} xx^T & x \\ x^T & 1 \end{bmatrix}$$

in (13.6) is positive semidefinite and has rank 1. As $x_i^2 = 1$ $(i = 1,\ldots,n)$, all the diagonal elements of this matrix equal 1. If we drop the rank 1 requirement, we can relax (13.6) to

$$\mathbf{Tr}\begin{bmatrix} A_i & b_i \\ b_i^T & c_i \end{bmatrix}\begin{bmatrix} X & y \\ y^T & 1 \end{bmatrix} \leq 0, \quad x_{jj} = 1, \quad j = 1,\ldots,n, \tag{13.7}$$

where X is now a symmetric $n \times n$ matrix such that

$$\begin{bmatrix} X & y \\ y^T & 1 \end{bmatrix} \succeq 0, \tag{13.8}$$

which is the same as

$$X - yy^T \succeq 0, \tag{13.9}$$

by the Schur complement theorem (Theorem A.9 on page 235). If we have a rank one solution of (13.7) and (13.8), then either y in (13.8) or $-y$ will be a feasible solution of the Boolean quadratic inequality (13.2). On the other hand, if the problem (13.7)–(13.8) is infeasible, then there can be no solution of (13.2).

We therefore have a general procedure by which we can relax a SAT problem to a set of linear matrix inequalities. The only non-mechanical step is to select which valid quadratic (in)equalities will be used to represent each clause.

Two SDP relaxations

In this chapter we consider the two SDP relaxations of SAT formulae:

- the so-called *gap relaxation* for k-SAT by De Klerk *et al.* [50] which uses the inequalities of the type (13.4) for k-clauses ($k \geq 3$);
- the relaxation due to Karloff and Zwick [99] for 3-SAT, which uses the first three of the seven inequalities in (13.3) for 3-clauses. We will refer to the resulting relaxation as the *K–Z relaxation*.

For the generic 2-clause

$$p_1 \vee p_2$$

both relaxations use the valid quadratic equality

$$(x_1 + x_2 - 1)^2 = 1.$$

Again, if p_1 is negated, then x_1 is replaced by $-x_1$ in (13.10), *etc.* Using $x_i^2 = 1$ as before, we get

$$x_1 x_2 - x_1 - x_2 = -1. \tag{13.10}$$

In both relaxations the generic 2-clause $p_1 \vee p_2$ corresponds to a 3×3 principal submatrix of the matrix in (13.8), namely

$$\begin{bmatrix} 1 & x_{12} & y_1 \\ x_{12} & 1 & y_2 \\ y_1 & y_2 & 1 \end{bmatrix},$$

and (13.10) implies

$$x_{12} - \sum_{i=1}^{2} y_i = -1.$$

Similarly, the generic 3-clause $p_1 \vee p_2 \vee p_3$ corresponds to a 4×4 principal submatrix of the matrix in (13.8), namely

$$\begin{bmatrix} 1 & x_{12} & x_{13} & y_1 \\ x_{12} & 1 & x_{23} & y_2 \\ x_{13} & x_{23} & 1 & y_3 \\ y_1 & y_2 & y_3 & 1 \end{bmatrix}.$$

For the gap relaxation, (13.5) implies

$$x_{12} + x_{13} + x_{23} - \sum_{i=1}^{3} y_i \leq 0. \tag{13.11}$$

Similarly, for the K–Z relaxation we use the first three inequalities in (13.3) to obtain:

$$\left.\begin{array}{rcl} x_{12} + x_{13} - y_2 - y_3 & \leq & 0 \\ x_{12} + x_{23} - y_1 - y_3 & \leq & 0 \\ x_{13} + x_{23} - y_1 - y_2 & \leq & 0 \end{array}\right\} \tag{13.12}$$

As before, the analogous inequalities for all other possible 3-clauses are obtained by replacing x_i by $-x_i$ in the first three inequalities of (13.3) if p_i appears negated in the clause.

Note that the identity matrix is always feasible for both the gap and K–Z relaxations of $\{3\}$-SAT formulae. This means that unsatisfiability can only be detected if the formula in question also has 2-clauses.

Recall that the gap relaxation is weaker than the K–Z relaxation for 3-SAT, because (13.5) is obtained by summing the three inequalities in (13.12). The reader may well ask why it is necessary to study both relaxations. The answer is simply that the gap relaxation is already strong enough to detect unsatisfiability for several interesting sub-classes of SAT problems, as we will show in the next section. On the other hand, the K–Z relaxation leads to better approximation algorithms for MAX-3-SAT; this is shown in Section 13.6.

The K–Z relaxation is only defined for 3-SAT. The gap relaxation is defined for general k-SAT as follows.

Definition 13.3 (Gap relaxation) *We define the gap relaxation of a given CNF formula in terms of the clause-variable matrix; the parameters in (13.7) become*

$$A_i = a_i a_i^T, \quad b_i = -a_i, \quad c_i = -\ell(\mathbf{C}_i)\left(\ell(\mathbf{C}_i) - 2\right), \quad i = 1, \ldots, m.$$

Moreover, for 2*-clauses, the inequality sign in (13.7) becomes equality.*

13.4 DETECTING UNSATISFIABILITY OF SOME CLASSES OF FORMULAE

In this section we look at some classes of formulae where unsatisfiability can be detected using the gap relaxation. (Recall that if the gap relaxation of a given formula is infeasible, then we obtain a certificate of unsatisfiability of the formula.)

A CLASS OF INFEASIBLE ASSIGNMENT PROBLEMS

Consider a set of n propositional variables. Let $\{S_1, \ldots, S_N\}$ and $\{T_1, \ldots, T_M\}$ denote an N-partition and an M-partition of this set of variables respectively. Furthermore, let us assume that $M < N$. We now define a class of unsatisfiable CNF formulae that we will call $\mathbf{F}_{CP}$.

$$\bigvee_{i \in S_k} p_i, \qquad 1 \leq k \leq N, \tag{13.13}$$

$$\neg p_i \vee \neg p_j, \quad i, j \in T_k, \ i \neq j, \ 1 \leq k \leq M. \tag{13.14}$$

We give two famous examples of problems that fit in this format.

Example 13.3 (Pigeonhole formulae) *Given $h + 1$ pigeons and h pigeonholes, decide whether it is possible to assign each pigeon to a hole in such a way that no two pigeons share the same hole.*

To obtain a satisfiability problem we introduce the logical variables

$$p_{i,j} = \begin{cases} \textit{TRUE} & \textit{if pigeon } i \textit{ is assigned to hole } j \\ \textit{FALSE} & \textit{otherwise.} \end{cases}$$

Each pigeon must have a pigeonhole:

$$p_{i,1} \vee p_{i,2} \vee \ldots \vee p_{i,h}, \quad i = 1, \ldots, h+1. \tag{13.15}$$

No two pigeons may share a pigeonhole:

$$\neg p_{i,k} \vee \neg p_{j,k} \quad i, j \in \{1, \ldots, h+1\} \; (i \neq j), \textit{ and } k \in \{1, \ldots, h\}.$$

We now show that this encoding of the pigeonhole problem fits the format $\mathbf{F}_{CP}$*: for each hole there is a set of 2-clauses, giving rise to* $M = h$ *separate sets* T_k*, each of size* $h+1$*, namely*

$$T_k := \{p_{1,k}, \ldots, p_{h+1,k}\}, \quad k = 1, \ldots, h.$$

Similarly, we have $N = h+1$ *disjoint sets corresponding to the clauses (13.15), namely*

$$S_k := \{p_{k,1}, \ldots, p_{k,h}\}, \quad k = 1, \ldots, h+1.$$

It is easy to see that each logical variable occurs both exactly once in the sets S_k *and exactly once in the sets* T_k*, so that* $\{T_1, \ldots, T_M\}$ *forms an* M*-partition of the set of logical variables, and* $\{S_1, \ldots, S_N\}$ *forms an* N*-partition, as required.* □

Example 13.4 (Mutilated chess boards) *Consider a chess board of size* $2s \times 2s$ *squares with the two opposing white corner squares removed (see Figure 13.1). Can*

Figure 13.1. A 'mutilated' chess board for $s = 3$.

the resulting 'mutilated' chess board be covered by rectangular dominoes of size 2×1 *(i.e. a single domino covers exactly two adjacent squares), such that each square is covered exactly once?*[3]

For the SAT encoding of this problem we introduce the logical variables

$$p_{i,j} = \begin{cases} \textit{TRUE} & \textit{if a domino covers squares } i \textit{ and } j \\ \textit{FALSE} & \textit{otherwise.} \end{cases}$$

Each square i must be covered:

$$p_{i,n_1} \vee p_{i,n_2} \vee \ldots \vee p_{i,n_t}, \tag{13.16}$$

where $n_1, \ldots, n_t$ are the t neighbouring squares of square i ($2 \leq t \leq 4$).

Each square must be covered only once:

$$\neg p_{i,k} \vee \neg p_{j,k} \tag{13.17}$$

where i, j are neighbouring squares of square k.

Thus we have $4(2s^2 - s - 1)$ logical variables.

We will only take a subset of the clauses (13.16) and (13.17) to obtain a formula of the form $\mathbf{F}_{CP}$. For all the black *squares we keep the clauses (13.16), while for all the* white *squares we only use the clauses (13.17). The first set corresponds to (13.13) and the second set to (13.14). Again, it is easy to see that each logical variable occurs in exactly one of the $N = 2s^2$ positive clauses, and in exactly one of the $M = 2s^2 - 2$ sets of negative clauses.* □

Both of the above problems are easily shown to be unsatisfiable by using suitable *cutting planes* (see *e.g.* Cook *et al.* [36]). Surprisingly, other techniques often exhibit exponential running times to solve formulas of this type. Indeed, Haken [75] proves that the unsatisfiability of the pigeonhole problem cannot be detected in polynomial time by the resolution algorithm. It is strongly conjectured that the resolution algorithm requires exponential running time on the mutilated chess board problem as well (Urquhart [177]).

[3]This is impossible — a single domino always covers one white and one black square, while the mutilated chess board has more black squares than white squares.

Let us consider the gap relaxation of $\mathbf{F}_{CP}$.

$$(\mathrm{SD}_{CP})\qquad \begin{array}{rl} \text{Find} & X \in \mathcal{S}_n,\ y \in \mathbf{R}^n \\ \text{s.t.} & e_{S_k}^T X e_{S_k} - 2e_{S_k}^T y \le |S_k|(|S_k| - 2),\ 1 \le k \le N \\ & e_{ij}^T X e_{ij} + 2e_{ij}^T y = 0,\ i, j, i \ne j \in T_k,\ 1 \le k \le M \\ & \mathrm{diag}(X) = e, \\ & X \succeq yy^T, \end{array}$$

where, e is the all-1 vector in $\mathbf{R}^n$, e_{S_k} is the incidence vector of the set S_k, *i.e.*

$$(e_{S_k})_i = \begin{cases} 1 & \text{if } p_i \in S_k \\ 0 & \text{otherwise,} \end{cases}$$

and e_{ij} is the $0 - 1$ vector with entries i and j equal to 1 and all other components zero.

We can now prove the following.

Theorem 13.1 *The gap relaxation (SD_{CP}) of $\mathbf{F}_{CP}$ is infeasible.*

Proof:

Note that from (13.9) it follows that $a^T X a \ge (a^T y)^2$ for any $a \in \mathbf{R}^n$. Thus we have

$$\left(e_{S_k}^T y\right)^2 - 2e_{S_k}^T y \le e_{S_k}^T X e_{S_k} - 2e_{S_k}^T y \le |S_k|(|S_k| - 2).$$

This implies that

$$2 - |S_k| \le e_{S_k}^T y \le |S_k|,\ 1 \le k \le N. \tag{13.18}$$

Now we consider the inequalities corresponding to the sets T_k. Summing over all the equalities corresponding to a set T_k for fixed k, we find that

$$e_{T_k}^T X e_{T_k} + (|T_k| - 2)e_{T_k}^T \mathrm{diag}(X) + 2(|T_k| - 1)e_{T_k}^T y = 0.$$

To verify this, note that each diagonal element x_{ii}, $i \in T_k$, occurs in exactly $|T_k| - 1$ inequalities; similarly, each linear term y_i, $i \in T_k$, occurs in exactly $|T_k| - 1$ inequalities as well. Simplifying this expression using that $\mathrm{diag}(X) = e$, we obtain

$$e_{T_k}^T X e_{T_k} + 2(|T_k| - 1)e_{T_k}^T y = -|T_k|(|T_k| - 2).$$

Using the semidefinite constraint again, we conclude that

$$\left(e_{T_k}^T y\right)^2 + 2(|T_k| - 1)e_{T_k}^T y \le e_{T_k}^T X e_{T_k} + 2(|T_k| - 1)e_{T_k}^T y \le -|T_k|(|T_k| - 2).$$

The outermost inequality implies

$$\left(e_{T_k}^T y\right)^2 + 2(|T_k| - 1)e_{T_k}^T y + |T_k|(|T_k| - 2) \le 0,$$

which in turn implies[4]

$$-|T_k| \leq e_{T_k}^T y \leq 2 - |T_k|, \ 1 \leq k \leq M.$$

Summing these inequalities we find that $-n \leq e^T y \leq 2M - n$, while from (13.18) we have that $2N - n \leq e^T y \leq n$, implying that $2N \leq e^T y + n \leq 2M$. Thus we conclude that (SD$_{CP}$) is infeasible because $N > M$. □

As a consequence, the gap relaxation can be used to detect unsatisfiability of both the pigeonhole (Example 13.3) and the mutilated chess board (Example 13.4) problems.

We now consider another example.

Example 13.5 *Let $G = (V, E)$ be a clique on n vertices. Decide whether one can legally colour G using t colours.*

For the SAT encoding of this problem we introduce the logical variables

$$p_{i,j} = \begin{cases} \textit{TRUE} & \textit{vertex } i \textit{ has colour } j \\ \textit{FALSE} & \textit{otherwise} \end{cases}$$

Each vertex i must be coloured by at least one of the t colours:

$$p_{i,1} \vee p_{i,2} \vee \ldots \vee p_{i,t}, \quad i = 1, \ldots, n. \tag{13.19}$$

Adjacent vertices may not be assigned the same colour:

$$\neg p_{ik} \vee \neg p_{jk} \quad (i, j) \in E, \ k = 1, \ldots, t. \tag{13.20}$$

Since G is a clique, the set of 2-clauses can be written as

$$\neg p_{ik} \vee \neg p_{jk} \quad i, j = 1, \ldots, n \ (i \neq j), \ k = 1, \ldots, t. \tag{13.21}$$

Given a truth assignment for this formula, then we can easily obtain a legal colouring of G. It may happen that a vertex is assigned more than one colour in a truth assignment, but then the excess colours can simply be removed (since there are no defect edges).

Note that the formula defined by (13.19) and (13.21) is now exactly a pigeonhole formula with $h = n - 1$ if we allow only $t = n - 1$ colours. The gap relaxation

---

[4]Here we use the inequality: $x^2 + bx + c \leq 0$ if and only if

$$\frac{-b - \sqrt{b^2 - 4c}}{2} \leq x \leq \frac{-b + \sqrt{b^2 - 4c}}{2}.$$

can therefore be used to prove that a clique of size n cannot be coloured with $n-1$ colours. □

For any given graph $G = (V, E)$ we can find the smallest integer value of t, say $t^*(G)$, such that the gap relaxation of the formula defined by (13.19) and (13.20) is feasible. By the last example, we will have $t^*(G) \geq \omega(G)$, where $\omega(G)$ denotes the clique number of G, as before. Moreover, $t^*(G)$ will be upper bounded by the chromatic number $\chi(G)$ of G, since $t^*(G)$ is obtained by relaxing the colouring problem. We therefore have

$$\omega(G) \leq t^*(G) \leq \chi(G).$$

This reminds one of the sandwich theorem on page 158, and suggests a link between $t^*(G)$ and $\vartheta(\bar{G})$. De Klerk *et al.* [42] have shown that

$$\omega(G) \leq t^*(G) \leq \lceil \vartheta(\bar{G}) \rceil \leq \chi(G).$$

It is conjectured (but not proven) that $t^*(G) = \lceil \vartheta(\bar{G}) \rceil$; in fact, it is shown in [42] that equality is obtained when suitable valid inequalities are added to the gap relaxation.

To quote Goemans [65]: 'It seems all roads lead to ϑ'!

The gap relaxation for other classes of formulae One can show that the gap relaxation of a 2-SAT formula is feasible if and only if the formula is satisfiable (see Feige and Goemans [56] and De Klerk *et al.* [50]).

The gap relaxation can also be used to solve *doubly balanced formulae* by introducing a suitable objective function. The reader is referred to De Klerk *et al.* [50] for details.

We stress that all the classes of formulae that have been described in this section are known to be polynomially solvable.

13.5 ROUNDING PROCEDURES

If the gap or K–Z relaxation of a given 3-SAT formula is feasible, then one can 'round' a feasible solution of the relaxation in an attempt to generate a truth assignment for the formula (if one exists). In other words, we can formulate approximation algorithms for MAX-3-SAT in this way, for the subclass of 3-SAT problems where the gap or K–Z relaxation is feasible. This subclass includes all satisfiable 3-SAT instances, of course. Moreover, it is possible to establish approximation guarantees for such approximation algorithms.

We will show here that by rounding a feasible solution of the K–Z relaxation of a given 3-SAT formula in a suitable way, we satisfy at least $7/8$ of the 3-clauses and a fraction 0.91226 of the 2-clauses (in expectation); this result is due to Karloff and Zwick [99].

DETERMINISTIC ROUNDING

Recall from page 216 that the vector y in (13.8) gives a truth assignment if X, y are feasible in (13.7) and X is a rank 1 matrix.

In general X will not be a rank 1 solution, but one can still check whether the rounded vector sign(y) yields a truth assignment.

We will refer to this heuristic as *deterministic rounding*. Deterministic rounding satisfies each 2-clause, say $\mathbf{C} = p_i \vee p_j$, for which $y_i \neq 0$ or $y_j \neq 0$ in the solution of the gap relaxation. The unresolved 2-clauses can subsequently be satisfied in a trivial way (see Feige and Goemans [56] and De Klerk *et al.* [50]).

RANDOMIZED ROUNDING WITH HYPERPLANES

Recall that the entry x_{ij} in the matrix in (13.8) corresponds to the product $x_i x_j$ of logical variables. In fact, we may write

$$x_{ij} = v_i^T v_j$$

where v_i and v_j are columns from the Choleski decomposition of the matrix in (13.8).

This shows that the product $x_i x_j$ is in fact relaxed to the inner product $v_i^T v_j$, *i.e.* we associate a vector v_i with each literal p_i.

The vector y in (13.8) can similarly be seen as a vector of inner products:

$$y_i = v_T^T v_i$$

where one can attach a special interpretation to the vector v_T as 'truth' vector: in a rank 1 solution, if $v_i = v_T$ then p_i is TRUE, and if $v_i = -v_T$ then p_i is FALSE.

This interpretation suggests a rounding scheme, introduced by Karloff and Zwick [99] as an extension of the ideas of Goemans and Williamson [66]:

1. Take the Choleski factorization of the matrix in (13.8);
2. Choose a random hyperplane through the origin;
3. If v_i lies on the same side of the hyperplane as v_T, then set p_i to TRUE; otherwise set p_i to FALSE.

We will refer to this procedure as *randomized rounding*; this heuristic can be de–randomized using the techniques in Mahajan and Ramesh [120].

13.6 APPROXIMATION GUARANTEES FOR THE ROUNDING SCHEMES

In this section we give a review of the analysis required to establish performance guarantees for the randomized rounding procedure. The analysis is very similar to that of Section 11.6 for the MAX-3-CUT problem; the methodology is largely due to Karloff and Zwick [99].

RANDOMIZED ROUNDING FOR **2**-*CLAUSES*

We again only consider the 2-clause $p_1 \vee p_2$ without loss of generality.

Let the vectors v_1, v_2 be associated with the literals of $p_1 \vee p_2$.

The randomized rounding procedure will fail to satisfy this clause if all three vectors $v_1, v_2, -v_T$ lie on the same side of the random hyperplane (see Figure 13.2).

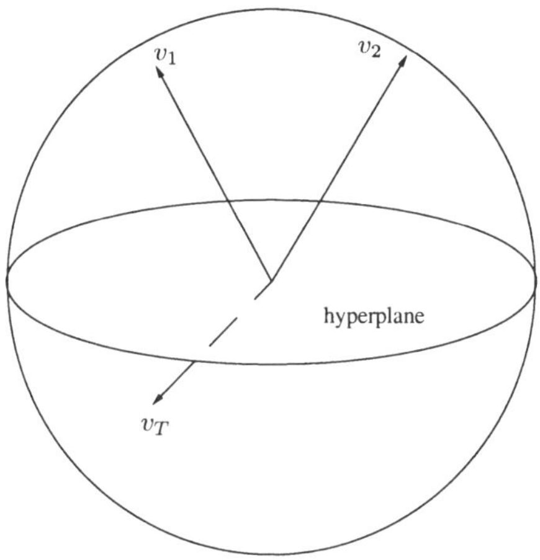

Figure 13.2. The situation where v_1 and v_2 are separated from v_T by a random hyperplane

In general, we want to know the probability that a given set of vectors lie on the same side of a random hyperplane. The probability that two given unit vectors v_1, v_2 lie on the same side of a random hyperplane is easy to determine: it only depends on the angle $\arccos(v_1^T v_2)$ between these vectors and is given by $1 - \arccos(v_1^T v_2)/\pi$ (see Goemans and Williamson [66]). One can use this observation to treat the three vector case using inclusion–exclusion; this is done by Karloff and Zwick [99]. We present a different derivation here which can be generalized to more vectors. The key is to consider a normal vector r to the random hyperplane. The clause will not be satisfied if the three vectors all have a positive (or all have a negative) inner product with r.

Note that the Gram matrix of $v_1, v_2, -v_T$ is the following matrix:

$$\bar{X}_2 := \begin{bmatrix} 1 & x_{12} & -y_1 \\ x_{12} & 1 & -y_2 \\ -y_1 & -y_2 & 1 \end{bmatrix},$$

and the K–Z and gap relaxations require

$$x_{12} - \sum_{i=1}^{2} y_i = -1. \qquad (13.22)$$

The vectors $v_1, v_2, -v_T$ can be viewed as three points on the unit sphere

$$\mathcal{S}^2 := \left\{x \in \mathbb{R}^3 \mid \|x\| = 1\right\},$$

and thus define a so-called *spherical triangle* (say $\mathbf{S}$) in the space $\mathcal{S}^2$.

The associated dual spherical triangle is defined as

$$\mathbf{S}^* := \left\{r \in \mathbb{R}^3 \ : \ r^T v_i \geq 0 \ (i = 1, 2), \ r^T v_T \geq 0\right\}$$

which, together with $-\mathbf{S}^*$ form the set of normal vectors for which all three vectors lie on the same side of the associated plane.

The probability that the clause is not satisfied is therefore given by:

$$p^{(2)} = 2\frac{\text{area}\,(\mathbf{S}^*)}{\text{area}\,(\mathcal{S}^2)} = \frac{\text{area}\,(\mathbf{S}^*)}{2\pi}.$$

It is well known that the area of a spherical triangle is given by its *angular excess*[5] (see *e.g.* Alekseevskij *et al.* [2]). The dihedral angels of $\mathbf{S}^*$ are given in terms of the edge lengths of $\mathbf{S}$ and equal $(\pi - \arccos(x_{12}))$, $(\pi - \arccos(-y_1))$, and $(\pi - \arccos(-y_2))$, respectively. It follows that the angular excess (*i.e.* area) of $\mathbf{S}^*$ is given by:

$$\begin{aligned} \text{area}\,(\mathbf{S}^*) &= (\pi - \arccos(x_{12}) + \pi - \arccos(-y_1) + \pi - \arccos(-y_2)) - \pi \\ &= 2\pi - \arccos(x_{12}) - \arccos(-y_1) - \arccos(-y_2), \end{aligned}$$

so that

$$p^{(2)} = 1 - \frac{1}{2\pi}\left(\arccos(x_{12}) + \arccos(-y_1) + \arccos(-y_2)\right). \tag{13.23}$$

We are therefore interested in the optimization problem:

$$\max_{\bar{X}_2} p^{(2)}$$

subject to $\bar{X}_2 \succeq 0$ and (13.22).

Since $p^{(2)}$ is a strongly quasi-concave function of $\bar{X}_2$, and the feasible region is convex, this optimization problem can be solved to global optimality, because each local optimum is also global in this case (see Theorem 3.5.9 in Bazaraa *et al.* [16]). The optimal solution is given by

$$\bar{X}_2 = \begin{bmatrix} 1 & -\frac{1}{3} & -\frac{1}{3} \\ -\frac{1}{3} & 1 & -\frac{1}{3} \\ -\frac{1}{3} & -\frac{1}{3} & 1 \end{bmatrix},$$

[5] The angular excess is the difference between the sum of the (dihedral) angles of the spherical triangle and π.

in which case the clause is satisfied by randomized rounding with probability

$$\frac{3}{2\pi}\arccos(-1/3) \approx 0.91226,$$

by (13.23).

RANDOMIZED ROUNDING FOR 3*-CLAUSES*

This analysis is perfectly analogous to the analysis of the 2-clause case, the only complication being that the probability function does not have a closed form representation.

Let the vectors v_1, v_2, v_3 be associated with the literals of $p_1 \vee p_2 \vee p_3$. As before, the randomized rounding procedure will fail to satisfy this clause if all four vectors $v_1, v_2, v_3, -v_T$ lie on the same side of the random hyperplane.

What is the probability of this event? This question has been answered by Karloff and Zwick [99]. Once again, we consider the normal vector r to the random hyperplane. The clause will not be satisfied if the four vectors all have a positive (or negative) inner product with r.

Note that the Gram matrix of $v_1, v_2, v_3, -v_T$ is the following matrix:

$$\bar{X}_3 := \begin{bmatrix} 1 & x_{12} & x_{13} & -y_1 \\ x_{12} & 1 & x_{23} & -y_2 \\ x_{13} & x_{23} & 1 & -y_3 \\ -y_1 & -y_2 & -y_3 & 1 \end{bmatrix}, \tag{13.24}$$

and that the K–Z relaxation requires

$$\begin{aligned} x_{12} + x_{13} - y_2 - y_3 &\leq 0 \\ x_{12} + x_{23} - y_1 - y_3 &\leq 0 \\ x_{13} + x_{23} - y_1 - y_2 &\leq 0. \end{aligned} \tag{13.25}$$

The vectors $v_1, v_2, v_3, -v_T$ can be viewed as four points on the unit hypersphere

$$\mathcal{S}^3 := \left\{x \in \mathbf{R}^4 \mid \|x\| = 1\right\},$$

and thus define a *spherical tetrahedron* (say $\mathbf{S}$) in the space $\mathcal{S}^3$.

The associated dual spherical tetrahedron is defined as

$$\mathbf{S}^* := \left\{r \in \mathbf{R}^4 \ : \ r^T v_i \geq 0 \ (i = 1, \ldots, 3), \ r^T v_T \geq 0\right\}$$

which, together with $-\mathbf{S}^*$ form the set of normal vectors for which all four vectors lie on the same side of the hyperplane.

The probability that the clause is not satisfied is therefore given by:

$$p^{(3)} = 2\frac{\text{volume}\,(\mathbf{S}^*)}{\text{volume}\,(\mathcal{S}^3)}. \tag{13.26}$$

The relative volume as a function of $\bar{X}_3$ is given by the following integral (see Section 11.6):

$$\frac{\text{volume}(\mathbf{S}^*)}{\text{volume}\,(\mathcal{S}^3)} = \frac{1}{\sqrt{\det(\bar{X}_3)\pi^4}} \int_0^\infty \cdots \int_0^\infty e^{-y^T \bar{X}_3^{-1} y} dy_1 \ldots dy_4.$$

In order to establish the worst-case performance guarantee for the randomized rounding scheme, we have to solve the optimization problem

$$\max_{\bar{X}_3} p^{(3)}$$

subject to $\bar{X}_3 \succeq 0$ and (13.25).

The volume function cannot be written in closed form, but can be simplified to a one dimensional integral (see Hsiang [87]) for spherical tetrahedrons. Surprisingly, the gradient of the volume function is explicitly known (see Karloff and Zwick [99]), even though the volume function itself is not.

The optimization problem we consider has a convex feasible region with nonempty interior, but the objective function (to be maximized) is not concave. It is therefore difficult to find the global maximum, but it can be done due to the small problem size. In particular, the KKT conditions are necessary conditions for optimality in this case (see Appendix B), and one can find and evaluate all KKT points numerically (see Karloff and Zwick [99] for details). After following this procedure, we find that the optimal solution is given by the identity matrix $\bar{X}_3 = I_4$. In this case the spherical tetrahedron $\mathbf{S}^*$ is simply the intersection of an orthant in $\mathbf{R}^4$ with the unit hypersphere $\mathcal{S}^3$. The relative volume of $\mathbf{S}^*$ is therefore $1/16$, and the 3-clause is therefore satisfied with probability

$$1 - p^{(3)} = 1 - 2\frac{1}{16} = \frac{7}{8}.$$

We can mention that the gap relaxation with randomized rounding only gives a $2/3$ approximation guarantee for 3-clauses (see De Klerk and Van Maaren [49]).

In summary, we have the following result.

Theorem 13.2 (Karloff and Zwick [99]) *Let a 3-SAT formula with a fraction p of 3-clauses and $(1-p)$ of 2-clauses be given, for which the K-Z relaxation is feasible. The deterministic rounding scheme satisfies a fraction of at least $1-p$ of the clauses. The randomized rounding scheme satisfies a fraction of at least*

$$\frac{7}{8}p + 0.91226(1-p)$$

of the clauses, in expectation.

Appendix A
Properties of positive (semi)definite matrices

In this appendix we list some well-known properties of positive (semi)definite matrices which are used in this monograph. The proofs which are omitted here may be found in [85]. A more detailed review of the matrix analysis which is relevant for SDP is given by Jarre in [94].

A.1 CHARACTERIZATIONS OF POSITIVE (SEMI)DEFINITENESS

Theorem A.1 *Let $X \in \mathcal{S}_n$. The following statements are equivalent:*

- $X \in \mathcal{S}_n^+$ *or* $X \succeq 0$ *(X is positive semidefinite);*
- $z^T X z \geq 0 \quad \forall z \in \mathbb{R}^n$;
- $\lambda_{\min}(X) \geq 0$;
- *All principal minors of X are nonnegative;*
- $X = LL^T$ *for some* $L \in \mathbb{R}^{n \times n}$.

We can replace 'positive semidefinite' by 'positive definite' in the statement of the theorem by changing the respective nonnegativity requirements to positivity, and by requiring that the matrix L in the last item be nonsingular. If X is positive definite ($X \in \mathcal{S}_n^{++}$ or $X \succ 0$), the matrix L can be chosen to be lower triangular, in which case we call $X = LL^T$ the *Choleski factorization* of X.

NB: In this monograph positive (semi)definite matrices are necessarily symmetric, *i.e.* we will use 'positive (semi)definite' instead of 'symmetric positive (semi)definite'.[1]

[1] In the literature a matrix $X \in \mathbb{R}^{n \times n}$ is sometimes called positive (semi)definite if its symmetric part $\frac{1}{2}(X + X^T)$ is positive (semi)definite.

As an immediate consequence of the second item in Theorem A.1 we have that

$$X \in \mathcal{S}_n^+ \Leftrightarrow AXA^T \in \mathcal{S}_n^+$$

for any given, nonsingular $A \in \mathbf{R}^{n\times n}$.

Another implication is that a block diagonal matrix is positive (semi)definite if and only if each of its diagonal blocks is positive (semi)definite.

A.2 SPECTRAL PROPERTIES

The characterization of positive (semi)definiteness in terms of nonnegative eigenvalues follows from the Raleigh–Ritz theorem.

Theorem A.2 (Raleigh–Ritz) *Let $A \in \mathcal{S}_n$. Then*

$$\lambda_{\min}(A) = \min_{z\in\mathbf{R}^n} \left\{z^T Az \mid \|z\| = 1\right\}. \tag{A.1}$$

It is well known that a symmetric matrix has an orthonormal set of eigenvectors, which implies the following result.

Theorem A.3 (Spectral decomposition) *Let $A \in \mathcal{S}_n$. Now A can be decomposed as*

$$A = Q^T\Lambda Q = \sum_{i=1}^{n} \lambda_i(A) q_i q_i^T$$

where Λ is a diagonal matrix with the eigenvalues $\lambda_i(A)$ $(i = 1, \dots, n)$ of A on the diagonal, and Q is an orthogonal matrix with a corresponding set of orthonormal eigenvectors $q_1, \dots, q_n$ of A as columns.

Since $\lambda_i(X) \geq 0$ $(i = 1, \dots, n)$ if $X \in \mathcal{S}_n^+$, we can define the *symmetric square root factorization* of $X \in \mathcal{S}_n^+$:

$$X^{\frac{1}{2}} = \sum_{i=1}^{n} \sqrt{\lambda_i(X)} q_i q_i^T \quad \left(X \in \mathcal{S}_n^+\right).$$

Note that $X^{\frac{1}{2}}X^{\frac{1}{2}} = X$; $X^{\frac{1}{2}}$ is the only matrix with this property.

Theorem A.4 *Let $X \in \mathcal{S}_n^{++}$ and $S \in \mathcal{S}_n^{++}$. Then all the eigenvalues of XS are real and positive.*

Proof:

The proof is immediate by noting that

$$XS \sim \left(X^{\frac{1}{2}}\right)^{-1} XS \left(X^{\frac{1}{2}}\right) = X^{\frac{1}{2}} S X^{\frac{1}{2}} \succ 0.$$

□

We will often use the notation $X^{-\frac{1}{2}} := \left(X^{\frac{1}{2}}\right)^{-1}$.

The eigenvalues of a symmetric matrix can be viewed as smooth functions on $\mathcal{S}_n$ in a sense made precise by the following theorem.

Theorem A.5 (Rellich) *Let an interval $(a, b) \subset \mathbf{R}$ be given. If $A : (a, b) \mapsto \mathcal{S}_n$ is a continuously differentiable function, then there exist n continuously differentiable functions $\lambda_i : (a, b) \mapsto \mathbf{R}$ $(i = 1, \ldots, n)$ such that $\lambda_1(t), \ldots, \lambda_n(t)$ give the values of the eigenvalues of $A(t)$ for each $t \in (a, b)$.*

The next lemma shows what happens to the spectrum of a positive semidefinite matrix if a skew symmetric matrix is added to it, in the case where the eigenvalues of the sum of the two matrices remain real numbers.

Lemma A.1 *Let $Q \in \mathcal{S}_n^{++}$, and let $M \in \mathbf{R}^{n \times n}$ be skew–symmetric ($M = -M^T$). One has $\det(Q + M) > 0$. Moreover, if $\lambda_i(Q + M) \in \mathbf{R}$ $(i = 1, \ldots, n)$, then*

$$0 < \lambda_{\min}(Q) \leq \lambda_{\min}(Q + M) \leq \lambda_{\max}(Q + M) \leq \lambda_{\max}(Q).$$

Proof:

First note that $Q + M$ is nonsingular since for all nonzero $x \in \mathbf{R}^n$:

$$x^T(Q + M)x = x^T Q x > 0,$$

using the skew symmetry of M. We therefore know that

$$\psi(t) := \det[Q + tM] \neq 0 \quad \forall t \in \mathbf{R},$$

since tM remains skew–symmetric. One now has that ψ is a continuous function of t which is nowhere zero and strictly positive for $t = 0$ as $\det(Q) > 0$. This shows $\det(Q + M) > 0$.

To prove the second part of the lemma, assume $\lambda > 0$ is such that $\lambda > \lambda_{\max}(Q)$. It then follows that $Q - \lambda I \prec 0$. By the same argument as above we then have $(Q + M) - \lambda I$ nonsingular, or

$$\det\left((Q + M) - \lambda I\right) \neq 0.$$

This implies that λ cannot be an eigenvalue of $Q + M$. Similarly, $Q + M$ cannot have an eigenvalue smaller than $\lambda_{\min}(Q)$. This gives the required result. □

The *spectral norm* $\|\cdot\|_2$ on $\mathbb{R}^{n\times n}$ is the norm induced by the Euclidean vector norm:

$$\|A\|_2 := \max_{x\in\mathbb{R}^n} \frac{\|Ax\|}{\|x\|} \quad (A \in \mathbb{R}^{n\times n}).$$

By the Raleigh–Ritz theorem, the spectral norm and spectral radius $\rho(\cdot)$ coincide for symmetric matrices:

$$\|A\|_2 = \rho(A) := \max_i |\lambda_i(A)| \quad \forall A \in \mathcal{S}_n.$$

The location of the eigenvalues of a matrix is bounded by the famous Gerschgorin theorem. For symmetric matrices the theorem states that

$$\lambda_k(A) \in \bigcup_{i=1}^{n} \left\{ t \;\middle|\; |t - a_{ii}| \le \sum_{j\neq i} |a_{ij}| \right\}, \quad k = 1, \ldots, n, \quad A \in \mathcal{S}_n.$$

As a consequence we find that the so-called diagonally dominant matrices are positive semi-definite.

Theorem A.6 (Diagonally dominant matrix is PSD) *A matrix $A \in \mathcal{S}_n$ is called diagonally dominant if*

$$a_{ii} \ge \sum_{j\neq i} |a_{ij}|, \quad i = 1, \ldots, n.$$

If A is diagonally dominant, then $A \in \mathcal{S}_n^+$.

A.3 THE TRACE OPERATOR AND THE FROBENIUS NORM

The trace of an $n \times n$ matrix A is defined as

$$\mathbf{Tr}\,(A) = \sum_{i=1}^{n} a_{ii}.$$

The trace is clearly a linear operator and has the following properties.

Theorem A.7 *Let $A \in \mathbb{R}^{n\times n}$ and $B \in \mathbb{R}^{n\times n}$. Then the following holds:*

- $\mathbf{Tr}\,(A) = \sum_{i=1}^{n} \lambda_i(A)$;
- $\mathbf{Tr}\,(A) = \mathbf{Tr}\,\left(A^T\right)$;
- $\mathbf{Tr}\,(AB) = \mathbf{Tr}\,(BA)$
- $\mathbf{Tr}\,(AB^T) = \mathbf{vec}(A)^T\mathbf{vec}(B) = \sum_{i,j=1}^{n} a_{ij}b_{ij}$.

The last item shows that we can view the usual Euclidean inner product on $\mathbb{R}^{n^2}$ as an inner product on $\mathbb{R}^{n\times n}$:

$$\langle A, B\rangle := \mathbf{Tr}\,\left(AB^T\right) = \mathbf{Tr}\,\left(B^TA\right) = \mathbf{Tr}\,\left(A^TB\right) = \mathbf{Tr}\,\left(B^TA\right). \tag{A.2}$$

The inner product in (A.2) induces the so-called *Frobenius* (or Euclidean) norm on $\mathbb{R}^{n\times n}$:

$$\|A\|^2 := \langle A, A\rangle = \mathbf{Tr}\left(AA^T\right) = \sum_{i,j=1}^{n} a_{ij}^2.$$

It now follows from the first item in Theorem (A.7) that

$$\|C\|^2 = \sum_{i=1}^{n} \lambda_i(C)^2 \text{ if } C \in \mathcal{S}_n.$$

The Frobenius and spectral norms are sub-multiplicative, *i.e.*

$$\|AB\| \le \|A\|\|B\|, \ \|AB\|_2 \le \|A\|_2\|B\|_2 \quad \forall A \in \mathbb{R}^{n\times n}, \ B \in \mathbb{R}^{n\times n},$$

and $\|A\|_2 \le \|A\|$ for all $A \in \mathbb{R}^{n\times n}$.

One can easily prove the useful inequality:

$$\mathbf{Tr}\,(AB) \le \lambda_{\max}(A)\mathbf{Tr}\,(B), \quad \text{for } A, B \succeq 0, \tag{A.3}$$

that is equivalent to

$$\|AB\| \le \|A\|_2\|B\|, \quad \text{for } A, B \in \mathbb{R}^{n\times n}. \tag{A.4}$$

Inequality (A.4) follows from (A.3) by replacing A by AA^T and B by BB^T in (A.3). Conversely, (A.3) follows from (A.4) by replacing A by $A^{\frac{1}{2}}$ and B by $B^{\frac{1}{2}}$ in (A.4).

Theorem A.8 (Fejer) *A matrix $A \in \mathcal{S}_n$ is positive semidefinite if and only if $\langle A, B\rangle \ge 0$ for all $B \in \mathcal{S}_n^+$. In other words, the cone $\mathcal{S}_n^+$ is self-dual.*

Proof:

Let $A \in \mathcal{S}_n^+$ and $B \in \mathcal{S}_n^+$; then

$$\langle A, B\rangle = \mathbf{Tr}\left(A^{\frac{1}{2}}A^{\frac{1}{2}}B^{\frac{1}{2}}B^{\frac{1}{2}}\right) = \mathbf{Tr}\left(A^{\frac{1}{2}}B^{\frac{1}{2}}B^{\frac{1}{2}}A^{\frac{1}{2}}\right) = \left\|A^{\frac{1}{2}}B^{\frac{1}{2}}\right\|^2 \ge 0.$$

Conversely, if $A \in \mathcal{S}_n$ and $\langle A, B\rangle \ge 0$ for all $B \in \mathcal{S}_n^+$, then let $x \in \mathbb{R}^n$ be given and set $B = xx^T \in \mathcal{S}_n^+$. Now

$$0 \le \langle A, B\rangle = \mathbf{Tr}\left(Axx^T\right) = \sum_{i,j=1}^{n} a_{ij}x_ix_j = x^TAx.$$

$\square$

For positive semidefinite matrices, the trace dominates the Frobenius norm, *i.e.*

$$\mathbf{Tr}\,(X) \ge \|X\| \quad \forall X \in \mathcal{S}_n^+.$$

This follows by applying the inequality

$$\sum_{i=1}^{n} x_i \geq \sqrt{\sum_{i=1}^{n} x_i^2} \quad \forall x \in \mathbf{R}_n^+$$

to the nonnegative eigenvalues of X. Similarly, one can apply the arithmetic-geometric mean inequality

$$\left(\prod_{i=1}^{n} x_i\right)^{1/n} \leq \frac{1}{n}\sum_{i=1}^{n} x_i \quad \forall x \in \mathbf{R}_n^+$$

to the eigenvalues of $X \in \mathcal{S}_n^+$ to obtain the inequality

$$\frac{1}{n}\mathbf{Tr}\,(X) \geq (\det(X))^{1/n} \quad \forall X \in \mathcal{S}_n^+,$$

where we have used the fact that $\det(A) = \prod_i \lambda_i(A)$ for any $A \in \mathbf{R}^{n\times n}$.

Lemma A.2 *If $X \in \mathcal{S}_n^+$ and $S \in \mathcal{S}_n^+$ and $\mathbf{Tr}\,(XS) = 0$, then $XS = SX = 0$.*

Proof:

By the properties of the trace operator

$$\mathbf{Tr}\,(XS) = \mathbf{Tr}\left(X^{\frac{1}{2}}X^{\frac{1}{2}}S^{\frac{1}{2}}S^{\frac{1}{2}}\right) = \mathbf{Tr}\left(S^{\frac{1}{2}}X^{\frac{1}{2}}X^{\frac{1}{2}}S^{\frac{1}{2}}\right) = \left\|S^{\frac{1}{2}}X^{\frac{1}{2}}\right\|^2.$$

Thus if $\mathbf{Tr}\,(XS) = 0$, it follows that $S^{\frac{1}{2}}X^{\frac{1}{2}} = 0$. Pre-multiplying by $S^{\frac{1}{2}}$ and post-multiplying by $X^{\frac{1}{2}}$ yields $SX = 0$, which in turn implies $(SX)^T = XS = 0$. □

The following lemma is used to prove that the search directions for the interior point methods described in this monograph are well defined. The proof given here is based on a proof given by Faybusovich [54].

Lemma A.3 *Let $A_i \in \mathcal{S}_n$ $(i = 1, \ldots, m)$ be linearly independent, and let $0 \neq Y \in \mathcal{S}_n^+$, $Z \in \mathcal{S}_n^{++}$. The matrix $M \in \mathcal{S}_m$ with entries*

$$m_{ij} := \mathbf{Tr}\,(A_i Z A_j Y), \quad i, j = 1, \ldots, m$$

is positive definite.

Proof:

We prove that the quadratic form

$$q(x) = x^T M x = \sum_{i,j=1}^{m} m_{ij} x_i x_j$$

is strictly positive for all nonzero $x \in \mathbf{R}^n$. To this end, note that for given $x \neq 0$,

$$q(x) = \mathbf{Tr}\left(\left(\sum_{i=1}^{m} x_i A_i\right) Z \left(\sum_{j=1}^{m} x_j A_j\right) Y\right).$$

Denoting $A(x) := \sum_{i=1}^{m} x_i A_i$ (which is nonzero by the linear independence of the $A_i's$), one has:

$$q(x) = \mathbf{Tr}\ (A(x)ZA(x)Y) > 0,$$

where the inequality follows from $0 \neq A(x)ZA(x) \succeq 0$ and $Y \succ 0$. □

A.4 THE LÖWNER PARTIAL ORDER AND THE SCHUR COMPLEMENT THEOREM

We define a partial ordering on $\mathcal{S}_n$ via:

$$A \succeq B \iff A - B \in \mathcal{S}_n^+.$$

This partial ordering is called the *Löwner partial order* on $\mathcal{S}^n$. (It motivates the alternative notation $X \succeq 0$ instead of $X \in \mathcal{S}_n^+$.) It follows immediately that

$$A \succeq B \iff C^T AC \succeq C^T BC \quad \forall C \in \mathbf{R}^{n\times n}.$$

One also has

$$A \succeq B \iff B^{-1} \succeq A^{-1} \quad (A \in \mathcal{S}_N^{++},\ B \in \mathcal{S}_n^{++}).$$

The Schur complement theorem gives us useful ways to express positive semidefiniteness of matrices with a block structure.

Theorem A.9 (Schur complement) *If*

$$M = \begin{bmatrix} A & B \\ B^T & C \end{bmatrix}$$

where A is positive definite and C is symmetric, then the matrix

$$C - B^T A^{-1} B$$

is called the Schur complement of A in X. The following are equivalent:

- *M is positive (semi)definite;*
- *$C - B^T A^{-1} B$ is positive (semi)definite.*

Proof:

The result follows by setting $D = -A^{-1}B$, and noting that

$$\begin{bmatrix} I & 0 \\ D^T & I \end{bmatrix} \begin{bmatrix} A & B \\ B^T & C \end{bmatrix} \begin{bmatrix} I & D \\ 0 & I \end{bmatrix} = \begin{bmatrix} A & 0 \\ 0 & C - B^T A^{-1} B \end{bmatrix}.$$

Since a block diagonal matrix is positive (semi)definite if and only if its diagonal blocks are positive (semi)definite, the proof is complete. □

Appendix B
Background material on convex optimization

In this Appendix we give some background material on convex analysis, convex optimization and nonlinear programming. All proofs are omitted here but may be found in the books by Rockafellar [160] (convex analysis) and Bazaraa *et al.* [16] (nonlinear programming).

B.1 CONVEX ANALYSIS

Definition B.1 (Convex set) *Let two points $x^1, x^2 \in \mathbb{R}^n$ and $0 \leq \lambda \leq 1$ be given. Then the point*

$$x = \lambda x^1 + (1 - \lambda)x^2$$

is a convex combination *of the two points x^1, x^2.*

The set $\mathcal{C} \subset \mathbb{R}^n$ is called convex, *if all convex combinations of any two points $x^1, x^2 \in \mathcal{C}$ are again in $\mathcal{C}$.*

Definition B.2 (Convex function) *A function $f : \mathcal{C} \to R$ defined on a convex set $\mathcal{C}$ is called* convex *if for all $x^1, x^2 \in \mathcal{C}$ and $0 \leq \lambda \leq 1$ one has*

$$f(\lambda x^1 + (1 - \lambda)x^2) \leq \lambda f(x^1) + (1 - \lambda)f(x^2).$$

The function is called strictly convex *if the last inequality is strict.*

A function is convex if and only if its epigraph is convex.

Definition B.3 (Epigraph) *The epigraph of a function $f : \mathcal{C} \to R$ is the $(n + 1)$-dimensional set*

$$\{(x, \tau) \,:\, f(x) \leq \tau, x \in \mathcal{C},\ \tau \in \mathbb{R}\}.$$

Theorem B.1 *A twice differentiable function f is convex (resp. strictly convex) on an open set $\mathcal{C}$ if and only if its Hessian $\nabla^2 f$ is positive semidefinite (resp. positive definite) on $\mathcal{C}$.*

Example B.1 *The function $f : \mathcal{S}_n^{++} \mapsto \mathbf{R}$ defined by $f(X) = -\log(\det(X))$ has a positive definite Hessian and is therefore strictly convex. (This is proven in Appendix C.)*

□

Strictly convex functions are useful for proving 'uniqueness properties', due to the following result.

Theorem B.2 *If a strictly convex function has a minimizer over a convex set, then this minimizer is unique.*

Definition B.4 (Convex cone) *The set $\mathcal{K} \subset \mathbf{R}^n$ is a* convex cone *if it is a convex set and for all $x \in \mathcal{K}$ and $0 < \lambda$ one has $\lambda x \in \mathcal{K}$.*

Example B.2 *Four examples of convex cones in $\mathcal{S}_n$ are:*

- *the symmetric positive semidefinite cone:*

$$\mathcal{S}_n^+ := \left\{ A \in \mathcal{S}_n \;\middle|\; x^T X x \geq 0 \quad \text{for all } x \in \mathbf{R}^n \right\};$$

- *the copositive cone:*

$$\mathcal{C}_n := \left\{ A \in \mathcal{S}_n \;\middle|\; x^T A x \geq 0 \quad \text{for all } x \in \mathbf{R}_+^n \right\};$$

- *the cone of completely positive matrices:*

$$\mathcal{C}_n^* = \left\{ A \in \mathcal{S}_n \;\middle|\; A = \sum_{i=1}^k x_i x_i^T,\ x_i \in \mathbf{R}_+^n \ \ (i = 1, \ldots, k) \ \text{ for any } k \right\};$$

- *the cone of nonnegative matrices:*

$$\mathcal{N}_n = \{ A \in \mathcal{S}_n \mid a_{ij} \geq 0 \ \ (i, j = 1, \ldots, n) \}.$$

□

Definition B.5 (Face (of a cone)) *A subset $\mathcal{F}$ of a convex cone $\mathcal{K}$ is called a* face *of $\mathcal{K}$ if for all $x \in \mathcal{F}$ and $y, z \in \mathcal{K}$ one has $x = y + z$ if and only if $y \in \mathcal{F}$ and $z \in \mathcal{F}$.*

Example B.3 *An example of a face of the cone of positive semidefinite matrices $\mathcal{S}_n^+$ is*

$$\mathcal{F} := \left\{ \left[\begin{array}{cc} U & 0_{r\times(n-r)} \\ 0_{(n-r)\times r} & 0_{(n-r)\times(n-r)} \end{array} \right] \;\middle|\; U \in \mathcal{S}_r^+ \right\}.$$

Note that if $A \in \mathcal{S}_n^+$ and $B \in \mathcal{S}_n^+$, then $A + B \in \mathcal{F}$ if and only if $A \in \mathcal{F}$ and $B \in \mathcal{F}$. □

Definition B.6 (Extreme ray (of a cone)) *A subset of a convex cone $\mathcal{K}$ is called an* extreme ray *of $\mathcal{K}$ if it is a one dimensional face of $\mathcal{K}$,* i.e. *a face that is a half line emanating from the origin.*

Example B.4 *Any $0 \neq x \in \mathbf{R}^n$ defines an extreme ray of the cone of positive semidefinite matrices $\mathcal{S}_n^+$ via*

$$\left\{U \mid U = cxx^T,\ c > 0\right\}.$$

Similarly, any $y \in \mathbf{R}_+^n$ defines an extreme ray of the cone of completely positive semidefinite matrices $\mathcal{C}_n^*$ via

$$\left\{U \mid U = cyy^T,\ c > 0\right\}.$$

□

Definition B.7 *Let a convex set $\mathcal{C}$ be given. The point $x \in \mathcal{C}$ is in the* relative interior *of $\mathcal{C}$ if for all $\hat{x} \in \mathcal{C}$ there exists $\tilde{x} \in \mathcal{C}$ and $0 < \lambda < 1$ such that $x = \lambda\hat{x} + (1-\lambda)\tilde{x}$. The set of relative interior points of the set $\mathcal{C}$ will be denoted by* $\mathrm{ri}\,(\mathcal{C})$.

Theorem B.3 *Assume that $\mathcal{C}_1$ and $\mathcal{C}_2$ are nonempty convex sets and* $\mathrm{ri}\,(\mathcal{C}_1) \cap \mathrm{ri}\,(\mathcal{C}_2) \neq \emptyset$. *Then*

$$\mathrm{ri}\,(\mathcal{C}_1 \cap \mathcal{C}_2) = \mathrm{ri}\,(\mathcal{C}_1) \cap \mathrm{ri}\,(\mathcal{C}_2).$$

Example B.5 *Note that the interior of the cone of positive semidefinite matrices $\mathcal{S}_n^+$ is the cone of positive definite matrices $\mathcal{S}_n^{++}$. Let $\mathcal{A} \subset \mathcal{S}_n$ denote an affine space. Assume that there exists an $X \in \mathcal{A}$ that is also positive definite. Then, by the last theorem,*

$$\mathrm{ri}\,\left(\mathcal{A} \cap \mathcal{S}_n^+\right) = \mathrm{ri}\,(\mathcal{A}) \cap \mathrm{ri}\,\left(\mathcal{S}_n^+\right) = \mathcal{A} \cap \mathcal{S}_n^{++},$$

since $\mathrm{ri}\,(\mathcal{A}) = \mathcal{A}$. □

Theorem B.4 (Separation theorem for convex sets) *Let $\mathcal{C}_1$ and $\mathcal{C}_2$ be nonempty convex sets in $\mathbb{R}^k$. There exists a $r \in \mathbb{R}^k$ such that*

$$\sup_{x\in\mathcal{C}_1} r^T x \le \inf_{y\in\mathcal{C}_2} r^T y$$

and

$$\inf_{x\in\mathcal{C}_1} r^T x < \sup_{y\in\mathcal{C}_2} r^T y$$

if and only if the relative interiors of $\mathcal{C}_1$ and $\mathcal{C}_2$ are disjoint.

The second inequality merely excludes the uninteresting case where the separating hyperplane contains both $\mathcal{C}_1$ and $\mathcal{C}_2$, *i.e.* it ensures so-called *proper separation.*

B.2 DUALITY IN CONVEX OPTIMIZATION

We consider the generic convex optimization problem:

$$\begin{array}{lll} (CO) & p^* = \inf_x & f(x) \\ & \text{s.t.} & g_j(x) \le 0, \quad j = 1, \cdots, m \\ & & x \in \mathcal{C}, \end{array} \tag{B.1}$$

where $\mathcal{C} \subseteq \mathbb{R}^n$ is a convex set and $f, g_1, \cdots, g_m$ are differentiable convex functions on $\mathcal{C}$ (or on an open set that contains the set $\mathcal{C}$).

For the convex optimization problem (CO) one defines the Lagrange function (or Lagrangian)

$$L(x,y) := f(x) + \sum_{j=1}^{m} y_j g_j(x) \tag{B.2}$$

where $x \in \mathcal{C}$ and $y \ge 0$.

Definition B.8 *A vector pair $(\bar{x}, \bar{y}) \in \mathbb{R}^{n+m}$, $\bar{x} \in \mathcal{C}$ and $\bar{y} \ge 0$ is called a saddle point of the Lagrange function L if*

$$L(\bar{x}, y) \le L(\bar{x}, \bar{y}) \le L(x, \bar{y}) \tag{B.3}$$

for all $x \in \mathcal{C}$ and $y \ge 0$.

One easily sees that (B.3) is equivalent with

$$L(\bar{x}, y) \le L(x, \bar{y}) \quad \text{for all} \quad x \in \mathcal{C}, \quad y \ge 0.$$

Lemma B.1 *The vector $(\bar{x}, \bar{y}) \in \mathbb{R}^{n+m}$, $\bar{x} \in \mathcal{C}$ and $\bar{y} \ge 0$ is a saddle point of $L(x, y)$ if and only if*

$$\inf_{x\in\mathcal{C}} \sup_{y\ge 0} L(x,y) = L(\bar{x}, \bar{y}) = \sup_{y\ge 0} \inf_{x\in\mathcal{C}} L(x,y). \tag{B.4}$$

Since we can reformulate (CO) as

$$p^* = \inf_{x \in \mathcal{C}} \sup_{y \geq 0} \left\{ f(x) + \sum_{j=1}^{m} y_j g_j(x) \right\},$$

it follows that $\bar{x}$ is an optimal solution of problem (CO) if there exists a $\bar{y} \geq 0$ such that $(\bar{x}, \bar{y})$ is a saddle point of the Lagrangian.

To ensure the existence of a saddle point of the Lagrangian, it is sufficient to require the so-called Slater regularity condition (Slater constraint qualification).

Assumption B.1 (Slater regularity) *There exists an $x \in \mathrm{ri}\,(\mathcal{C})$ such that*

- $g_j(x) < 0$ *if g_j not linear or affine;*
- $g_j(x) \leq 0$ *if g_j is linear or affine.*

Under the Slater regularity assumption, we therefore have a one-to-one correspondence between a saddle point of the Lagrangian and an optimal solution of (CO).

Theorem B.5 (Karush–Kuhn–Tucker) *The convex optimization problem (CO) is given. Assume that the Slater regularity condition is satisfied. The vector $\bar{x}$ is an optimal solution of (CO) if and only if there is a vector $\bar{y}$ such that $(\bar{x}, \bar{y})$ is a saddle point of the Lagrange function L.*

The formulation of the saddle point condition in Lemma (B.1) motivates the concept of the Lagrangian dual.

Definition B.9 (Lagrangian dual) *Denote*

$$\psi(y) = \inf_{x \in \mathcal{C}} \{ f(x) + \sum_{j=1}^{m} y_j g_j(x) \}.$$

The problem

$$\sup \psi(y)$$
$$y \geq 0$$

is called the Lagrangian dual of the convex optimization problem (CO).

It is straightforward to show the so-called weak duality property.

Theorem B.6 (Weak duality) *If $\bar{x}$ is a feasible solution of (CO) and $\bar{y} \geq 0$, then*

$$\psi(\bar{y}) \leq f(\bar{x})$$

and equality holds if and only if $\inf_{x \in C} \{f(x) + \sum_{j=1}^{m} \bar{y}_j g_j(x)\} = f(\bar{x})$.

Under the Slater regularity assumption we have a stronger duality result, by the Karush–Kuhn–Tucker theorem and Lemma (B.1).

Theorem B.7 (Strong duality) *Assume that (CO) satisfies the Slater regularity condition, and let $\bar{x}$ be a feasible solution of (CO). Now the vector $\bar{x}$ is an optimal solution of (CO) if and only if there exists a $\bar{y} \geq 0$ such that $\bar{y}$ is an optimal solution of the Lagrangian dual problem and*

$$\psi(\bar{y}) = f(\bar{x}).$$

B.3 THE KKT OPTIMALITY CONDITIONS

We now state the Karush–Kuhn–Tucker (KKT) necessary and sufficient optimality conditions for problem (CO). First we define the notion of a KKT point.

Definition B.10 (KKT point) *The vector $(\hat{x}, \bar{y}) \in C \times \mathbf{R}^m$ is called a Karush–Kuhn–Tucker (KKT) point of (CO) if*

$$\begin{array}{ll} (i) & g_j(\hat{x}) \leq 0, \textit{for all } j = 1, \dots, m \\ (ii) & 0 = \nabla f(\hat{x}) + \sum_{j=1}^{m} \bar{y}_j \nabla g_j(\hat{x}) \\ (iii) & \sum_{j=1}^{m} \bar{y}_j g_j(\hat{x}) = 0 \\ (iv) & \bar{y} \geq 0. \end{array}$$

A KKT point is a saddle point of the Lagrangian L of (CO). Conversely, a saddle point of L is a KKT point of (CO). This leads us to the following result.

Theorem B.8 (KKT conditions) *If $(\bar{x}, \bar{y})$ is a KKT point, then $\bar{x}$ is an optimal solution of (CO). Conversely — under the Slater regularity assumption — a feasible solution $\bar{x}$ of (CO) is optimal if there exists a $\bar{y} \in \mathbf{R}^m$ such that $(\bar{x}, \bar{y})$ is a KKT point.*

We say that $x \in C$ meets the KKT conditions if there exists a $y \geq 0$ such that (x, y) is a KKT point of (CO).

If we drop the convexity requirements on f and g_i in the statement of (CO), then the KKT conditions remain *necessary* optimality conditions under the Slater regularity assumption.

Appendix C
The function $\log\det(X)$

In this appendix we develop the matrix calculus needed to derive the gradient and Hessian of the function $\log\det(X)$, and show that it is a strictly concave function.

Lemma C.1 *Let* $f : \mathrm{int}\,(\mathcal{S}_n^+) \mapsto \mathbb{R}$ *be given by*

$$f(X) = \log\det X,$$

Denoting

$$\nabla f(X) := \begin{bmatrix} \frac{\partial f(X)}{\partial x_{11}} & \cdots & \frac{\partial f(X)}{\partial x_{1n}} \\ \vdots & \ddots & \vdots \\ \frac{\partial f(X)}{\partial x_{n1}} & \cdots & \frac{\partial f(X)}{\partial x_{nn}} \end{bmatrix},$$

one has $\nabla f(X) = X^{-1}$.

Proof:

Let $X \in \mathrm{int}\,(\mathcal{S}_n^+)$ be given and let $H \in \mathcal{S}_n$ be such that $X + H \in \mathrm{int}(\mathcal{S}_n^+)$. One has

$$\begin{aligned} f(X+H) - f(X) &= \log\det(X+H) - \log\det(X) \\ &= \log\det\left(X^{-1}(X+H)\right) \\ &= \log\det\left(I + X^{-\frac{1}{2}}HX^{-\frac{1}{2}}\right). \end{aligned}$$

By the arithmetic-geometric inequality applied to the eigenvalues of $X^{-\frac{1}{2}}HX^{-\frac{1}{2}}$ one has

$$\begin{aligned} \log\det\left(I + X^{-\frac{1}{2}}HX^{-\frac{1}{2}}\right) &\le \log\left(\frac{1}{n}\mathbf{Tr}\left(I + X^{-\frac{1}{2}}HX^{-\frac{1}{2}}\right)\right)^n \\ &= n\log\left(\frac{1}{n}\mathbf{Tr}\left(I + X^{-\frac{1}{2}}HX^{-\frac{1}{2}}\right)\right) \end{aligned}$$

$$= n \log \left(1 + \frac{1}{n}\mathbf{Tr}\left(X^{-\frac{1}{2}}HX^{-\frac{1}{2}}\right)\right).$$

Using the well-known inequality $\log(1+t) \leq t$ we arrive at

$$f(X+H) - f(X) \leq \mathbf{Tr}\left(X^{-\frac{1}{2}}HX^{-\frac{1}{2}}\right) = \langle X^{-1}, H\rangle.$$

This shows that X^{-1} is a subgradient of f at X. Since f is assumed differentiable, the subgradient is unique and equals the gradient $\nabla f(X)$. □

The proof of the next result is trivial.

Lemma C.2 *Let $f : \text{int}\,(\mathcal{S}_n^+) \mapsto \mathbf{R}$ be given by*

$$f(X) = \mathbf{Tr}\,(CX),$$

where $C \in \mathcal{S}_n$. One has $\nabla f(X) = C$.

The following result is used to derive the Hessian of the log-barrier function $f_{bar}(X) = -\log\det(X)$.

Lemma C.3 *Let $f : \mathcal{S}_n^{++} \mapsto \mathbf{R}$ be given by*

$$f(X) = \log\det X.$$

If $\nabla^2 f$ denotes the derivative of $\nabla f : X \mapsto X^{-1}$ with respect to X, then $\nabla^2 f(X)$ is the linear operator which satisfies

$$\nabla^2 f(X)H = -X^{-1}HX^{-1}, \quad \forall H \in \mathcal{S}_n,$$

for a given invertible X.

Proof:

Let $\mathrm{L}(\mathcal{S}_n, \mathcal{S}_n)$ denote the space of linear operators which map $\mathcal{S}_n$ to $\mathcal{S}_n$. The Frechet derivative of ∇f is defined as the (unique) function $\nabla^2 f : \mathcal{S}_n \mapsto \mathrm{L}(\mathcal{S}_n, \mathcal{S}_n)$ such that

$$\lim_{\|H\|\to 0} \frac{\left\|\nabla f(X+H) - \nabla f(X) - \nabla^2 f(X)H\right\|}{\|H\|} = 0. \tag{C.1}$$

We show that $\nabla^2 f(X)H := -X^{-1}HX^{-1}$ satisfies (C.1). To this end, let $H \in \mathcal{S}_n$ be such that $(X+H)$ is invertible, and consider

$$\begin{aligned}
& \left\|\nabla f(X+H) - \nabla f(X) - \nabla^2 f(X)H\right\| \\
= \; & \left\|(X+H)^{-1} - X^{-1} + X^{-1}HX^{-1}\right\| \\
= \; & \left\|(X+H)^{-1}\left(I - (X+H)X^{-1} + (X+H)X^{-1}HX^{-1}\right)\right\| \\
= \; & \left\|(X+H)^{-1}\left(HX^{-1}HX^{-1}\right)\right\| \\
\leq \; & \left\|(X+H)^{-1}\right\| \|H\| \left\|X^{-1}HX^{-1}\right\|,
\end{aligned}$$

which shows that (C.1) indeed holds. □

By Lemma A.3, the Hessian of the function $f(X) = -\log\det(X)$ is a positive definite operator which implies that f is strictly convex on $\mathcal{S}_n^{++}$. We state this observation as a theorem.

Theorem C.1 *The function $f : \mathcal{S}_n^{++} \mapsto \mathbf{R}$ defined by*

$$f(X) = -\log\det(X)$$

is strictly convex.

An alternative proof of this theorem is given in [85] (Theorem 7.6.6).

Appendix D
Real analytic functions

This appendix gives some elementary properties of analytic functions which are used in this monograph. It is based on notes by M. Halická [76].

Definition D.1 *A function $f(x) : \mathbf{R} \to \mathbf{R}$ is said to be analytic at $x = a$ if there exist $r > 0$ and $\{a_n\}$ such that*

$$f(x) = \sum_{n=0}^{\infty} a_n (x-a)^n, \quad \forall x : |x - a| < r. \tag{D.1}$$

Remark D.1 *Taking the nth derivative in (D.1), it is easy to see that $a_n = \frac{f^{(n)}(a)}{n!}$. Hence the series in (D.1) is the Taylor series of $f(x)$ at $x = a$.*

Remark D.2 *A function $f(x)$ that is analytic at $x = a$ is infinitely differentiable in some neighborhood of a and the corresponding Taylor series converges to $f(x)$ in some neighborhood of a (sometimes this is used as the definition of an analytic function). If a function is infinitely differentiable, then it is not necessarily analytic. A well-known example is the Cauchy function $f(x) = e^{-1/x^2}$ for $x \neq 0$ and $f(x) = 0$ for $x = 0$. At $x = 0$ all derivatives are zero and hence the corresponding Taylor series converges to the zero function.*

Definition D.2 *Let $I \subset \mathbf{R}$ be an open interval. Then f is analytic on I if it is analytic at any point $a \in I$.*

We can extend this definition to closed intervals by making the following changes to the above.

(a) if $f(x)$ is defined for all $x \geq a$, then $f(x)$ is defined to be analytic at a if there exist an $r > 0$ and a series of coefficients $\{a_n\}$ such that the equality in (D.1) holds for all $x \geq a$.

(b) if $f(x)$ is defined for all $x > a$, then we say that $f(x)$ can be analytically extended to a if there exist an $r > 0$ and a series of coefficients $\{a_n\}$ such that the equality in (D.1) holds for all $x \geq a$.

The following result is used in the proof of Theorem 3.6 (that the central path has a unique limit point in the optimal set).

Theorem D.1 *Let $f : [0, 1) \rightarrow \mathbf{R}$ be analytic on $[0, 1)$. Let $f(0) = 0$ and $x = 0$ be an accumulation point of all x such that $f(x) = 0$. Then $f(x) = 0$ for all $x \in [0, 1)$.*

Proof:

Since f is analytic at $x = 0$, there exist $r > 0$ and a_n such that

$$f(x) = \sum_{n=0}^{\infty} a_n x^n, \quad \forall x : 0 \leq x < r. \tag{D.2}$$

We show that $a_n = 0, \forall n$. Let a_k be the first non-zero coefficient in (D.2). Hence

$$f(x) = x^k(a_k + a_{k+1}x + \ldots), \quad \forall x : 0 \leq x < r. \tag{D.3}$$

Let $0 < \rho < r$. The series in (D.3) converges at $x = \rho$ and hence the numbers $a_n \rho^n$ are bounded. Let $|a_n|\rho^n < K$, i.e., $|a_n| < K/\rho^n, \forall n$. Then

$$|f(x)| \geq |x^k|(|a_k| - \frac{K|x|}{\rho^{k+1}} - \frac{K|x|^2}{\rho^{k+2}} - \ldots) = |x^k| \left(|a_k| - \frac{K|x|}{\rho^k(\rho - |x|)} \right), \tag{D.4}$$

for all $0 < x < \rho$. The last expression in (D.4) is positive for all sufficiently small x. This is in contradiction with the assumption that f has roots arbitrarily close to $x = 0$. Hence $a_n = 0, \forall n$ and $f(x) = 0$ in some right neighborhood of 0. Using the analyticity in $(0, 1)$ it can be shown that it is zero on the whole domain. □

Appendix E
The (symmetric) Kronecker product

This appendix contains various results about the Kronecker and symmetric Kronecker product. The part on the symmetric Kronecker product is based on Appendix D in Aarts [1].

We will use the **vec** and **svec** notation frequently in this appendix, and restate the definitions here for convenience.

Definition E.1 *For any symmetric $n \times n$ matrix U, the vector* $\mathbf{svec}(U) \in \mathbf{R}^{\frac{1}{2}n(n+1)}$ *is defined as*

$$\mathbf{svec}(U) = \left(u_{11}, \sqrt{2}u_{21}, \ldots, \sqrt{2}u_{n1}, u_{22}, \sqrt{2}u_{32}, \ldots, \sqrt{2}u_{n2}, \ldots, u_{nn}\right)^T,$$

such that

$$\mathbf{svec}(U)^T \mathbf{svec}(U) = \mathbf{Tr}\,(U^T U) = \mathbf{vec}(U)^T \mathbf{vec}(U),$$

where

$$\mathbf{vec}(U) = (u_{11}, u_{21}, \ldots, u_{n1}, u_{12}, \ldots, u_{nn})^T.$$

The inverse map of **svec** is denoted by **smat**.

E.1 THE KRONECKER PRODUCT

Definition E.2 *Let $G \in \mathbf{R}^{m\times n}$ and $K \in \mathbf{R}^{r\times s}$. Then $G\otimes K$ is defined as the $mr \times ns$ matrix with block structure*

$$G \otimes K = [g_{ij}K] \quad i = 1, \ldots, m, \ j = 1, \ldots, n,$$

i.e. *the block in position ij is given by $g_{ij}K$.*

The following identities are proven in Horn and Johnson [86]. (We assume that the sizes of the matrices are such that the relations are defined and that the inverses exist where referenced.)

Theorem E.1 *Let K, L, G and H be real matrices.*

- $G \otimes K\mathbf{vec}(H) = \mathbf{vec}\left(KHG^T\right)$;
- $(G \otimes K)^T = G^T \otimes K^T$;
- $(G \otimes K)^{-1} = G^{-1} \otimes K^{-1}$;
- $(G \otimes K)(H \otimes L) = (GH) \otimes (KL)$;
- *The eigenvalues of $G \otimes K$ are given by $\lambda_i(G)\lambda_j(K)\ \forall\ i,j = 1,\dots,n$. As a consequence, if G and K are positive (semi)definite, then so is $G \otimes K$;*
- *If $Gx_i = \lambda_i(G)x_i$ and $Ky_j = \lambda_j(K)y_j$, then $\mathbf{vec}\left(y_j x_i^T\right)$ is an eigenvector of $G \otimes K$ with corresponding eigenvalue $\lambda_i(G)\lambda_j(K)$.*

As an application of the Kronecker product we can analyse the solutions of so-called Lyapunov equations.

Theorem E.2 *Let $A \in \mathbf{R}^{n\times n}$ and $B \in \mathcal{S}_n$. The Lyapunov equation*

$$AX + XA^T = B \tag{E.1}$$

has a unique symmetric solution if A and $-A$ have no eigenvalues in common.

Proof:

Using the first item in Theorem E.1, we can rewrite (E.1) as

$$\mathbf{vec}\left(AX + XA^T\right) = \mathbf{vec}\left(AXI + IXA^T\right) = (I \otimes A + A \otimes I)\,\mathbf{vec}(X) = \mathbf{vec}(B).$$

By using the fourth item of Theorem E.1 we have

$$(I \otimes A)(A \otimes I) = (IA) \otimes (AI) = (AI) \otimes (IA) = (A \otimes I)(I \otimes A).$$

In other words, the matrices $A \otimes I$ and $I \otimes A$ commute and therefore share a set of eigenvectors. Moreover, by the fifth item of Theorem E.1 we know that the eigenvalues of $A\otimes I$ and $I\otimes A$ are obtained by taking n copies of the spectrum of A. Therefore each eigenvalue of $(I \otimes A + A \otimes I)$ is given by $\lambda_i(A) + \lambda_j(A)$ for some i, j. It follows that the matrix $(I \otimes A + A \otimes I)$ is nonsingular if A and $-A$ have no eigenvalues in common. In this case equation (E.1) has a unique solution. We must still verify that this solution is symmetric. To this end, note that if X is a solution of (E.1), then so is X^T. This completes the proof. □

E.2 THE SYMMETRIC KRONECKER PRODUCT

Definition E.3 *The symmetric Kronecker product of any two $n \times n$ matrices G and K (not necessarily symmetric), is a mapping on a vector $u = \mathbf{svec}(U)$ where U is a symmetric $n \times n$ matrix and is defined as*

$$(G \otimes_s K)u = \frac{1}{2}\mathbf{svec}(KUG^T + GUK^T). \qquad \text{(E.2)}$$

Note that the linear operator $G \otimes_s K$ is defined implicitly in (E.2). We can give a matrix representation of $G \otimes_s K$ by introducing the orthogonal $\frac{1}{2}n(n+1) \times n$ matrix Q (*i.e.* $QQ^T = I_{\frac{1}{2}n(n+1)}$), with the property that

$$Q\mathbf{vec}(U) = \mathbf{svec}(U) \text{ and } Q^T\mathbf{svec}(U) = \mathbf{vec}(U) \quad \forall U \in \mathcal{S}_n. \qquad \text{(E.3)}$$

Theorem E.3 *Let Q be the unique orthogonal $\frac{1}{2}n(n+1) \times n$ matrix that satisfies (E.3). For any $G \in \mathbb{R}^{n\times n}$ and $K \in \mathbb{R}^{n\times n}$ one has*

$$G \otimes_s K = \frac{1}{2}Q\left(G \otimes K + K \otimes G\right)Q^T.$$

Proof:

Let $U \in \mathcal{S}_n$ be given and note that

$$\begin{aligned}
\frac{1}{2}Q\left(G \otimes K + K \otimes G\right)Q^T\mathbf{svec}(U) &= \frac{1}{2}Q\left(G \otimes K + K \otimes G\right)\mathbf{vec}(U) \\
&= \frac{1}{2}Q\left((G \otimes K)\mathbf{vec}(U) + (K \otimes G)\mathbf{vec}(U)\right) \\
&= \frac{1}{2}Q\left(\mathbf{vec}\left(KUG^T\right) + \mathbf{vec}\left(GUK^T\right)\right) \\
&= \frac{1}{2}Q\left(\mathbf{vec}\left(KUG^T + \left(KUG^T\right)^T\right)\right) \\
&= \frac{1}{2}\mathbf{svec}\left(KUG^T + GUK^T\right) \\
&= (G \otimes_s K)\mathbf{svec}(U),
\end{aligned}$$

where we have used the first identity in Theorem E.1 to obtain the third equality. □

Definition E.4 *If for every vector $u = \mathbf{svec}(U)$ where U is a symmetric nonzero matrix,*

$$u^T(G \otimes_s K)u > 0,$$

then $(G \otimes_s K)$ is called positive definite.

Lemma E.1 *The symmetric Kronecker product has the following properties.*

1. $(G \otimes_s K) = (K \otimes_s G)$;
2. $(G \otimes_s K)(H \otimes_s L) = \frac{1}{2}(GH \otimes_s KL + GL \otimes_s KH)$;
3. $\mathbf{svec}(U)^T\mathbf{svec}(V) = \mathbf{Tr}\,(UV) = \mathbf{Tr}\,(VU)$ *for two symmetric matrices U and V;*
4. *If G and K are symmetric positive definite, then $(G \otimes_s K)$ is positive definite;*
5. $(G \otimes_s K)^T = (G^T \otimes_s K^T)$.

Proof:

1. This directly follows from Definition E.3.
2. Let U be a symmetric matrix, then

$$\begin{aligned}
& (G \otimes_s K)(H \otimes_s L)\mathbf{svec}(U) \\
= \;& \frac{1}{2}(G \otimes_s K)\mathbf{svec}(HUL^T + LUH^T) \\
= \;& \frac{1}{4}\mathbf{svec}\big(GHUL^TK^T + KHUL^TG^T + GLUH^TK^T + KLUH^TG^T\big) \\
= \;& \frac{1}{4}\mathbf{svec}\big(GHU(KL)^T + KLU(GH)^T + KHU(GL)^T + GLU(KH)^T\big) \\
= \;& \frac{1}{2}(GH \otimes_s KL + GL \otimes_s KH)\mathbf{svec}(U).
\end{aligned}$$

3. See Definition E.1.
4. For every symmetric nonzero matrix U we have to prove that

$$u^T(G \otimes_s K)u > 0,$$

where $u = \mathbf{svec}(U)$. Now

$$u^T(G \otimes_s K)u = \frac{1}{2}u^T\mathbf{svec}(GUK + KUG) = \frac{1}{2}\mathbf{svec}(U)^T\mathbf{svec}(GUK + KUG),$$

since G and K are symmetric. By using Property 3 we derive

$$\frac{1}{2}(\mathbf{svec}(U))^T\mathbf{svec}(GUK + KUG) = \frac{1}{2}(\mathbf{Tr}\,(UGUK) + \mathbf{Tr}\,(UKUG)).$$

Since U is nonzero and K and G are symmetric positive definite and therefore nonsingular, we obtain that $K^{\frac{1}{2}}UG^{\frac{1}{2}} \neq 0$ and thus

$$\frac{1}{2}(\mathbf{Tr}\,(UGUK) + \mathbf{Tr}\,(UKUG)) = \mathbf{Tr}\,(UKUG) = \left\|K^{\frac{1}{2}}UG^{\frac{1}{2}}\right\|^2 > 0.$$

5. Define $u = \mathbf{svec}(U)$ and $\ell = \mathbf{svec}(L)$ for arbitrary symmetric matrices U and L and $m = (G \otimes_s K)u$. Now

$$\ell^T m = (\mathbf{svec}(L))^T (G \otimes_s K)u = \frac{1}{2}(\mathbf{svec}(L))^T \mathbf{svec}(GUK^T + KUG^T)$$

and by using Property 3 it follows that

$$\begin{aligned} & \frac{1}{2}(\mathbf{svec}(L))^T \mathbf{svec}(GUK^T + KUG^T) \\ = & \frac{1}{2}\mathbf{Tr}\,(LGUK^T + LKUG^T) \\ = & \frac{1}{2}\mathbf{Tr}\,(UK^T LG + UG^T LK) \\ = & \frac{1}{2}(\mathbf{svec}(U))^T \mathbf{svec}(K^T LG + G^T LK) \\ = & (\mathbf{svec}(U))^T (G^T \otimes_s K^T)\mathbf{svec}(L) \\ = & (\mathbf{svec}(U))^T (G^T \otimes_s K^T)\ell. \end{aligned}$$

Since $\ell^T m = m^T \ell$, we obtain

$$m^T = (\mathbf{svec}(U))^T (G^T \otimes_s K^T)$$

and from the definition of m we derive

$$m^T = ((G \otimes_s K)\mathbf{svec}(U))^T = \mathbf{svec}(U)^T (G \otimes_s K)^T$$

and thus

$$(G \otimes_s K)^T = (G^T \otimes_s K^T).$$

□

Definition E.5 *For every nonsingular matrix P we define the operators $\mathcal{E}$ and $\mathcal{F}$ as*

$$\mathcal{E} = (P \otimes_s P^{-T} S), \qquad \mathcal{F} = (PX \otimes_s P^{-T}).$$

By using Property 2 in Lemma E.1 we obtain

$$\mathcal{E} = (P \otimes_s P^{-T} S) = (I \otimes_s P^{-T} S P^{-1})(P \otimes_s P)$$

and

$$\mathcal{F} = (PX \otimes_s P^{-T}) = (PXP^T \otimes_s I)(P^{-T} \otimes_s P^{-T}).$$

We define the operator H_P for any nonsingular $n \times n$ matrix P as

$$H_P(Q) = \frac{1}{2}(PQP^{-1} + P^{-T} Q^T P^T).$$

The next theorems are based on Theorem 3.1 and Theorem 3.2 in Todd *et al.* [173].

Theorem E.4 *If the matrices X and S are positive definite and the matrix $H_P(XS)$ is symmetric positive semidefinite, then $\mathcal{E}^{-1}\mathcal{F}$ is positive definite.*

Proof:

Consider an arbitrary nonzero vector $g \in \mathbf{R}^{\frac{n(n+1)}{2}}$. We now prove that $g^T\mathcal{E}^{-1}\mathcal{F}g > 0$. Defining $k := \mathcal{E}^{-T}g$ and $K = \mathbf{smat}(k)$, where k is also nonzero since $\mathcal{E}^{-T}$ exists, it follows by using Property 5 in Lemma E.1 that

$$g^T\mathcal{E}^{-1}\mathcal{F}g = k^T\mathcal{F}\mathcal{E}^Tk = k^T(PX \otimes_s P^{-T})(P^T \otimes_s SP^{-1})k.$$

By using Property 2 in Lemma E.1 we derive

$$\begin{aligned} k^T(PX \otimes_s P^{-T})(P^T \otimes_s SP^{-1})k &= \frac{1}{2}k^T(PXP^T \otimes_s P^{-T}SP^{-1})k \\ &\quad +\frac{1}{2}k^T(PXSP^{-1} \otimes_s I)k. \end{aligned}$$

Since X and S are symmetric positive definite, PXP^T and $P^{-T}SP^{-1}$ are symmetric positive definite and by using Property 4 in Lemma E.1 it follows that $(PXP^T \otimes_s P^{-T}SP^{-1})$ is positive definite. Therefore

$$\begin{aligned} & \frac{1}{2}k^T(PXP^T \otimes_s P^{-T}SP^{-1})k + \frac{1}{2}k^T(PXSP^{-1} \otimes_s I)k \\ > \; & \frac{1}{2}k^T(PXSP^{-1} \otimes_s I)k \\ = \; & \frac{1}{4}(\mathbf{svec}(K))^T\mathbf{svec}(PXSP^{-1}K + KP^{-T}SXP^T). \end{aligned}$$

By using Property 3 in Lemma E.1 we now obtain

$$\begin{aligned} & \frac{1}{4}\mathbf{svec}(K)^T\mathbf{svec}\left(PXSP^{-1}K + KP^{-T}SXP^T\right) \\ = \; & \frac{1}{4}\mathbf{Tr}\left(K\left(PXSP^{-1} + P^{-T}SXP^T\right)K\right) \\ = \; & \frac{1}{2}\mathbf{Tr}\left(KH_P(XS)K\right). \end{aligned}$$

Since we assumed that $H_P(XS)$ is positive semidefinite and the matrix K is symmetric, it follows that the matrix $KH_P(XS)K$ is positive semidefinite. Thus

$$\frac{1}{2}\mathbf{Tr}\left(KH_P(XS)K\right) \geq 0.$$

□

Theorem E.5 *If the matrices X and S are symmetric positive definite and the matrices PXP^T and $P^{-T}SP^{-1}$ commute, then $H_P(XS)$ is symmetric positive definite.*

Proof:

If PXP^T and $P^{-T}SP^{-1}$ commute, it follows that

$$PXSP^{-1} = (PXP^T)(P^{-T}SP^{-1}) = (P^{-T}SP^{-1})(PXP^T) = P^{-T}SXP^T.$$

Therefore the matrix $PXSP^{-1}$ is symmetric and

$$H_P(XS) = \frac{1}{2}(PXSP^{-1} + P^{-T}SXP^T) = PXSP^{-1}.$$

From Theorem A.4 we know that XS — and therefore $PXSP^{-1}$ — have positive eigenvalues. Since $H_P(XS)$ is symmetric it is therefore positive definite. □

Appendix F
Search directions for the embedding problem

Conditions for the existence and uniqueness of several search directions for the self–dual embedding problem of Chapter 4 (Section 4.2) are derived here.

The feasible directions of interior point methods for the embedding problem (4.4) can be computed from the following generic linear system:

$$\begin{array}{llllllll} & \mathbf{Tr}\,(A_i\Delta X) & -\Delta\tau b_i & +\Delta\theta\bar{b}_i & & & & = 0 \\ -\sum_{i=1}^m \Delta y_i A_i & & +\Delta\tau C & -\Delta\theta\bar{C} & -\Delta S & & & = 0 \\ b^T\Delta y & -\mathbf{Tr}\,(C\Delta X) & & +\Delta\theta\alpha & & -\Delta\rho & & = 0 \\ -\bar{b}^T\Delta y & +\mathbf{Tr}\,(\bar{C}\Delta X) & -\Delta\tau\alpha & & & & -\Delta\nu & = 0 \end{array} \tag{F.1}$$

where $i = 1, \ldots, m$, and

$$\left.\begin{array}{rcl} H_P(\Delta XS + X\Delta S) & = & \mu I - H_P(XS), \\ \rho\Delta\tau + \tau\Delta\rho & = & \mu - \tau\rho \\ \nu\Delta\theta + \theta\Delta\nu & = & \mu - \theta\nu, \end{array}\right\} \tag{F.2}$$

where H_P is the linear transformation given by

$$H_P(M) := \frac{1}{2}\left[PMP^{-1} + P^{-T}M^TP^T\right],$$

for any matrix M, and where the *scaling matrix* P determines the symmetrization strategy. The best-known choices of P from the literature are listed in Table F.1. We will now prove (or derive sufficient conditions) for existence and uniqueness of the search directions corresponding to each of the choices of P in Table F.1. To this end,

P	Reference
$\left[X^{\frac{1}{2}}\left(X^{\frac{1}{2}}SX^{\frac{1}{2}}\right)^{-\frac{1}{2}}X^{\frac{1}{2}}\right]^{\frac{1}{2}}$	Nesterov and Todd [138];
$X^{-\frac{1}{2}}$	Monteiro [124], Kojima *et al.* [108];
$S^{\frac{1}{2}}$	Monteiro [124], Helmberg *et al.* [84], Kojima *et al.* [108];
I	Alizadeh, Haeberley and Overton [6];

Table F.1. Choices for the scaling matrix P.

we will write the equations (F.1) and (F.2) as a single linear system and show that the coefficient matrix of this system is nonsingular.[1]

We will use the notation

$$\tilde{A}_i := \begin{bmatrix} A_i & & \\ & -b_i & \\ & & \bar{b}_i \end{bmatrix} \quad (i = 1, \ldots, m),$$

and define $\tilde{P}$ by replacing X by $\tilde{X}$ and S by $\tilde{S}$ in Table F.1. We will rewrite (F.2) by using the *symmetric Kronecker product*. For a detailed review of the Kronecker product see Appendix E; we only restate the relevant definitions here for convenience.

- **svec**$(X) := [X_{11}, \sqrt{2}X_{12}, \ldots \sqrt{2}X_{1n}, X_{22}, \sqrt{2}X_{23}, \ldots, X_{nn}]^T$;
- The *symmetric Kronecker product* $G \otimes_s K$ of $G, K \in \mathbb{R}^{n\times n}$ is implicitly defined via

$$(G \otimes_s K)\,\mathbf{svec}(H) := \frac{1}{2}\mathbf{svec}\left(KHG^T + GHK^T\right) \ (\forall\ H \in \mathcal{S}^n).$$

Using the symmetric Kronecker notation, we can combine (F.1) and (F.2) as

$$\begin{bmatrix} 0 & \tilde{\mathcal{A}} & 0 \\ -\tilde{\mathcal{A}}^T & \mathcal{S}_{kew} & I \\ 0 & E & F \end{bmatrix} \begin{bmatrix} \Delta y \\ \mathbf{svec}\left(\Delta\tilde{X}\right) \\ \mathbf{svec}\left(\Delta\tilde{S}\right) \end{bmatrix} = \begin{bmatrix} 0 \\ 0 \\ \mathbf{svec}\left(\mu I - H_{\tilde{P}}(\tilde{X}\tilde{S})\right) \end{bmatrix} \tag{F.3}$$

[1] The approach used here is a straightforward extension of the analysis by Todd *et al.* in [173], where this result was proved for SDP problems in the standard form (P) and (D).

where

$$\begin{aligned}
\tilde{\mathcal{A}} &:= [\mathbf{svec}(\tilde{A}_1) \;\ldots\; \mathbf{svec}(\tilde{A}_m)]^T \\
\mathcal{S}_{kew} &:= \begin{bmatrix} 0 & \mathbf{svec}(C) & -\mathbf{svec}(\bar{C}) \\ -\mathbf{svec}(C)^T & 0 & \alpha \\ \mathbf{svec}(\bar{C})^T & -\alpha & 0 \end{bmatrix} \\
E &:= \tilde{P} \otimes_s (\tilde{P}^{-T}\tilde{S}), \quad F := (\tilde{P}\tilde{X}) \otimes_s \tilde{P}^{-T}.
\end{aligned}$$

By Theorem E.4 in Appendix E we have the following result.

Lemma F.1 (Toh *et al.* [175]) *Let $\tilde{P}$ be invertible and $\tilde{X}$ and $\tilde{S}$ symmetric positive definite. Then the matrices E and F are invertible. If one also has $H_{\tilde{P}}(\tilde{X}\tilde{S}) \succ 0$, then the symmetric part of $E^{-1}F$ is also positive definite.*

We are now in a position to prove a sufficient condition for uniqueness of the search direction.

Theorem F.1 *The linear system (F.3) has a unique solution if $H_{\tilde{P}}(\tilde{X}\tilde{S}) \succ 0$.*

Proof:

We consider the homogeneous system

$$\begin{bmatrix} 0 & \tilde{\mathcal{A}} & 0 \\ -\tilde{\mathcal{A}}^T & \mathcal{S}_{kew} & I \\ 0 & E & F \end{bmatrix} \begin{bmatrix} \Delta y \\ \mathbf{svec}\left(\Delta\tilde{X}\right) \\ \mathbf{svec}\left(\Delta\tilde{S}\right) \end{bmatrix} = \begin{bmatrix} 0 \\ 0 \\ 0 \end{bmatrix}, \tag{F.4}$$

and prove that it has only the zero vector as solution.

From (F.4) we have

$$\mathbf{svec}\left(\Delta\tilde{S}\right) = \tilde{\mathcal{A}}^T \Delta y - \mathcal{S}_{kew}\mathbf{svec}\left(\Delta\tilde{X}\right)$$

and

$$\mathbf{svec}\left(\Delta\tilde{S}\right) = -\left(F^{-1}E\right)\mathbf{svec}\left(\Delta\tilde{X}\right). \tag{F.5}$$

Eliminating $\mathbf{svec}\left(\Delta\tilde{S}\right)$ from the last two equations gives

$$\tilde{\mathcal{A}}^T \Delta y - \mathcal{S}_{kew}\mathbf{svec}\left(\Delta\tilde{X}\right) + \left(F^{-1}E\right)\mathbf{svec}\left(\Delta\tilde{X}\right) = 0. \tag{F.6}$$

System (F.4) also implies

$$\tilde{\mathcal{A}}\mathbf{svec}\left(\Delta\tilde{X}\right) = 0. \tag{F.7}$$

From (F.6) we have

$$\mathbf{svec}\left(\Delta\tilde{X}\right)^T \tilde{\mathcal{A}}^T \Delta y - \mathbf{svec}\left(\Delta\tilde{X}\right)^T \mathcal{S}_{kew}\mathbf{svec}\left(\Delta\tilde{X}\right) + \mathbf{svec}\left(\Delta\tilde{X}\right)^T \left(F^{-1}E\right)\mathbf{svec}\left(\Delta\tilde{X}\right) = 0.$$

The first term on the left-hand side is zero, by (F.7), and the second term is zero by the skew-symmetry of $\mathcal{S}_{kew}$. We therefore have

$$\mathbf{svec}\left(\Delta\tilde{X}\right)^T \left(F^{-1}E\right)\mathbf{svec}\left(\Delta\tilde{X}\right) = 0,$$

which shows that $\Delta\tilde{X} = 0$, since EF^{-1} is assumed to be (non-symmetric) positive definite. It follows that $\Delta\tilde{S} = 0$ by (F.5). Furthermore, $\Delta y = 0$ by (F.6), since $\tilde{\mathcal{A}}$ has full rank (the matrices A_i $(i = 1, \ldots, m)$ are linearly independent). □

All that remains is to analyze the condition

$$H_{\tilde{P}}(\tilde{X}\tilde{S}) \succ 0 \qquad \text{(F.8)}$$

in the theorem. For the first three choices of $\tilde{P}$ in Table F.1, condition (F.8) always holds (by Theorem E.5 in Appendix E). For $\tilde{P} = I$ (the so-called AHO direction), (F.8) becomes the condition $\tilde{X}\tilde{S} + \tilde{S}\tilde{X} \succ 0$.

An alternative sufficient condition for existence of the AHO direction was derived by Monteiro and Zanjácomo [126], namely

$$\left\| \frac{1}{\mu}\tilde{X}^{\frac{1}{2}}\tilde{S}\tilde{X}^{\frac{1}{2}} - I \right\| \leq \frac{1}{2},$$

where $\mu = \mathbf{Tr}\left(\tilde{X}\tilde{S}\right)/\tilde{n}$.

Appendix G
Regularized duals

In this appendix we review some strong duality results for SDP problems that do not satisfy the Slater condition.

The duals in question are obtained through a procedure called regularization. Although a detailed treatment of regularization is beyond the scope of this monograph, the underlying idea is quite simple:[1] if the problem

$$(D): \quad d^* := \sup_{y,S} \left\{ b^T y \;\middle|\; \sum_{i=1}^{m} y_i A_i + S = C, \; S \succeq 0, \; y \in \mathbf{R}^m \right\}$$

is feasible, but not strictly feasible, we can obtain a 'strictly feasible reformulation' by replacing the semidefinite cone by a suitable lower dimensional face (say $\mathcal{F} \subset \mathcal{S}_n^+$) of it, such that the new problem

$$(\bar{D}): \quad d^* := \sup_{y,S} \left\{ b^T y \;\middle|\; \sum_{i=1}^{m} y_i A_i + S = C, \; S \in \mathcal{F}, \; y \in \mathbf{R}^m \right\}$$

is strictly feasible in the sense that there exists a pair (y, S) such that $S \in \mathrm{ri}\,(\mathcal{F})$ and $\sum_{i=1}^{m} y_i A_i + S = C$. Problem $(\bar{D})$ will have a perfect dual, by the conic duality theorem (Theorem 2.3). The main point is to find an explicit expression of the dual of the face $\mathcal{F}$. The resulting dual problem now takes the form:

$$(P') \quad \inf_{X} \left\{ \mathbf{Tr}\,(CX) \mid \mathbf{Tr}\,(A_i X) = b_i \; (i = 1, \ldots, m), \; X \in \mathcal{F}^* \right\},$$

where $\mathcal{F}^*$ denotes the dual cone of $\mathcal{F}$. In the SDP case $\mathcal{F}^*$ can be described by a system of linear matrix inequalities. There is more than one way to do this and one can obtain different dual problems within this framework (see Pataki [144] for details).

[1] The idea of regularization was introduced by Borwein and Wolkowicz [30]. For a more recent (and simplified) treatment the reader is referred to the excellent exposition by Pataki [144].

Ramana [153] first obtained a regularized dual for (D);[2] the so-called gap-free (or extended Lagrange–Slater) primal dual (P_{gf}) of (D) takes the form:

$$p^*_{gf} := \inf \mathbf{Tr}\ (C(U_0 + W_L))$$

subject to

$$\begin{array}{rcl} \mathbf{Tr}\ (A_k(U_0 + W_L)) & = & b_k, \quad k = 1, \ldots, m \\ \mathbf{Tr}\ (C(U_i + W_{i-1})) & = & 0, \quad i = 1, \ldots, L \\ \mathbf{Tr}\ (A_k(U_i + W_{i-1})) & = & 0, \quad i = 1, \ldots, L, \ k = 1, \ldots, m \\ W_0 & = & 0 \\ \begin{bmatrix} I & W_i^T \\ W_i & U_i \end{bmatrix} & \succeq & 0, \quad i = 1, \ldots, L \\ U_0 & \succeq & 0, \end{array}$$

where the variables are $U_i \succeq 0$ and $W_i \in \mathbf{R}^{n \times n}$, $i = 0, \ldots, L$, and

$$L = \min \left\{ n, \ \frac{1}{2} n(n+1) - m - 1 \right\}.$$

Note that the gap-free primal problem is easily cast in the standard primal form. Moreover, its size is polynomial in the size of (D). Unlike the Lagrangian dual (P) of (D), (P_{gf}) has the following desirable features:

- (Weak duality) If $(y, S) \in \mathcal{D}$ and (U_i, W_i) $(i = 0, \ldots, m)$ is feasible for (P_{gf}), then
$$b^T y \leq \mathbf{Tr}\ (C(U_0 + W_m)).$$
- (Dual boundedness) If (D) is feasible, its optimal value is finite if and only if (P_{gf}) is feasible.
- (Zero duality gap) The optimal value p^*_{gf} of (P_{gf}) equals the optimal value of (D) if and only if both (P_{gf}) and (D) are feasible.
- (Attainment) If the optimal value of (D) is finite, then it is attained by (P_{gf}).

The standard (Lagrangian) dual problem associated with (P_{gf}) is called the *corrected dual* (D_{cor}). The pair (P_{gf}) and (D_{cor}) are now in perfect duality; see Ramana and Freund [154].

Moreover, a feasible solution to (D) can be extracted from a feasible solution to (D_{cor}). The only problem is that (D_{cor}) does not necessarily attain its supremum, even if (D) does.

[2] In fact, Ramana did not derive this dual via regularization initially; it was only shown subsequently that it can be derived in this way in Ramana *et al.* [155].

Example G.1 *It is readily verified that the weakly infeasible problem* (D) *in Example 2.2 has a weakly infeasible corrected problem* (D_{cor}).

The possible duality relations are listed in Table G.1. The optimal value of D_{cor} is denoted by d^*_{cor}.

Status of (D)	Status of (P_{gf})	Status of (D_{cor})
$d^* < \infty$	$p^*_{gf} = d^*$	$d^*_{cor} = d^*$
unbounded	infeasible	unbounded
infeasible	unbounded	infeasible

Table G.1. Duality relations for a given problem (D), its gap-free dual (P_{gf}) and its corrected problem (D_{cor}).

References

[1] L.P. Aarts. Primal-dual search directions in semidefinite optimization. Master's thesis, Delft University of Technology, Faculty of Information Technology and Systems, Delft, The Netherlands, 1999.

[2] D.V. Alekseevskij, E.B. Vinberg, and A.S. Solodovnikov. Geometry of spaces of constant curvature. In E.B. Vinberg, editor, *Geometry II*, volume 29 of *Encyclopedia of Mathematical Sciences*. Springer-Verlag, 1993.

[3] F. Alizadeh. *Combinatorial optimization with interior point methods and semi–definite matrices*. PhD thesis, University of Minnesota, Minneapolis, USA, 1991.

[4] F. Alizadeh. Interior point methods in semidefinite programming with applications to combinatorial optimization. *SIAM Journal on Optimization*, 5:13–51, 1995.

[5] F. Alizadeh, J.-P.A. Haeberley, and M.L. Overton. Complementarity and nondegeneracy in semidefinite programming. *Mathematical Programming, Series B*, 77(2):129–162, 1997.

[6] F. Alizadeh, J.-P.A. Haeberley, and M.L. Overton. Primal-dual methods for semidefinite programming: convergence rates, stability and numerical results. *SIAM Journal on Optimization*, 8(3):746–768, 1998.

[7] F. Alizadeh and S. Schmieta. Symmetric cones, potential reduction methods. In H. Wolkowicz, R. Saigal, and L. Vandenberghe, editors, *Handbook of semidefinite programming*, pages 195–233. Kluwer Academic Publishers, Norwell, MA, 2000.

[8] E.D. Andersen. Finding all linearly dependent rows in large-scale linear programming. *Optimization Methods and Software*, 6(3):219–227, 1995.

[9] E.D. Andersen and K.D. Andersen. The MOSEK interior point optimizer for linear programming: an implementation of the homogeneous algorithm. In H. Frenk, K. Roos, T. Terlaky, and S. Zhang, editors, *High performance optimization*, pages 197–232. Kluwer Academic Publishers, 2000.

[10] E.D. Andersen, J. Gondzio, C. Mészáros, and X. Xu. Implementation of interior-point methods for large scale linear programs. In T. Terlaky, editor, *Interior point methods of mathematical programming*, pages 189–252. Kluwer, Dordrecht, The Netherlands, 1996.

[11] M.F. Anjos and H. Wolkowicz. A strengthened SDP relaxation via a second lifting for the Max-Cut problem. *Special Issue of Discrete Applied Mathematics devoted to Foundations of Heuristics in Combinatorial Optimization*, to appear.

[12] K.M. Anstreicher and M. Fampa. A long–step path following algorithm for semidefinite programming problems. In P. M. Pardalos and H. Wolkowicz, editors, *Topics in Semidefinite and Interior-Point Methods*, volume 18 of *Fields Institute Communications Series*, pages 181–196. American Mathematical Society, 1998.

[13] V. Balakrishnan and F. Wang. Sdp in systems and control theory. In H. Wolkowicz, R. Saigal, and L. Vandenberghe, editors, *Handbook of semidefinite programming*, pages 421–442. Kluwer Academic Publishers, Norwell, MA, 2000.

[14] G.P. Barker and D. Carlson. Cones of diagonally dominant matrices. *Pacific Journal of Mathematics*, 57:15–32, 1975.

[15] E.R. Barnes. A variation on Karmarkar's algorithm for solving linear programming problems. *Mathematical Programming*, 36:174–182, 1986.

[16] M.S. Bazarraa, H.D. Sherali, and C.M. Shetty. *Nonlinear Programming: Theory and Algorithms*. John Wiley and Sons, New York, 1993.

[17] R. Beigel and D. Eppstein. 3-Coloring in time $O(1.3446^n)$: a no-MIS algorithm. In *Proc. 36th IEEE Symp. Foundations of Comp. Sci.*, pages 444–453, 1995.

[18] M. Bellare and P. Rogaway. The complexity of approximating a nonlinear program. *Mathematical Programming*, 69:429–441, 1995.

[19] R. Bellman and K. Fan. On systems of linear inequalities in hermitian matrix variables. In V.L. Klee, editor, *Convexity*, Vol. 7 Proc. Symposia in Pure Mathematics, pages 1–11. Amer. Math. Soc., Providence, RI, 1963.

[20] A. Ben-Tal, L. El Ghaoui, and A.S. Nemirovski. Robustness. In H. Wolkowicz, R. Saigal, and L. Vandenberghe, editors, *Handbook of semidefinite programming*, pages 139–162. Kluwer Academic Publishers, Norwell, MA, 2000.

[21] A. Ben-Tal and A.S. Nemirovski. Structural design. In H. Wolkowicz, R. Saigal, and L. Vandenberghe, editors, *Handbook of semidefinite programming*, pages 443–467. Kluwer Academic Publishers, Norwell, MA, 2000.

[22] S.J. Benson and Y. Ye. DSDP3: Dual scaling algorithm for general positive semidefinite programming. Working paper, Computational Optimization Lab, Dept. of Management Science, University of Iowa, Iowa City, USA, 2001.

[23] S.J. Benson, Y. Ye, and X. Zhang. Solving large-scale sparse semidefinite programs for combinatorial optimization. *SIAM Journal on Optimization*, 10:443–461, 2000.

[24] Lenore Blum, Mike Shub, and Steve Smale. On a theory of computation and complexity over the real numbers: NP-completeness, recursive functions and universal machines. *Bull. Amer. Math. Soc. (N.S.)*, 21(1):1–46, 1989.

[25] I.M. Bomze. On standard quadratic optimization problems. *Journal of Global Optimization*, 13:369–387, 1998.

[26] I.M. Bomze, M. Budinich, P.M. Pardalos, and M. Pelillo. The maximum clique problem. In D.-Z. Du and P.M. Pardalos, editors, *Handbook of Combinatorial Optimization*, volume suppl. Vol. A, pages 1–74. Kluwer, Dordrecht, 1999.

[27] I.M. Bomze and E. de Klerk. Solving standard quadratic optimization problems via linear, semidefinite and copositive programming. Technical Report TR 2001–03, Insitut für Statistik und Decision Support Systems, Universität Wien, 2001.

[28] I.M. Bomze, M. Dür, E. de Klerk, C. Roos, A. Quist, and T. Terlaky. On copositive programming and standard quadratic optimization problems. *Journal of Global Optimization*, 18:301–320, 2000.

[29] R. Boppana and M.M. Halldórsson. Approximating maximum independent sets by excluding subgraphs. *Bit*, 32:180–196, 1992.

[30] J.M. Borwein and H. Wolkowicz. Regularizing the abstract convex program. *J. Math. Anal. Appl.*, 83(2):495–530, 1981.

[31] S.E. Boyd, L. El Ghaoui, E. Feron, and V. Balakrishnan. *Linear matrix inequalities in system and control theory.* SIAM Studies in Applied Mathematics, Vol. 15. SIAM, Philadelphia, USA, 1994.

[32] S. Burer and R.D.C. Monteiro. A Projected Gradient Algorithm for Solving the Maxcut SDP Relaxation. *Optimization Methods and Software*, 15:175–200, 2001.

[33] J.-D. Cho and M. Sarrafzadeh. Fast approximation algorithms on Maxcut, k-Coloring and k-Color ordering for VLSI applications. *IEEE Transactions on Computers*, 47(11):1253–1266, 1998.

[34] C. Choi and Y. Ye. Solving sparse semidefinite programs using the dual scaling algorithm with an iterative solver. Working paper, Computational Optimization Lab, Dept. of Management Science, University of Iowa, Iowa City, USA, 2000.

[35] S.A. Cook. The complexity of theorem proving procedures. In *Proceedings of the 3rd annual ACM symposium on the Theory of Computing*, pages 151–158, 1971.

[36] W. Cook, C.R. Coullard, and G. Turan. On the complexity of cutting plane proofs. *Discrete Applied Mathematics*, 18:25–38, 1987.

[37] B. Craven and B. Mond. Linear programming with matrix variables. *Linear Algebra Appl.*, 38:73–80, 1981.

[38] D. Cvetković, M. Cangalović, and V. Kovačevič-Vujcić. Semidefinite programming methods for the traveling salesman problem. In *Proceedings of the 7th International IPCO conference*, pages 126–136, 1999.

[39] E. de Klerk. *Interior Point Methods for Semidefinite Programming*. PhD thesis, Delft University of Technology, Delft, The Netherlands, 1997.

[40] E. de Klerk and D.V. Pasechnik. On the performance guarantee of MAX-3-CUT approximation algorithms. Manuscript, March, 2001.

[41] E. de Klerk and D.V. Pasechnik. Approximation of the stability number of a graph via copositive programming. *SIAM Journal on Optimization*, to appear.

[42] E. de Klerk, D.V. Pasechnik, and J.P. Warners. On approximate graph colouring and MAX-k-CUT algorithms based on the ϑ-function. Submitted to *Journal of Combinatorial Optimization*, 2000.

[43] E. de Klerk, J. Peng, C. Roos, and T. Terlaky. A scaled Gauss-Newton primal–dual search direction for semidefinite optimization. *SIAM Journal on Optimization*, 11:870–888, 2001.

[44] E. de Klerk, C. Roos, and T. Terlaky. Initialization in semidefinite programming via a self–dual, skew–symmetric embedding. *OR Letters*, 20:213–221, 1997.

[45] E. de Klerk, C. Roos, and T. Terlaky. A short survey on semidefinite programming. In W.K. Klein Haneveld, O.J. Vrieze, and L.C.M. Kallenberg, editors, *Ten years LNMB*. Stichting Mathematisch Centrum, Amsterdam, The Netherlands, 1997.

[46] E. de Klerk, C. Roos, and T. Terlaky. Infeasible–start semidefinite programming algorithms via self–dual embeddings. In P. M. Pardalos and H. Wolkowicz, editors, *Topics in Semidefinite and Interior-Point Methods*, volume 18 of *Fields Institute Communications Series*, pages 215–236. American Mathematical Society, 1998.

[47] E. de Klerk, C. Roos, and T. Terlaky. On primal–dual path–following algorithms in semidefinite programming. In F. Giannessi, S. Komlósi, and T. Rapcsák, editors, *New trends in mathematical programming*, pages 137–157. Kluwer Academic Publishers, Dordrecht, The Netherlands, 1998.

[48] E. de Klerk, C. Roos, and T. Terlaky. Polynomial primal–dual affine scaling algorithms in semidefinite programming. *Journal of Combinatorial Optimization*, 2:51–69, 1998.

[49] E. de Klerk and H. van Maaren. On semidefinite programming relaxations of $(2 + p)$-SAT. Submitted to *Annals of Mathematics of Artificial Intelligence*, 2000.

[50] E. de Klerk, H. van Maaren, and J.P. Warners. Relaxations of the satisfiability problem using semidefinite programming. *Journal of automated reasoning*, 24:37–65, 2000.

[51] D. den Hertog, E. de Klerk, and C. Roos. On convex quadratic approximation. CentER Discussion Paper 2000-47, CentER for Economic Research, Tilburg University, Tilburg, The Netherlands, 2000.

[52] J. Dieudonné. *Foundations of Modern Analysis*. Academic Press, New York, 1960.

[53] I.I. Dikin. Iterative solution of problems of linear and quadratic programming. *Doklady Akademii Nauk SSSR*, 174:747–748, 1967. (Translated in: *Soviet Mathematics Doklady*, 8:674–675, 1967).

[54] L. Faybusovich. On a matrix generalization of affine–scaling vector fields. *SIAM J. Matrix Anal. Appl.*, 16:886–897, 1995.

[55] L. Faybusovich. Semi–definite programming: a path–following algorithm for a linear–quadratic functional. *SIAM Journal on Optimization*, 6(4):1007–1024, 1996.

[56] U. Feige and M. Goemans. Approximating the value of two prover proof systems with applications to MAX 2SAT and MAX DICUT. In *Proc. Third Israel Symposium on Theory of Computing and Systems*, pages 182–189, 1995.

[57] U. Feige and J. Kilian. Zero knowledge and the chromatic number. *J. Comput. System Sci.*, 57:187–199, 1998.

[58] A.V. Fiacco and G.P. McCormick. *Nonlinear programming: sequential unconstrained minimization techniques.* John Wiley & Sons, New York, 1968. (Reprint: Volume 4 of *SIAM Classics in Applied Mathematics*, SIAM Publications, Philadelphia, USA, 1990).

[59] A. Frieze and M. Jerrum. Improved approximation algorithms for MAX k-cut and MAX BISECTION. In *Proc. 4th IPCO conference*, pages 1–13, 1995.

[60] K. Fujisawa, M. Kojima, and K. Nakata. Exploiting sparsity in primal–dual interior–point methods for semidefinite programming. *Mathematical Programming*, 79:235–253, 1997.

[61] M.R. Garey and D.S. Johnson. *Computers and intractability: a guide to the theory of NP–completeness*. W.H. Freeman and Company, Publishers, San Fransisco, USA, 1979.

[62] A. Genz. Numerical computation of multivariate normal probabilities. *J. Comp. Graph. Stat.*, 1:141–149, 1992.

[63] P.E. Gill, W. Murray, M.A. Saunders, J.A. Tomlin, and M.H. Wright. On projected Newton barrier methods for linear programming and an equivalence to Karmarkar's projective method. *Mathematical Programming*, 36:183–209, 1986.

[64] M. Goemans and F. Rendl. Combinatorial optimization. In H. Wolkowicz, R. Saigal, and L. Vandenberghe, editors, *Handbook of semidefinite programming*, pages 343–360. Kluwer Academic Publishers, Norwell, MA, 2000.

[65] M.X. Goemans. Semidefinite programming in combinatorial optimization. *Math. Programming*, 79(1-3, Ser. B):143–161, 1997. Lectures on mathematical programming (ismp97).

[66] M.X. Goemans and D.P. Williamson. Improved approximation algorithms for maximum cut and satisfiability problems using semidefinite programming. *Journal of the ACM*, 42(6):1115–1145, 1995.

[67] M.X. Goemans and D.P. Williamson. Approximation algorithms for MAX 3-CUT and other problems via complex semidefinite programming. In *Proceedings of the 33rd annual symposium on theory of computing (STOC '01)*, pages 443–452. ACM, 2001.

[68] D. Goldfarb and K. Scheinberg. Interior point trajectories in semidefinite programming. *SIAM Journal on Optimization*, 8(4):871–886, 1998.

[69] A.J. Goldman and A.W. Tucker. Theory of linear programming. In H.W. Kuhn and A.W. Tucker, editors, *Linear inequalities and related systems, Annals of Mathematical Studies, No. 38*, pages 53–97. Princeton University Press, Princeton, New Jersey, 1956.

[70] C.C. Gonzaga. Path following methods for linear programming. *SIAM Review*, 34:167–227, 1992.

[71] L.M. Graña Drummond and Y. Peterzil. The central path in smooth convex semidefinite programs. *Optimization*, to appear.

[72] M. Grötschel, L. Lovász, and A. Schrijver. *Geometric Algorithms and Combinatorial Optimization*. Springer–Verlag, Berlin, 1988.

[73] M. Grötschel, L.A. Lovász, and A. Schrijver. *Geometric algorithms and combinatorial optimization*. Springer Verlag, Berlin, 1988.

[74] J. Gu, P.W. Purdom, J. Franco, and B.W. Wah. Algorithms for the satisfiability (SAT) problem: a survey. In D. Du, J. Gu, and P.M. Pardalos, editors, *Satisfiability problem: Theory and applications*, volume 35 of *DIMACS series in Discrete Mathematics and Computer Science*. American Mathematical Society, 1997.

[75] A. Haken. The intractability of resolution. *Theoretical Computer Science*, 39:297–308, 1985.

[76] M. Halická. Private communication, 2001.

[77] M. Halická. Analyticity of the central path at the boundary point in semidefinite programming. *European Journal of Operations Research*, to appear.

[78] M. Halická, E. de Klerk, and C. Roos. On the convergence of the central path in semidefinite optimization. *SIAM Journal on Optimization*, to appear.

[79] E. Halperin and U. Zwick. Approximation algorithms for MAX 4-SAT and rounding procedures for semidefinite programs. In *Proc. 7th IPCO*, pages 202–217, 1999.

[80] J. Håstad. Some optimal inapproximability results. In *Proc. 29th Ann. ACM Symp. on Theory of Comp.*, pages 1–10, 1997.

[81] J. Håstad. Clique is hard to approximate within $|V^{1-\epsilon}|$. *Acta Mathematica*, 182:105–142, 1999.

[82] B. He, E. de Klerk, C. Roos, and T. Terlaky. Method of approximate centers for semi–definite programming. *Optimization Methods and Software*, 7:291–309, 1997.

[83] C. Helmberg and F. Oustry. Bundle methods and eigenvalue functions. In H. Wolkowicz, R. Saigal, and L. Vandenberghe, editors, *Handbook of semidefinite programming*, pages 307–337. Kluwer Academic Publishers, Norwell, MA, 2000.

[84] C. Helmberg, F. Rendl, R.J. Vanderbei, and H. Wolkowicz. An interior-point method for semidefinite programming. *SIAM Journal on Optimization*, 6:342–361, 1996.

[85] R.A. Horn and C.R. Johnson. *Matrix Analysis*. Cambridge University Press, 1985.

[86] R.A. Horn and C.R. Johnson. *Topics in Matrix Analysis*. Cambridge University Press, 1991.

[87] W.-Y. Hsiang. On infinitesimal symmetrization and volume formula for spherical or hyperbolic tetrahedrons. *Quart. J. Math. Oxford*, 39(2):463–468, 1988.

[88] B. Jansen. *Interior Point Techniques in Optimization*. PhD thesis, Delft University of Technology, Delft, The Netherlands, 1995.

[89] B. Jansen, C. Roos, and T. Terlaky. The theory of linear programming : Skew symmetric self–dual problems and the central path. *Optimization*, 29:225–233, 1994.

[90] B. Jansen, C. Roos, and T. Terlaky. Interior point methods: a decade after Karmarkar. A survey, with application to the smallest eigenvalue problem. *Statistica Neerlandica*, 50, 1995.

[91] B. Jansen, C. Roos, and T. Terlaky. A polynomial primal–dual Dikin–type algorithm for linear programming. *Mathematics of Operations Research*, 21:431–353, 1996.

[92] B. Jansen, C. Roos, and T. Terlaky. A family of polynomial affine scaling algorithms for positive semi–definite linear complementarity problems. *SIAM Journal on Optimization*, 7(1):126–140, 1997.

[93] B. Jansen, C. Roos, T. Terlaky, and J.-Ph. Vial. Primal–dual algorithms for linear programming based on the logarithmic barrier method. *Journal of Optimization Theory and Applications*, 83:1–26, 1994.

[94] F. Jarre. Convex analysis on symmetric matrices. In H. Wolkowicz, R. Saigal, and L. Vandenberghe, editors, *Handbook of semidefinite programming*, pages 13–27. Kluwer Academic Publishers, Norwell, MA, 2000.

[95] Jr. J.E. Dennis and H. Wolkowicz. Sizing and least change secant methods. *SIGNUM*, 10:1291–1314, 1993.

[96] J. Jiang. A long step primal–dual path following method for semidefinite programming. *OR Letters*, 23(1,2):53–62, 1998.

[97] V. Kann, S. Khanna, J. Lagergren, and A. Panconesi. On the hardness of approximating Max k-Cut and its dual. *Chicago Journal of Theoretical Computer Science*, 1997(2), June 1997.

[98] D. Karger, R. Motwani, and M. Sudan. Approximate graph coloring by semidefinite programming. In *35th Symposium on Foundations of Computer Science*, pages 2–13. IEEE Computer Society Press, 1994.

[99] H. Karloff and U. Zwick. A 7/8-approximation algorithm for MAX 3SAT? In *Proc. 38th FOCS*, pages 406–415, 1997.

[100] N.K. Karmarkar. A new polynomial–time algorithm for linear programming. *Combinatorica*, 4:373–395, 1984.

[101] L. Khachiyan. A polynomial time algorithm in linear programming. *Soviet Mathematics Doklady*, 20:191–194, 1979.

[102] S. Khanna, N. Linial, and S. Safra. On the hardness of approximating the chromatic number. *Combinatorica*, 20:393–415, 2000.

[103] D.E. Knuth. The sandwich theorem. *The Electronic Journal of Combinatorics*, 1:1–48, 1994.

[104] M. Kojima, N. Megiddo, T. Noma, and A. Yoshise. *A unified approach to interior point algorithms for linear complementarity problems*, volume 538 of *Lecture Notes in Computer Science*. Springer Verlag, Berlin, Germany, 1991.

[105] M. Kojima, M. Shida, and S. Shindoh. Local Convergence of Predictor-Corrector Infeasible-Interior-Point Algorithms for SDP's and SDLCP's. *Mathematical Programming*, 80:129–160, 1998.

[106] M. Kojima, M. Shida, and S. Shindoh. A note on the Nesterov-Todd and the Kojima-Shindoh-Hara search directions in semidefinite programming. *Optimization Methods and Software*, 11-12:47–52, 1999.

[107] M. Kojima, M. Shidah, and S. Shindoh. A predictor-corrector interior point algorithm for the semidefinite linear complementarity problem using the Alizadeh-Haeberley-Overton search direction. *SIAM Journal on Optimization*, 9(2):444–465, 1999.

[108] M. Kojima, S. Shindoh, and S. Hara. Interior point methods for the monotone semidefinite linear complementarity problem in symmetric matrices. *SIAM Journal on Optimization*, 7(1):88–125, 1997.

[109] S. Kruk, M. Muramatsu, F. Rendl, R.J. Vanderbei, and H. Wolkowicz. The Gauss-Newton direction in linear and semidefinite programming. *Optimization Methods and Software*, 15(1):1–27, 2001.

[110] J.B. Lasserre. An explicit exact SDP relaxation for nonlinear 0-1 programs. In *Proc. IPCO VIII*, pages 293–303, 2001.

[111] J.B. Lasserre. Global optimization with polynomials and the problem of moments. *SIAM J. Optimization*, 11:796–817, 2001.

[112] M. Laurent. A comparison of the Sherali-Adams, Lovász-Schrijver and Lasserre relaxations for 0-1 programming. Technical report pna-r0108, CWI, Amsterdam, The Netherlands, 2001.

[113] M. Laurent. Tighter linear and semidefinite relaxations for max-cut based on the Lovász-Schrijver lift-and-project procedure. *SIAM Journal on Optimization*, to appear.

[114] A.S. Lewis and M.L. Overton. Eigenvalue optimization. *Acta Numerica*, 5:149–190, 1996.

[115] L. Lovász. On the Shannon capacity of a graph. *IEEE Trans. on Information Theory*, 25:1–7, 1979.

[116] L. Lovász and A. Schrijver. Cones of matrices and set–functions and 0–1 optimization. *SIAM Journal on Optimization*, 1(2):166–190, 1991.

[117] Z.-Q. Luo, J.F. Sturm, and S. Zhang. Superlinear convergence of a Symmetric primal–dual path following algorithm for semidefinite programming. *SIAM Journal on Optimization*, 8(1):59–81, 1998.

[118] Z.-Q. Luo, J.F. Sturm, and S. Zhang. Conic convex programming and self-dual embedding. *Optimization Methods and Software*, 14(3):196–218, 2000.

[119] I.J. Lustig, R.E. Marsten, and D.F. Shanno. Interior point methods : Computational state of the art. *ORSA Journal on Computing*, 6:1–15, 1994.

[120] S. Mahajan and H. Ramesh. Derandomizing semidefinite programming based approximation algorithms. In *Proc. of the 36th Annual IEEE Symposium on Foundations of Computer Science*, pages 162–169. IEEE, 1995.

[121] S. Mehrotra. On the implementation of a (primal–dual) interior point method. *SIAM Journal on Optimization*, 2:575–601, 1992.

[122] J. Milnor. *Singular points of complex hypersurfaces*. Annals of mathematics studies. Princeton University Press, Princeton, New Jersey, 1968.

[123] S. Mizuno, M.J. Todd, and Y. Ye. On adaptive step primal–dual interior–point algorithms for linear programming. *Mathematics of Operations Research*, 18:964–981, 1993.

[124] R.D.C. Monteiro. Primal-dual path-following algorithms for semidefinite programming. *SIAM Journal on Optimization*, 7(3):663–678, 1997.

[125] R.D.C. Monteiro, I. Adler, and M.G.C. Resende. A polynomial–time primal–dual affine scaling algorithm for linear and convex quadratic programming and its power series extension. *Mathematics of Operations Research*, 15:191–214, 1990.

[126] R.D.C. Monteiro and P.R. Zanjácomo. A note on the existence of the Alizadeh-Haeberley-Overton direction in semidefinite programming. *Mathematical Programming*, 78(3):393–397, 1997.

[127] R.D.C. Monteiro and P.R. Zanjácomo. Implementation of primal-dual methods for semidefinite programming based on Monteiro and Tsuchiya directions and their variants. *Optimization Methods and Software*, 11/12:91–140, 1999.

[128] R. Motwani and P. Raghavan. *Randomized Algorithms*. Cambridge University Press, 1995.

[129] T.S. Motzkin and E.G. Straus. Maxima for graphs and a new proof of a theorem of Turán. *Canadian J. Math.*, 17:533–540, 1965.

[130] M. Muramatsu. Affine scaling algorithm fails for semidefinite programming. *Mathematical Programming*, 83(3):393–406, 1998.

[131] M. Muramatsu and R.J. Vanderbei. Primal-dual affine-scaling algorithm fails for semidefinite programming. *Mathematics of Operations Research*, 24:149–175, 1999.

[132] K.G. Murty and S.N. Kabadi. Some NP-complete problems in quadratic and linear programming. *Mathematical Programming*, 39:117–129, 1987.

[133] A. Nemirovskii and P. Gahinet. The projective method for solving linear matrix inequalities. *Mathematical Programming, Series B*, 77(2):163–190, 1997.

[134] Y.E. Nesterov. Global quadratic optimization on the sets with simplex structure. Discussion paper 9915, CORE, Catholic University of Louvain, Belgium, 1999.

[135] Yu. Nesterov. Long–step strategies in interior point potential reduction methods. *Mathematical Programming, Series B*, 76(1):47–94, 1997.

[136] Yu. Nesterov. Quality of semidefinite relaxation for nonconvex quadratic optimization. Discussion paper 9719, CORE, Catholic University of Louvain, Belgium, March 1997.

[137] Yu. Nesterov and A.S. Nemirovski. *Interior point polynomial algorithms in convex programming*. SIAM Studies in Applied Mathematics, Vol. 13. SIAM, Philadelphia, USA, 1994.

[138] Yu. Nesterov and M.J. Todd. Self–scaled barriers and interior–point methods for convex programming. *Mathematics of Operations Research*, 22(1):1–42, 1997.

[139] Yu. Nesterov, H. Wolkowicz, and Y. Ye. Semidefinite programming relaxations of nonconvex quadratic optimization. In H. Wolkowicz, R. Saigal, and L. Vandenberghe, editors, *Handbook of semidefinite programming*, pages 361–419. Kluwer Academic Publishers, Norwell, MA, 2000.

[140] J.A. Olkin. Using semi–definite programming for controller design in active noise control. *SIAG/OPT Views and News*, 8:1–5, Fall 1996.

[141] P.A. Parillo. *Structured Semidefinite Programs and Semi-algebraic Geometry Methods in Robustness and Optimization.* PhD thesis, California Institute of Technology, Pasadena, California, USA, 2000. Available at: http://www.cds.caltech.edu/~pablo/.

[142] D.V. Pasechnik. Bipartite sandwiches: bounding the size of a maximum biclique. ArXiv e-print, 1999. Available online at http://xxx.lanl.gov/abs/math.CO/9907109.

[143] G. Pataki. On the rank of extreme matrices in semidefinite programs and the multiplicity of optimal eigenvalues. *Mathematics of Operations Research*, 23(2):339–358, 1998.

[144] G. Pataki. A Simple Derivation of a Facial Reduction Algorithm, and Extended Dual Systems. Technical Report, Columbia University, New York, USA, 2000.

[145] G. Pataki. The geometry of semidefinite programming. In H. Wolkowicz, R. Saigal, and L. Vandenberghe, editors, *Handbook of semidefinite programming*, pages 29–66. Kluwer Academic Publishers, Norwell, MA, 2000.

[146] G. Pólya. Über positive Darstellung von Polynomen. *Vierteljschr. Naturforsch. Ges. Zürich*, 73:141–145, 1928. (also Collected Papers, Vol. 2, 309–313, MIT Press, Cambridge, Mass., London 1974).

[147] L. Porkolab and L. Khachiyan. On the complexity of semidefinite programs. *Journal of Global Optimization*, 10:351–365, 1997.

[148] F.A. Potra and R. Sheng. A superlinearly convergent primal–dual infeasible–interior–point algorithm for semidefinite programming. *SIAM Journal on Optimization*, 8(4):1007–1028, 1998.

[149] F.A. Potra and R. Sheng. On homogeneous interior–point algorithms for semidefinite programming. *Optimization methods and software*, 9:161–184, 1998.

[150] V. Powers and B. Reznick. A new bound for Pólya's Theorem with applications to polynomials positive on polyhedra. In *Proceedings of MEGA 2000*, 2000. To appear in *J. Pure and Applied Algebra*.

[151] W.R. Pulleyblank. The combinatorial optimization top 10 list. Plenary talk at the ISMP2000 conference, Atlanta, August 7-11, 2000.

[152] A.J. Quist, E. de Klerk, C. Roos, and T. Terlaky. Copositive relaxation for general quadratic programming. *Optimization Methods and Software*, 9:185–209, 1998.

[153] M. Ramana. An exact duality theory for semidefinite programming and its complexity implications. *Mathematical Programming, Series B*, 77(2):129–162, 1997.

[154] M. Ramana and R. Freund. On the ELSD duality theory for SDP. Technical report, Center of Applied Optimization, University of Florida, Gainsville, Florida, USA, 1996.

[155] M. Ramana, L. Tunçel, and H. Wolkowicz. Strong duality for semidefinite programming. *SIAM Journal on Optimization*, 7(3):641–662, 1997.

[156] M.V. Ramana and P.M. Pardalos. Semidefinite programming. In T. Terlaky, editor, *Interior point methods of mathematical programming*, pages 369–398. Kluwer, Dordrecht, The Netherlands, 1996.

[157] J. Renegar. Condition numbers, the barrier method, and the conjugate-gradient method. *SIAM Journal on Optimization*, 6(4):879–912, 1996.

[158] J. Renegar. *A mathematical view of interior-point methods in convex optimization*. SIAM, Philadelphia, PA, 2001.

[159] Bruce Reznick. Some Concrete Aspects of Hilbert's 17th Problem. Preprint 98-002, Department of Mathematics, University of Illinois at Urbana-Champaign, Urbana, Illinois, USA, 1998.

[160] R.T. Rockafellar. *Convex analysis*. Princeton University Press, Princeton, New Jersey, 1970.

[161] C. Roos, T. Terlaky, and J.-Ph. Vial. *Theory and Algorithms for Linear Optimization: An interior point approach*. John Wiley & Sons, New York, 1997.

[162] C. Roos and J.-Ph. Vial. A polynomial method of approximate centers for linear programming. *Mathematical Programming*, 54:295–305, 1992.

[163] Alexander Schrijver. A comparison of the Delsarte and Lovász bounds. *IEEE Trans. Inform. Theory*, 25(4):425–429, 1979.

[164] H.D. Sherali and W.P. Adams. A hierarchy of relaxations between the continuous and convex hull representations for 0–1 programming problems. *SIAM Journal on Discrete Mathematics*, 3(3):411–430, 1990.

[165] M. Shida, S. Shindoh, and M. Kojima. Existence of search directions in interior–point algorithms for the SDP and the monotone SDLCP. *SIAM Journal on Optimization*, 8(2):387–396, 1998.

[166] N.Z. Shor. Quadratic optimization problems. *Soviet Journal of Computer and System Sciences*, 25:1–11, 1987.

[167] J.F. Sturm. Using SeDuMi 1.02, a MATLAB toolbox for optimization over symmetric cones. *Optimization Methods and Software*, 11-12:625–653, 1999.

[168] J.F. Sturm. Error bounds for linear matrix inequalities. *SIAM Journal on Optimization*, 10(4):1228–1248, 2000.

[169] J.F. Sturm and S. Zhang. On the long step path–following method for semidefinite programming. *OR Letters*, 22:145–150, 1998.

[170] J.F. Sturm and S. Zhang. Symmetric primal–dual path following algorithms for semidefinite programming. *Applied Numerical Mathematics*, 29:301–316, 1999.

[171] K. Tanabe. Centered Newton method for mathematical programming. In *System Modelling and Optimization: Proceedings of the 13th IFIP–Conference, Berlin, August/September 1987*, volume 113 of *Lecture Notes in Control and Information Sciences*, pages 197–206. Springer-Verlag, New York, 1988.

[172] M.J. Todd. A study of search directions in primal-dual interior-point methods for semidefinite programming. *Optimization Methods and Software*, 11:1–46, 1999.

[173] M.J. Todd, K.C. Toh, and R.H. Tütüncü. On the Nesterov-Todd direction in semidefinite programming. *SIAM Journal on Optimization*, 8(3):769–796, 1998.

[174] M.J. Todd and Y. Ye. A centered projective algorithm for linear programming. *Mathematics of Operations Research*, 15:508–529, 1990.

[175] K.C. Toh, M.J. Todd, and R.H. Tütüncü. SDPT3 — a Matlab software package for semidefinite programming. *Optimization Methods and Software*, 11:545–581, 1999.

[176] L. Tunçel. Potential reduction and primal–dual methods. In H. Wolkowicz, R. Saigal, and L. Vandenberghe, editors, *Handbook of semidefinite programming*, pages 235–266. Kluwer Academic Publishers, Norwell, MA, 2000.

[177] A. Urquhart. Open problem posed at SAT'98 workshop, May 10–14, Paderborn, Germany, 1998.

[178] H. van Maaren. Elliptic approximations of propositional formulae. *Discrete Applied Mathematics*, 96-97:223–244, 1999.

[179] L. Vandenberghe and S. Boyd. SP: Software for semidefinite programming. User's Guide, Stanford University, Stanford, USA, 1994.

[180] L. Vandenberghe and S. Boyd. A primal–dual potential reduction algorithm for problems involving matrix inequalities. *Math. Programming*, 69:205–236, 1995.

[181] L. Vandenberghe and S. Boyd. Semidefinite programming. *SIAM Review*, 38:49–95, 1996.

[182] R.J. Vanderbei, M.S. Meketon, and B.A. Freedman. A modification of Karmarkar's linear programming algorithm. *Algorithmica*, 1:395–407, 1986.

[183] H. Wolkowicz, R. Saigal, and L. Vandenberghe, editors. *Handbook of semidefinite programming*. Kluwer Academic Publishers, Norwell, MA, 2000.

[184] S.J. Wright. *Primal–dual interior point methods*. SIAM, Philadelphia, 1997.

[185] X. Xu, P.-F. Hung, and Y. Ye. A simplified homogeneous and self–dual linear programming algorithm and its implementation. *Annals of OR*, 62:151–171, 1996.

[186] H. Yang, Y. Ye, and J. Zhang. An approximation algorithm for the two-parallel machines scheduling problem with capacity constraints. Manuscript, Department of Management Sciences, Henry B. Tippie College of Business Administration, University of Iowa, 2000.

[187] Y. Ye. Toward probabilistic analysis of interior–point algorithms for linear programming. *SIAM Journal on Optimization*, 19:38–52, 1994.

[188] Y. Ye. *Interior point algorithms*. Discrete Mathematics and Optimization. Wiley-Interscience, New York, 1997.

[189] Y. Ye. Approximating quadratic programming with bound and quadratic constraints. *Mathematical Programming*, 84:219–226, 1999.

[190] Y. Ye. A .699 approximation algorithm for max-bisection. *Mathematical Programming*, 90:101–111, 2001.

[191] Y. Ye, M.J. Todd, and S. Mizuno. An $\mathcal{O}(\sqrt{n}L)$–iteration homogeneous and self–dual linear programming algorithm. *Mathematics of Operations Research*, 19:53–67, 1994.

[192] Y. Zhang. On extending some primal-dual interior point algorithms from linear programming to semidefinite programming. *SIAM Journal on Optimization*, 8(2):365–386, 1998.

[193] Q. Zhao, S.E. Karisch, F. Rendl, and H. Wolkowicz. Semidefinite programming relaxations for the quadratic assignment problem. *Journal of Combinatorial Optimization*, 2(1):71–109, 1998.

[194] U. Zwick. Outward rotations: a new tool for rounding solutions of semidefinite programming relaxations, with applications to MAX CUT and other problems. In *Proc. 31st STOC*, pages 679–687, 1999.

Index

Applied Optimization

1. D.-Z. Du and D.F. Hsu (eds.): *Combinatorial Network Theory.* 1996 ISBN 0-7923-3777-8
2. M.J. Panik: *Linear Programming: Mathematics, Theory and Algorithms.* 1996 ISBN 0-7923-3782-4
3. R.B. Kearfott and V. Kreinovich (eds.): *Applications of Interval Computations.* 1996 ISBN 0-7923-3847-2
4. N. Hritonenko and Y. Yatsenko: *Modeling and Optimimization of the Lifetime of Technology.* 1996 ISBN 0-7923-4014-0
5. T. Terlaky (ed.): *Interior Point Methods of Mathematical Programming.* 1996 ISBN 0-7923-4201-1
6. B. Jansen: *Interior Point Techniques in Optimization.* Complementarity, Sensitivity and Algorithms. 1997 ISBN 0-7923-4430-8
7. A. Migdalas, P.M. Pardalos and S. Storøy (eds.): *Parallel Computing in Optimization.* 1997 ISBN 0-7923-4583-5
8. F.A. Lootsma: *Fuzzy Logic for Planning and Decision Making.* 1997 ISBN 0-7923-4681-5
9. J.A. dos Santos Gromicho: *Quasiconvex Optimization and Location Theory.* 1998 ISBN 0-7923-4694-7
10. V. Kreinovich, A. Lakeyev, J. Rohn and P. Kahl: *Computational Complexity and Feasibility of Data Processing and Interval Computations.* 1998 ISBN 0-7923-4865-6
11. J. Gil-Aluja: *The Interactive Management of Human Resources in Uncertainty.* 1998 ISBN 0-7923-4886-9
12. C. Zopounidis and A.I. Dimitras: *Multicriteria Decision Aid Methods for the Prediction of Business Failure.* 1998 ISBN 0-7923-4900-8
13. F. Giannessi, S. Komlósi and T. Rapcsák (eds.): *New Trends in Mathematical Programming.* Homage to Steven Vajda. 1998 ISBN 0-7923-5036-7
14. Ya-xiang Yuan (ed.): *Advances in Nonlinear Programming.* Proceedings of the '96 International Conference on Nonlinear Programming. 1998 ISBN 0-7923-5053-7
15. W.W. Hager and P.M. Pardalos: *Optimal Control.* Theory, Algorithms, and Applications. 1998 ISBN 0-7923-5067-7
16. Gang Yu (ed.): *Industrial Applications of Combinatorial Optimization.* 1998 ISBN 0-7923-5073-1
17. D. Braha and O. Maimon (eds.): *A Mathematical Theory of Design: Foundations, Algorithms and Applications.* 1998 ISBN 0-7923-5079-0

Applied Optimization

18. O. Maimon, E. Khmelnitsky and K. Kogan: *Optimal Flow Control in Manufacturing.* Production Planning and Scheduling. 1998 ISBN 0-7923-5106-1
19. C. Zopounidis and P.M. Pardalos (eds.): *Managing in Uncertainty: Theory and Practice.* 1998 ISBN 0-7923-5110-X
20. A.S. Belenky: *Operations Research in Transportation Systems:* Ideas and Schemes of Optimization Methods for Strategic Planning and Operations Management. 1998 ISBN 0-7923-5157-6
21. J. Gil-Aluja: *Investment in Uncertainty.* 1999 ISBN 0-7923-5296-3
22. M. Fukushima and L. Qi (eds.): *Reformulation: Nonsmooth, Piecewise Smooth, Semismooth and Smooting Methods.* 1999 ISBN 0-7923-5320-X
23. M. Patriksson: *Nonlinear Programming and Variational Inequality Problems.* A Unified Approach. 1999 ISBN 0-7923-5455-9
24. R. De Leone, A. Murli, P.M. Pardalos and G. Toraldo (eds.): *High Performance Algorithms and Software in Nonlinear Optimization.* 1999 ISBN 0-7923-5483-4
25. A. Schöbel: *Locating Lines and Hyperplanes.* Theory and Algorithms. 1999 ISBN 0-7923-5559-8
26. R.B. Statnikov: *Multicriteria Design.* Optimization and Identification. 1999 ISBN 0-7923-5560-1
27. V. Tsurkov and A. Mironov: *Minimax under Transportation Constrains.* 1999 ISBN 0-7923-5609-8
28. V.I. Ivanov: *Model Development and Optimization.* 1999 ISBN 0-7923-5610-1
29. F.A. Lootsma: *Multi-Criteria Decision Analysis via Ratio and Difference Judgement.* 1999 ISBN 0-7923-5669-1
30. A. Eberhard, R. Hill, D. Ralph and B.M. Glover (eds.): *Progress in Optimization.* Contributions from Australasia. 1999 ISBN 0-7923-5733-7
31. T. Hürlimann: *Mathematical Modeling and Optimization.* An Essay for the Design of Computer-Based Modeling Tools. 1999 ISBN 0-7923-5927-5
32. J. Gil-Aluja: *Elements for a Theory of Decision in Uncertainty.* 1999 ISBN 0-7923-5987-9
33. H. Frenk, K. Roos, T. Terlaky and S. Zhang (eds.): *High Performance Optimization.* 1999 ISBN 0-7923-6013-3
34. N. Hritonenko and Y. Yatsenko: *Mathematical Modeling in Economics, Ecology and the Environment.* 1999 ISBN 0-7923-6015-X
35. J. Virant: *Design Considerations of Time in Fuzzy Systems.* 2000 ISBN 0-7923-6100-8

Applied Optimization

36. G. Di Pillo and F. Giannessi (eds.): *Nonlinear Optimization and Related Topics.* 2000 ISBN 0-7923-6109-1
37. V. Tsurkov: *Hierarchical Optimization and Mathematical Physics.* 2000 ISBN 0-7923-6175-X
38. C. Zopounidis and M. Doumpos: *Intelligent Decision Aiding Systems Based on Multiple Criteria for Financial Engineering.* 2000 ISBN 0-7923-6273-X
39. X. Yang, A.I. Mees, M. Fisher and L.Jennings (eds.): *Progress in Optimization.* Contributions from Australasia. 2000 ISBN 0-7923-6286-1
40. D. Butnariu and A.N. Iusem: *Totally Convex Functions for Fixed Points Computation and Infinite Dimensional Optimization.* 2000 ISBN 0-7923-6287-X
41. J. Mockus: *A Set of Examples of Global and Discrete Optimization.* Applications of Bayesian Heuristic Approach. 2000 ISBN 0-7923-6359-0
42. H. Neunzert and A.H. Siddiqi: *Topics in Industrial Mathematics.* Case Studies and Related Mathematical Methods. 2000 ISBN 0-7923-6417-1
43. K. Kogan and E. Khmelnitsky: *Scheduling: Control-Based Theory and Polynomial-Time Algorithms.* 2000 ISBN 0-7923-6486-4
44. E. Triantaphyllou: *Multi-Criteria Decision Making Methods.* A Comparative Study. 2000 ISBN 0-7923-6607-7
45. S.H. Zanakis, G. Doukidis and C. Zopounidis (eds.): *Decision Making: Recent Developments and Worldwide Applications.* 2000 ISBN 0-7923-6621-2
46. G.E. Stavroulakis: *Inverse and Crack Identification Problems in Engineering Mechanics.* 2000 ISBN 0-7923-6690-5
47. A. Rubinov and B. Glover (eds.): *Optimization and Related Topics.* 2001 ISBN 0-7923-6732-4
48. M. Pursula and J. Niittymäki (eds.): *Mathematical Methods on Optimization in Transportation Systems.* 2000 ISBN 0-7923-6774-X
49. E. Cascetta: *Transportation Systems Engineering: Theory and Methods.* 2001 ISBN 0-7923-6792-8
50. M.C. Ferris, O.L. Mangasarian and J.-S. Pang (eds.): *Complementarity: Applications, Algorithms and Extensions.* 2001 ISBN 0-7923-6816-9
51. V. Tsurkov: *Large-scale Optimization – Problems and Methods.* 2001 ISBN 0-7923-6817-7
52. X. Yang, K.L. Teo and L. Caccetta (eds.): *Optimization Methods and Applications.* 2001 ISBN 0-7923-6866-5
53. S.M. Stefanov: *Separable Programming Theory and Methods.* 2001 ISBN 0-7923-6882-7

Applied Optimization

54. S.P. Uryasev and P.M. Pardalos (eds.): *Stochastic Optimization: Algorithms and Applications.* 2001 ISBN 0-7923-6951-3
55. J. Gil-Aluja (ed.): *Handbook of Management under Uncertainty.* 2001 ISBN 0-7923-7025-2
56. B.-N. Vo, A. Cantoni and K.L. Teo: *Filter Design with Time Domain Mask Constraints: Theory and Applications.* 2001 ISBN 0-7923-7138-0
57. S. Zlobec: *Stable Parametric Programming.* 2001 ISBN 0-7923-7139-9
58. M.G. Nicholls, S. Clarke and B. Lehaney (eds.): *Mixed-Mode Modelling: Mixing Methodologies for Organisational Intervention.* 2001 ISBN 0-7923-7151-8
59. F. Giannessi, P.M. Pardalos and T. Rapcsák (eds.): *Optimization Theory. Recent Developments from Mátraháza.* 2001 ISBN 1-4020-0009-X
60. K.M. Hangos, R. Lakner and M. Gerzson: *Intelligent Control Systems.* An Introduction with Examples. 2001 ISBN 1-4020-0134-7
61. D. Gstach: *Estimating Output-Specific Efficiencies.* 2002 ISBN 1-4020-0483-4
62. J. Geunes, P.M. Pardalos and H.E. Romeijn (eds.): *Supply Chain Management: Models, Applications, and Research Directions.* 2002 ISBN 1-4020-0487-7
63. M. Gendreau and P. Marcotte (eds.): *Transportation and Network Analysis: Current Trends.* Miscellanea in Honor of Michael Florian. 2002 ISBN 1-4020-0488-5
64. M. Patriksson and M. Labbé (eds.): *Transportation Planning.* State of the Art. 2002 ISBN 1-4020-0546-6
65. E. de Klerk: *Aspects of Semidefinite Programming.* Interior Point Algorithms and Selected Applications. 2002 ISBN 1-4020-0547-4

KLUWER ACADEMIC PUBLISHERS – DORDRECHT / BOSTON / LONDON